Fuels and Petrochemicals Division 2013

Core Programming Area at the 2013 AIChE Annual Meeting: Global Challenges for Engineering a Sustainable Future

San Francisco, California, USA
3 – 8 November 2013

ISBN: 978-1-63439-043-9

Printed from e-media with permission by:

Curran Associates, Inc.
57 Morehouse Lane
Red Hook, NY 12571

Some format issues inherent in the e-media version may also appear in this print version.

Printed by Curran Associates, Inc. (2014)

For permission requests, please contact AIChE
at the address below.

AIChE
3 Park Avenue
New York, NY 10016-5991

Phone: (203) 702-7660
Fax: (203) 775-5177

www.aiche.org

Additional copies of this publication are available from:

Curran Associates, Inc.
57 Morehouse Lane
Red Hook, NY 12571 USA
Phone: 845-758-0400
Fax: 845-758-2634
Email: curran@proceedings.com
Web: www.proceedings.com

TABLE OF CONTENTS

(5a) Modeling of Electrochemical Energy Storage Devices for Electric Vehicles

Chandrasekaran, R., Georgia Institute of Technology & Center for Fuel Cell and Battery Technologies,Georgia Tech Research Institute

Originally presented on: 11/3/2013 15:30:00 - 15:45:00

Abstract: Electrochemical energy storage devices such as lithium-ion batteries for electric vehicles have to be designed and optimized for performance, life [[i],[ii],[iii]], safety and cost. Customers of plug-in hybrid and all-electric vehicles may also place a premium on the ability to quickly recharge the battery. Modeling and simulation enables detailed spatial-temporal analyses of responses of interest which may be impossible or difficult to be determined by experiments. This is also usually achieved at a fraction of time and cost compared to experiments.

Isothermal, physics-based model [[iv]] was used to simulate the galvanostatic charge and discharge performance [[v]] of dual lithium-ion insertion cell sandwich at various current densities. Modeling results were validated with experimental data and accuracy of parameters [[vi]] was analyzed. The limitations to cell performance at high charge [[vii]] and discharge rates were quantified in terms of their contributions to cell overpotential and further analyzed. Simple expressions were provided to quickly evaluate resistances and potential drop wherever feasible. Engineering solutions for averting bottlenecks to fast charge acceptance and high rate discharge performance and the corresponding improvements in performance were developed from the simulations. Ongoing university collaborations [[viii]], [[ix]], [[x]] to address these issues and other in-house efforts to understand [[xi]] and measure [[xii]] the input parameters for modeling supplier cell chemistry [[xiii]] and Li-Si negative electrode [[xiv],[xv]] will be briefly mentioned.

Acknowledgements: Andy Drews and Ted Miller are acknowledged for the support.

---

[i]. R. Chandrasekaran, W. Bi, T. F. Fuller, *J. Power Sources*, 182 (2008) 546–557.

[ii]. R. Chandrasekaran, J. Denis, M. Skinner, T. F. Fuller, # 457c, *AIChE Annual Meeting Proceedings*, November 2008.

[iii]. R. Chandrasekaran, G. Sikha, B. N. Popov, *J. Applied Electrochemistry*, 35(10), 1005 – 1013 (2005).

[iv]. Thomas F. Fuller, Marc Doyle, John Newman, *J. Electrochem. Soc.*, 141(1), 1994, 1-10.

[v]. R. Chandrasekaran, Y. Seong, C. Bae, J. Jung, K. Kim, K. Cheong, T. J. Miller, Talk # F4.05, *Materials*

Research Society (MRS) Online Proceedings, Spring 2013 Meeting (under review).

[vi]. R. Chandrasekaran, Talk # O6.9, *MRS Spring Meeting, April 2012.*

[vii]. R. Chandrasekaran, *Electrochemical Society (ECS) Fall Meeting*, San Francisco, Oct 27- Nov 1, 2013 (abstract submitted).

[viii]. R. Chandrasekaran, A. R. Drews, A. Shaligram, J. Sakamoto, *MRS Proceedings* (online), Volume 1440, 2012. DOI: http://dx.doi.org/10.1557/opl.2012.1287

[ix]. J. Sakamoto, E. Rangasamy, H. Lee, R. Chandrasekaran, A.R. Drews, T.J.Miller, Talk # J16.07, *MRS Fall Meeting*, Boston, 2012.

[x]. V. George and T. F. Fuller, *ECS Fall Meeting*, Abstract # 1215, Hawaii, 2012.

[xi]. D. M. Bernardi, R. Chandrasekaran, J. Y. Go, *J. Electrochem. Soc.*, 2013 (under review).

[xii]. H. Wen, A. R. Drews, T.J. Miller, M. Karulkar, V. Anandan, R. Chandrasekaran, *Battery Congress Proceedings*, April 2013.

[xiii]. D.M. Bernardi, J. Go, *J. Power Sources* 196 (2011) 412–427.

[xiv]. R. Chandrasekaran, A. Magasinski, G. Yushin, T.F. Fuller, *J. Electrochem.Soc.,* 157 (10), A1139-A1151 (2010).

[xv]. R. Chandrasekaran and T.F. Fuller, *J. Electrochem.Soc.*, 158 (8) A859-A871 (2011).

(5b) Characterisation of Thermal Response and Gas Generation Products of Li-Ion Cells Under Abusive Conditions

Hacker, V., Graz University of Technology
Voitic, G., Graz University of Technology
Golubkov, A.,
Originally presented on: 11/3/2013 15:45:00 - 16:00:00

The number of lithium-ion battery applications continues to increase rapidly. Commercially available lithium-ion batteries contain a multitude of harmful and flammable substances. Despite intense safety measures, incidents might occur. This demands a better understanding of potential risks of this technology that could arise under critical conditions. Studies and models have already given a good understanding of cell behaviour under abusive treatment [1,2,3].

Based on this knowledge a flexible test rig was constructed in the laboratory, which allows the investigation of thermal response and gas generation products of various commercially available consumer cell formats and cell types in an open reactor. In the reactor, commercial lithium-ion pouch cells are heated up far beyond their specified temperature rating until the self-reaction within the cells is initialised. The so called thermal runaway is then observed in the experiments. The generated heat is measured and the produced gases are analysed. The on-set temperature that indicates the beginning of the thermal runaway reactions, the self-heating rate of the battery and the maximum temperature are determined. The decomposition reactions that occur at the cathode and anode side are derived from the gas chromatographic analyses of the ventilated gases.

The gas compositions of each gas sample together with the observed temperature profiles give a better understanding of the influence of cathode material, cell format and state of charge on the thermal runaway characteristic.

Acknowledgment

The authors would like to acknowledge the financial support of the "COMET K2 - Competence Centres for Excellent Technologies Programme" of the Austrian Federal Ministry for Transport, Innovation and Technology (BMVIT), the Austrian Federal Ministry of Economy, Family and Youth (BMWFJ), the Austrian Research Promotion Agency (FFG), the Province of Styria and the Styrian Business Promotion Agency (SFG).
The project was supported by the following industrial partners: Magna Steyr Battery Systems GmbH & Co OG and by BASF SE.

Reference

[1] Can-Yong Jhu, et al., Thermal runaway potential of $LiCoO_2$ and $Li(Ni_{1/3}Co_{1/3}Mn_{1/3})O_2$ batteries determined with adiabatic calorimetry methodology, Applied Energy 100 (2012), 127–131.
[2] Kong W., et al., Gas evolution behaviors for several cathode materials in lithium-ion batteries, Journal of Power Sources 142 (2005), 285–291.
[3] Abraham D.P., et al., Diagnostic examination of thermally abused high-power lithium-ion cells, Journal of Power Sources 161 (2006), 648–657.

(5c) Transient Studies of a Sodium-Sulfur Cell

Caprio, S., West Virginia University
Bhattacharyya, D., West Virginia University
Originally presented on: 11/3/2013 16:00:00 - 16:15:00

In the modern grid era, multiple fossil-fueled power generation facilities and renewable energy sources are expected to interact actively. Power produced by the renewable energy sources is intermittent and variable. Thus, a strong penetration of the renewable energy sources can create a challenging situation not only for grid stability and reliability but also from the perspective of the fossil-fueled power generation facilities that have to cycle their load. An electrical storage can be used to improve the grid stability and reliability by storing energy when the electricity price is low and discharging it when needed. The sodium-sulfur battery has high potential for electrical storage at the grid level due to its high energy density, low cost of the reactants, and high open-circuit voltage. However, the use of the sodium-sulfur batteries at the grid level requires high current density operation that can cause cell deterioration leading to lower sulfur utilization and lower energy efficiency. In addition, it can result in undesired thermal runaway leading to potentially hazardous situation. A rigorous, dynamic model of a sodium-sulfur battery can be used to study these phenomena, design the battery for optimal transient performance, and develop mitigation strategies.

With this motivation, a first-principles dynamic model of a sodium-sulfur cell has been developed. The state of discharge (SOD) of a sodium-sulfur cell significantly affects the heat generation rate, rates of electrochemical reactions, and internal resistance. To capture these phenomena correctly, a fully coupled thermal-electrochemical model is developed considering the operation of the cell in 0-85% state of discharge. The thermal model considers heat generation due to Ohmic loss, Peltier heat, and heat due to the entropy change. Species conservation equations are written in the sulfur electrode for the chemical and ionic species by considering the phase transition and change in the composition depending on the SOD. The electrochemical reactions are modeled by using Arrhenius-type rate equations with temperature-dependent terms and varying species concentration depending on the SOD. In addition, the potential distribution, and the cell resistance for this spatially distributed system has been modeled. The physicochemical properties are considered to be temperature-dependent.

The PDE-based model is solved in Aspen Custom Modeler by using method of lines. A number of disturbances as would be expected in a grid-connected system is simulated and the key transient profiles are studied. Our work shows that an appropriate thermal management strategy is necessary for high current-density operation, especially in the case of high penetration of the renewable energy into the grid.

(5d) Cycle Life Prediction for Ncm-Composite/Graphite Lithium Ion Battery Cells

Purewal, J., HRL Laboratories, LLC
Wang, J., HRL Laboratories, LLC
Graetz, J., HRL Laboratories, LLC
Tataria, H., GM Vehicle Engineering
Verbrugge, M. W., General Motors Research and Development Center
Originally presented on: 11/3/2013 16:15:00 - 16:30:00

Secondary lithium-ion batteries used in electric vehicles must last 10 years. Predicting the capacity and power loss of the batteries over this span requires accelerated testing methods such as storing and cycling the individual cells at elevated temperatures. Physics-based models may be able to improve capacity fade prediction using a smaller test matrix, greatly reducing the testing burden. In this paper, we study capacity fade in NCM-composite batteries (1.5 Ah) containing a composite Li(Ni0.33Mn0.33Co0.33)O2 + LiMn2O4 positive electrode and a graphite negative electrode. The batteries were cycled at a wide range of conditions, including temperature (10°C to 46°C), discharge current (0.5C to 6.5C) and depth of discharge (10% to 90%). The measured capacity loss is described by a single particle model which calculates lithium loss from two related mechanisms: (1) mechanical degradation at the graphite negative electrode particle surface caused by diffusion-induced stresses; (2) chemical degradation caused by lithium loss to the solid-electrolyte interphase (SEI). Fatigue crack growth is modeled based upon the cyclic diffusion-induced stresses that are generated by repeated lithium insertion and de-insertion in the graphite particle. This simplified model is able to reproduce the dominant capacity loss features observed for the NCM/graphite cells. In particular, the increased capacity fade observed for cells at low temperature (10°C) and high discharge rates (5C–6.5C) is linked to crack propagation on graphite anode particles and subsequent formation of a Li-consuming SEI layer on newly exposed crack surfaces. Fading of rate and pulse power capability during cycling is also modeled.

(5e) Convection Battery - Elimination of Dendrite Failure

Suppes, G., University of Missouri - Columbia
Dornbusch, D., University of Missouri
Hilton, R., University of Missouri - Columbia
Originally presented on: 11/3/2013 16:30:00 - 16:55:00

The convection battery pumps liquid electrolyte between the electrode and counter-electrode. It is different from flow batteries in that the same electrolyte is used by both electrodes and the active reagents are stored as solids. This paper is on lithium-ion chemistry performance in the battery.

The convection battery allows for unique degrees of freedom to remove heat and to eliminate dendrite modes of failure. This paper describes experimental techniques to first generate dendrite mode of failure on a consistent basis and then to alter the performance using operation specific to the convection battery to eliminate the failure.

Major advantages of operation without dendrite mode of failure include: reduced battery costs, higher and more-consistent battery voltages, and higher energy densities. A typical carbon:lithium ration for intercalation is 6:1. By example, a lithium-plated carbon electrode has about one sixth the carbon and about one third the price of an electrode based on intercalation.

(5f) Convection Battery - Performance Improvement and Characterization

Suppes, G., University of Missouri â€" Columbia
Hilton, R., University of Missouri â€" Columbia
Gordon, M., University of Missouri-Columbia
Dornbusch, D., University of Missouri
Originally presented on: 11/3/2013 16:55:00 - 17:10:00

The convection battery pumps liquid electrolyte between the electrode and counter-electrode. It is different from flow batteries in that the same electrolyte is used by both electrodes and the active reagents are stored as solids. This paper is on lithium-ion chemistry performance in the battery.

A design constraint of the convection battery is that the electrodes and separator must be permeable to flow at thicknesses of 0.5 to 6 mm. This constraint has led to reduced energy densities in the electrode materials for initial materials used for studies. This paper is on the use of composite electrodes to increase energy density. The use of composite electrodes more than doubled the energy densities.

The use of reference electrodes to facilitate improved understandings of overpotential losses. These studies revealed that the anode (discharge mode) is a major source of voltage loss for batteries that are assembled with lithium in the discharged state.

(5g) Key Parameters for Electrothermal Dynamics and Control of 15-Ah Prismatic Li-Ion Batteries

Kim, S. U., University of Michigan
Secondo, L., Trine University
Monroe, C. W., University of Michigan
Siegel, J., University of Michigan
Stefanopoulou, A., University of Michigan
Originally presented on: 11/3/2013 17:10:00 - 17:25:00

Four dimensionless key parameters that describe the electrical and thermal response of prismatic lithium-ion batteries are presented. Understanding of the thermal response of lithium-ion cells is essential because it would allow the batteries to be operated more efficiently in extreme environments. Two important issues about the thermal response are "cold start" and "thermal runaway." Thermal instability can be leveraged for the fast cold start of military vehicles and robotic platforms. Thermal instability can be suppressed to prevent thermal runaway, making operation in very warm environments safer.[1]

To introduce the minimum number of essential parameters of the battery responses, the simplest models are carefully chosen. At least two physical models are necessary for describing an electrochemical system: an electrochemical model and a local energy balance. For the electrochemical model, we select Newman and Tobias model, which provides an analytic solution with an assumption that the ion concentration gradient is negligible.[2] The analytic solution describes how the reaction current is distributed in the porous electrodes.

For the energy balance[3,4], three different heat sources are necessary to match our experimental data: (1) Joule heating of electrode matrix and electrolyte (simple *IR* heating), (2) another Joule heating generated by the interfacial resistance between electrode and electrolyte, and (3) the reaction heating from reversible entropy change. The Joule heating is always exothermic; however, the reversible heat can be endothermic or exothermic depending on the electrochemical reaction.

Even though the physics are minimalized for simplicity, the number of physicochemical parameters in our models was more than ten; however, the number could be decreased to four after rigorous dimensional analysis. The analytic solutions in 1D geometry allow demonstration of how these four essential properties affect the system responses (see Fig. 1). Therefore, not only do these parameters suggest what are important factors for the battery response, but they also help to control the battery in efficient ways, since this test shows that many parameters have the same effect in the electrical/thermal response.

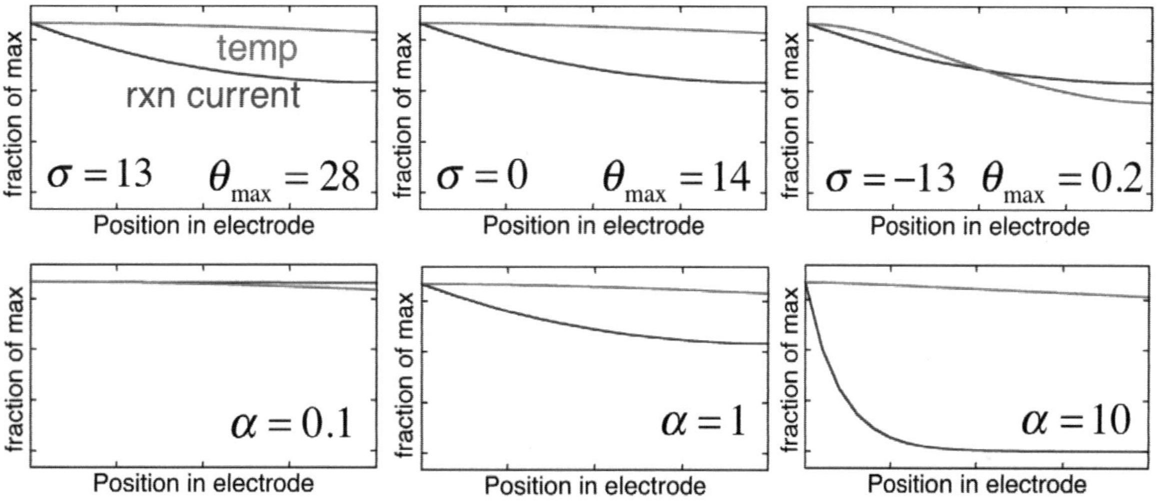

Fig. 1. The effect of reaction heat (1st row) and kinetic resistance (2nd row). Reaction heat (σ) changes the max temperature (θ_{max}) and its profile, and the large ratio of bulk and kinetic resistances (α) makes reaction distribution more concentrated on separator side.

[1] T. M. Bandhauer, S. Garimella, and T. F. Fuller, Journal of The Electrochemical Society **158**, R1 (2011).

[2] J. S. Newman and C. W. Tobias, **109**(12), 1183 (1962).

[3] D. Bernardi, E. Pawlikowski, and J. Newman, J. Electrochem. Soc. **132**, 5 (1985).

[4] J. Newman and K. E. Thomas-Alyea, *Electrochemical Systems,* 3rd ed. Wiley, 2004.

(5h) Inorganic Electrolyte Membranes for Redox Flow Batteries

Xu, Z., University of Cincinnati
Michos, I., University of Cincinnati
Yang, R., University of Cincinnati
Dong, J., University of Cincinnati
Originally presented on: 11/3/2013 17:25:00 - 17:40:00

Inorganic Electrolyte Membranes for Redox Flow Batteries

Zhi Xu, Ioannis Michos, Ruidong Yang,Junhang Dong

Department of Chemical Engineering, University of Cincinnati, Cincinnati, OH 45221, USA

Redox-flow batteries (RFBs) are emerging as useful means of electrical energy storage for distributed renewable power systems and large-scale grids because of a number of advantages such as low cost and small environmental footprint. The RFB operates by electrode reactions of dissolved metal ions constituting red-ox couples. The RFB operation is enabled by internal transfer of non-reactive ions, often protons, through an ion exchange membrane (IEM). Conventional IEMs are made of ionic polymers such as perfluorosulfonic polymers which have shortcomings of relatively low ion selectivity and long term material instability. This presentation reports our recent demonstration of purely inorganic nanoporous membranes as an alternative IEM in vanadium RFB (VRFB). We will discuss the operation principles of the new type of nanoporous inorganic IEM and the performance of the new IEM equipped RFB.

(5i) Slurry Electrode for An All-Iron Flow Battery for Low Cost Large-Scale Energy Storage

Petek, T. J., Case Western Reserve University

Wainright, J. S.,

Savinell, R. F., Case Western Reserve University

Originally presented on: 11/3/2013 17:40:00 - 17:55:00

Slurry Electrode for an All-Iron Flow Battery for Low Cost Large-Scale Energy Storage

Tyler J Petek, Jesse S Wainright, and Robert F Savinell

Case Western Reserve University

Department of Chemical Engineering

10900 Euclid Avenue, Cleveland, OH 44106 USA

The energy production market must change its output to match real time demand; making production very inefficient and restricting the grid's ability to rely upon intermittent renewable sources, i.e. solar and wind. Both of these issues can be alleviated by grid scale energy storage that could store energy when it is available and distribute energy as needed.

In typical batteries, all of the active materials are stored inside the battery. This directly couples the storage capacity and power so there is very little economic advantage to scaling, making standard batteries cost prohibitive for grid level energy storage. In redox flow batteries, the electrochemically active species are dissolved in an electrolyte and stored externally. During operation, the chemicals are pumped through the active cell where the reactions take place. Now, the energy storage capacity is dictated by the amount of chemicals stored and the power is dictated by the size of the cell. This allows redox flow batteries to have a significant advantage of scaling and drives down the cost significantly.

The all-iron flow battery is a potentially low cost energy storage device that utilizes domestically abundant materials (iron is the only active element), cheap aqueous electrolytes, and inexpensive separators. During charging, ferrous iron (Fe^{2+}) is oxidized at the positive electrode to ferric iron (Fe^{3+}) while ferrous iron is reduced to form iron metal at the negative electrode. These reactions are reversed during discharge. A schematic of the all-iron battery is shown in figure 1. This all-iron flow battery has a nominal cell potential of 1.2 Volts. [1] With conventional electrodes the energy storage capacity and power rating are coupled because the iron is plated in the cell. In order to decouple the energy and power ratings of the battery and regain the economic advantages of a flow battery without losing the advantages of the all-iron chemistry, a new electrode design needs to be developed. This project is investigating the use of slurry electrodes in order to decouple the all-iron battery.

A slurry electrode is made by flowing electrically conductive particles in an electrolyte containing the dissolved iron species. Iron is plated onto the particles in the negative side while charging. The particles can then carry the iron metal out of the cell to be stored in external reservoirs. This decouples the energy storage capacity and power rating, regaining the economic advantages of scaling inherent to true flow batteries. Some known applications of slurry electrodes include waste water treatment [2] and in zinc-air batteries. [3] A schematic of a slurry electrode is shown in figure 2.

Slurry electrodes have been tested utilizing a variety of particles including natural carbon micro-flakes, multiwall carbon nanotubes, carbon nanofibers, nickel micro-fibers, and copper micro-particles. All of the slurries tested thus far have exhibited shear thinning behavior as seen in figure 3. Based upon the viscosity characteristics, the flow of a slurry electrode was simulated in a 1000 cm^2 flow field in SolidWORKS. The parasitic pumping costs for a 5 kW flow battery is estimated to be less than 1 W at flow rates up to 23 times the stoichiometric flow rate based upon these simulations. The stoichiometric flow rate is the minimum flow rate required to sustain the applied current assuming 100% conversion of the reactant. For these calculations, it was assumed that the 5 kW battery was made up of 28 cells, each 1000 cm^2, and operated at 200 mA/cm^2.

Ideally, to control the current distribution and to ensure iron deposition occurs on the slurry particles and not the current collector, the electronic conductivity of the slurry electrodes needs to be greater than the ionic conductivity of the electrolyte. The conductivity of the base electrolyte at room temperature is about 120 mS/cm. When pressed into a dry pellet, the conductivity of most of the particles investigated is 10,000 ? 25,000 mS/cm. However, the electronic conductivity of the slurries while flowing has been found to be significantly lower; about 1 ? 40 mS/cm. Figure 4 shows the electronic conductivity of a variety of slurry electrodes. The three orders of magnitude decrease in conductivity is thought to be due to particle-to-particle resistances. The conductivity in figure 4 is shown against the weight percent of particles in the slurry normalized to the maximum weight percent. The maximum weight percent is when the particles are no longer wetted when mixed into the electrolyte and was found experimentally.

While the slurry electrodes investigated thus far have been shown to have appropriately low viscosities and pumping costs, their electronic conductivities are not high enough for flow battery electrode applications. Further work is being done in novel electrode designs and slurry compositions, as well as current collector designs, in an attempt to control the deposition of iron while keeping the overpotentials low.

Acknowledgements

This is an ARPA-E funded project, contract # DE-AR0000352. The authors would also like to acknowledge Nicholas Sinclair, Jason Pickering, Matthew Madler, Joseph Plazek, Mallory Miller, and Mirko Antloga.

References

[1] LW Hruska and RF Savinell, *J. Electrochem. Soc.*, **128**, 18-25 (1981).

[2] B Kastening, T Boinowitz, M Heins, *J. Appl. Electrochem.*, **27**, 147-152 (1997).

[3] PC Foller, *J. Appl. Electrochem.*, **16**, 527-543 (1986).

[4] CRC Handbook of Chemistry and Physics, 60th ed. CRC Press: Boca Raton, FL. 1980: D-237.

[5] YS Muzychka, J Edge, *J. Fluid. Eng.*, **130**, 111201-1 (2008)

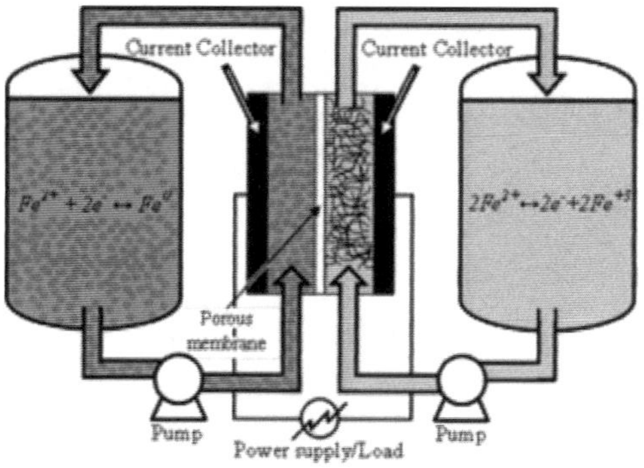

Figure 1: Schematic of the all-iron flow battery. This schematic shows a slurry electrode on the negative side and a felt electrode on the positive side.

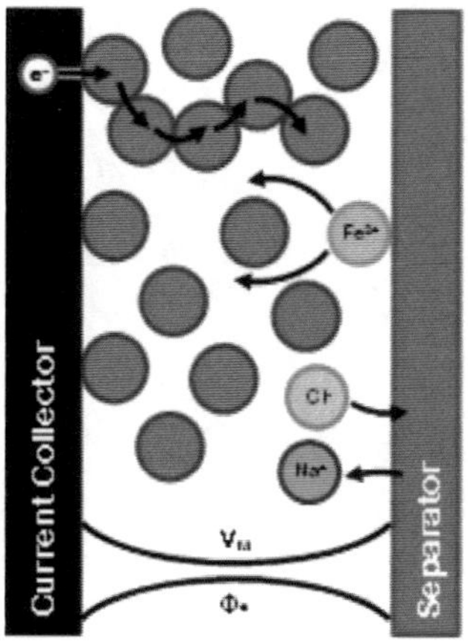

Figure 2: A schematic of a slurry electrode.

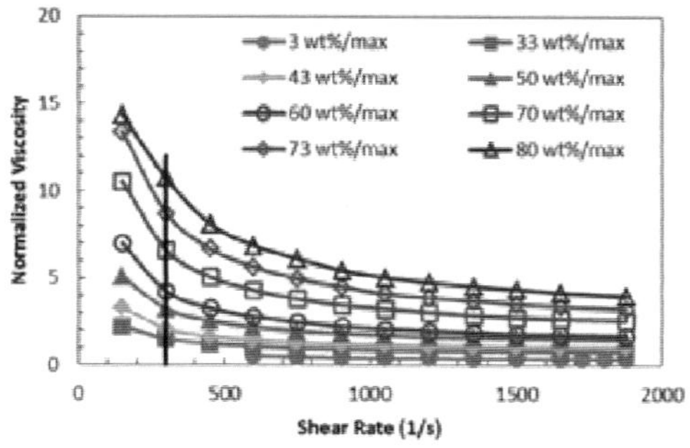

Figure 3: This figure shows the normalized viscosity plotted against shear rate for 20µm natural flake carbon particles at varying weight percent. The viscosity is normalized by the base electrolyte, 2M $FeCl_2$, which is about 4 cP at room temperature. [4] The vertical line at 300 s^{-1} indicates the shear rate associated with 5X stoichiometric flow rates in the 1000 cm^2 cells that were simulated. [5]

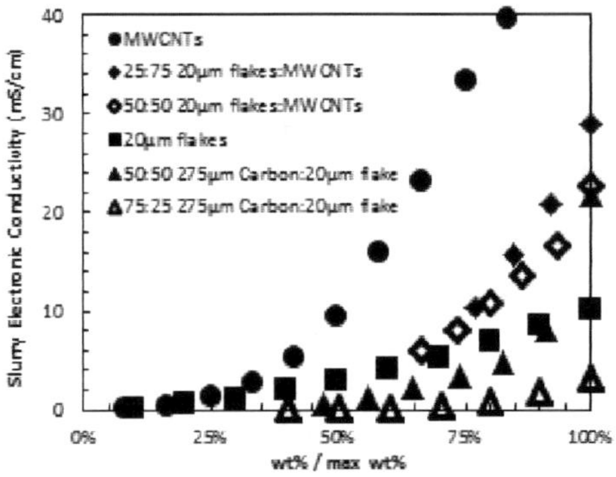

Figure 4: Electronic conductivity of varying carbon slurries. Slurry concentrations are normalized to the maximum concentrations for each type of particle. At the maximum weight percent, the MWCNTs exhibited an electronic conductivity of 85 mS/cm.

(5j) Acid and Alkaline Hydrogen-Bromine Fuel Cell Systems for Electrical Energy Storage

Nguyen, T. V., The University of Kansas
Yarlagadda, V. R., University of Kansas
Lin, G., TVN Systems, Inc
Weng, G., The University of Hong Kong
Li, V., The University of Hong Kong
Chan, K. Y., The University of Hong Kong
Originally presented on: 11/3/2013 17:55:00 - 18:10:00

Comparison of Acid and Alkaline Hydrogen-Bromine Fuel Cell Systems

Trung Van Nguyen[a*], Venkata Yarlagadda[a], Guangyu Lin[b], Guoming Weng[c], Vanessa Li[c], and Kwong-Yu Chan[c]

[a]Department of Chemical & Petroleum Engineering

The University of Kansas

Lawrence, KS, USA

[b]TVN Systems, Inc.

Lawrence, KS, USA

[c]Department of Chemistry

The University of Hong Kong

Hong Kong SAR, China

[*]Corresponding Author: cptvn@ku.edu

Abstract

The hydrogen bromine (H_2-Br_2) fuel cell system is an attractive system for electrical energy storage because of its high round-trip conversion efficiency, high power density capability, and anticipated low costs.

The hydrogen-bromine fuel cell system can be operated in the acid or alkaline modes. The charge and discharge electrode reactions in an acid H_2-Br_2 fuel cell system are as follows:

Bromine Electrode:

$$Br_{2\,(aq)} + 2e^- \leftrightarrow 2Br^-_{(aq)},\ \boldsymbol{E^o = +1.09\ V}$$

Hydrogen Electrode:

$$H_{2\,(g)} \leftrightarrow 2H^+_{(aq)} + 2e^-,\ \boldsymbol{E^o = +0.0\ V}$$

The H^+ ions migrate from the hydrogen side across a proton conducting membrane to the bromine side during discharge to combine with the Br^- ions to form hydrobromic acid.

Overall Reaction:

$$H_{2\,(g)} + Br_{2\,(aq)} \leftrightarrow 2HBr_{(aq)},\ \boldsymbol{E^o = +1.09\ V}$$

The charge and discharge electrode reactions in an alkaline H_2-Br_2 fuel cell system are as follows:

Bromine Electrode:

$$Br_{2\,(aq)} + 2e^- \leftrightarrow 2Br^-_{(aq)},\ \boldsymbol{E^o = +1.09\ V}$$

Hydrogen Electrode:

$$H_{2\,(g)} + 2OH^-_{(aq)} \leftrightarrow 2H_2O + 2e^-,\ \boldsymbol{E^o = +0.83\ V}$$

The cations (e.g., K^+), associated with the OH^- ions, migrate from the hydrogen electrode across a cation (K^+) conducting membrane to the bromine side and combine with the Br^- ions to form KBr as shown in the overall reaction.

Overall Reaction:

$$H_{2\,(g)} + Br_{2\,(aq)} + 2KOH_{(aq)} \leftrightarrow 2H_2O + 2KBr_{(aq)},\ \hat{A}\ \hat{A}\ \hat{A}\ \boldsymbol{E^o = +1.92\ V}$$

Based on the reactions shown above the alkaline system offers higher cell voltage, which is an advantage because of potentially higher power output. However, the hydrogen reactions in this system are two-phase reactions involving gaseous hydrogen and liquid-phase hydroxide ion reactants and will require more complex electrode structure and fuel cell design. The other advantages of this system include the fact that non-noble catalysts can be used for the hydrogen actions and lower corrosiveness.

This presentation will discuss the advantages and disadvantages of the alkaline and acid H_2-Br_2 fuel cell systems and compare the discharge and charge performance of both systems.

Acknowledgements

This work was funded in part by the National Science Foundation through grant number EFRI-1038234 and the Research Grants Council of Hong Kong through a General Research Fund (GRF HKU 700210P). A Visiting Professorship to Trung Nguyen was provided by the Initiative on Clean Energy and Environment (ICEE), University of Hong Kong.

(5k) Coupling Anion- and Cation-Exchange Membranes for Redox Flow Batteries With Mixed Ion Charges

Gu, S., University of Delaware
Gong, K.,
Yan, E.,
Yan, Y., University of Delaware
Originally presented on: 11/3/2013 18:10:00 - 18:25:00

Coupling Anion- and Cation-Exchange Membranes for Redox Flow Batteries with Mixed Ion Charges

Shuang Gu*, Ke Gong, Emily Z Yan and Yushan Yan*

Department of Chemical & Biomolecular Engineering, University of Delaware

5 Innovation Way, Newark, DE 19716, USA

**shgu@udel.edu and *yanys@udel.edu*

Electrochemical rechargeable batteries have shown remarkable improvements in power and energy density in recent decades and are finding increasingly broader applications in consumer electronics, electric vehicles, and electrical grid systems[1]. Their basic cell structure however has remained the same since their first discovery in 1859[2]: two redox pairs residing in the solid electrodes separated by a liquid electrolyte. In 1974, the first redox flow battery (RFB)[3] (i.e., chromium-iron RFB) was invented by introducing one ion-exchange membrane (IEM) into the cell. This IEM isolates negative electrode redox pair (negative pair) from positive electrode redox pair (positive pair) even when all of the four redox ions are freely dissolved in electrolytes; and at the same time it allows the passage of non-electroactive balancing ions. The liberation of redox pairs from the solid electrodes enables the transfer of energy storage function from electrodes to liquid electrolytes in external tanks, and this decoupling of energy storage from power delivery provides RFBs with unprecedented design flexibility and scalability. Significant efforts and progresses have been made in developing efficient and economical RFB systems in the last two decades mostly motivated by the needs of smart grid energy storage.

In the original chromium-iron RFB design, one anion-exchange membrane (AEM) was used to isolate the cation/cation negative pair (i.e.,Cr^{3+}/Cr^{2+}) from the cation/cation positive pair (Fe^{3+}/Fe^{2+}) to prevent self-discharge while allowing the conduction of the non-electroactive anions (i.e., Cl^{-} in HCl solution) to balance the electrode charges. Similarly, one cation-exchange membrane (CEM) can work for the cell with an anion/anion negative pair and an anion/anion positive pair. However, the single-IEM configuration is not able to effectively isolate redox pairs that have mixed ion charges, because neither one single AEM nor one single CEM can effectively prevent the mixing of electroactive redox ions, resulting in permanent coulombic efficiency loss. Beside low coulombic efficiency, the crossovered redox ions often interfere with the opposite electrode reaction, leading to voltage efficiency loss. As a result, the energy efficiency of those RFBs is significantly limited.

Here we present a universal design by using couple- or triple-IEMs in one cell that can accommodate any combination of ion charges for both the negative and the positive pairs. Composed of one AEM, one CEM, and a middle electrolyte in between, the couple-IEM cell provides a solution for redox pairs of all combinations of ion charges except two hybrid pairs (i.e., an anion-cation pair vs. an anion-cation pair). For two hybrid pairs, a triple-IEM cell with three membranes (CEM/AEM/CEM or AEM/CEM/AEM) and two middle electrolytes is needed. The couple- and triple-IEM designs bring unprecedented freedom in choosing redox pairs and electrolytes. Of particular importance, ultra-high cell voltage can be achieved in aqueous RFBs, e.g., 3.08 V (standard voltage hereinafter) of a zinc-cerium RFB [denoted as $Zn(OH)_4^{2-}/Zn||Ce_2O^{6+}/Ce^{3+}$, where $Zn(OH)_4^{2-}/Zn$ and Ce_2O^{6+}/Ce^{3+} are negative and positive pairs, respectively, and the double vertical lines are for the CEM and AEM couple]. The multi-IEM approach may also have implications in primary and other secondary battery designs.

References

[1] Z. G. Yang, J. L. Zhang, M. C. W. Kintner-Meyer, X. C. Lu, D. W. Choi, J. P. Lemmon, J. Liu, *Chem Rev* **2011**, *111*, 3577-3613; B. Dunn, H. Kamath, J. M. Tarascon, *Science* **2011**, *334*, 928-935.

[2] G. Plante, *Comptes Rendus de l'Academie des Sciences* **1860**, 640-642.

[3] L. H. Thaller, *The 9th Intersociety Energy Conversion Engineering Conference Proceedings* **1974**, 924-928.

Key words: Coupling, Ion exchange membranes, Mixed ion charges, Redox flow batteries, Energy storage.

(24a) Intrinsic Chemistry of Biomass Pyrolysis for Biofuels: Primary and Secondary Reactions of Cellulose Decomposition

Dauenhauer, P. J., University of Massachusetts Amherst
Originally presented on: 11/3/2013 15:30:00 - 15:50:00

The next generation of biomass pyrolysis models must be capable of predicting the molecular-level chemistry occurring during the fast pyrolysis process [1]. The first step in developing molecular prediction requires elucidating the primary and secondary reactions of biopolymers such as cellulose during pyrolysis. Experimental evidence has indicated that the reactive intermediates for carbohydrates, lignin, and lignocellulose at temperatures exceeding 500 °C exist within a short-lived intermediate liquid which subsequently evaporates to vapors and gases [2]. We utilize a novel 'thin-film pyrolysis' technique to elucidate the primary and secondary reaction pathways of cellulose to volatile products including anhydrosugars, pyrans, furans, and light oxygenates [3,4]. The distribution of products from carbohydrate polymers is shown to depend upon the chain-length, with longer chains producing more anhydrosugars such as levoglucosan. Finally, the secondary, condensed-phase pathways of levoglucosan are revealed using isotopically-labeled samples to depend upon the residence time within molten cellulose. Combination of these results provides the first insight into the chemical pathways which dictate the composition of bio-oil.

References

[1] M.S. Mettler, D.G. Vlachos, P.J. Dauenhauer, *Energy & Environmental Science* **2012**, 5, 7797-7809.

[2] A.R. Teixeira, K.G. Mooney, J.S. Kruger, C.L. Williams, W.J. Suszynski, L.D. Schmidt, D.P. Schmidt, P.J. Dauenhauer, *Energy & Environmental Science* **2011**, 4, 4306.

[3] M.S. Mettler, D.G. Vlachos, P.J. Dauenhauer, *Energy & Environmental Science* **2012**, 5, 7797-7809.

[4] M.S. Mettler, A.D. Paulsen, D.G. Vlachos, P.J. Dauenhauer, *Energy & Environmental Science* **2012**, 5, 7864-7868.

(24b) Cellulose-Hemicellulose, Cellulose-Lignin Interactions During Fast Pyrolysis

Zhang, J., Iowa State University
Choi, Y. S., Iowa State University
Brown, R. C., Iowa State University
Shanks, B. H., Iowa State University
Originally presented on: 11/3/2013 15:50:00 - 16:10:00

Cellulose, hemicellulose and lignin are the major biopolymers in biomass and during their fast pyrolysis, each component generates distinct species, which leads to a complex chemical composition of the pyrolytic bio-oil product. Given this complexity, complete chemical speciation with a high mass balance closure can provide insight into the underlying chemical mechanisms during fast pyrolysis. Previously, the pyrolysis behavior of the pure biopolymers was studied to explore their thermal deconstruction mechanisms. The pyrolysis behavior of cellulose with different degrees of crystallinity and polymerization was also examined and it has been found that similar product distributions were obtained for all of the celluloses under fast pyrolysis conditions as long as it was appropriately demineralized. In the current work, the interaction effects between cellulose-hemicellulose and cellulose-lignin under fast pyrolysis conditions were determined by comparing the pyrolysis products from their native mixtures, physical mixtures and superposition of individual components. A micropyrolyzer-GC-MS/FID in conjunction with a gas analyzer was used for online pyrolysis product analysis. It was found that no apparent interaction exists for either the physical mixture or the native cellulose-hemicellulose mixture. However, in the case of the native cellulose-lignin mixture, the interaction effects depended on the type of biomass pyrolyzed. Herbaceous biomass exhibited an apparent interaction effect, represented by a diminished yield of levoglucosan and enhanced yield of low molecular weight compounds and furans. However, these interaction effects were not found for woody biomass. It is speculated that these results were due to a different degree of covalent linkage between cellulose and lignin in these biomass samples.

(24c) Pyrolysis Chemistry of Lignin Model Compounds; β-O-4 and α-O-4 Linkages

Choi, Y. S., Iowa State University
Zhang, J., Iowa State University
Shanks, B. H., Iowa State University
Originally presented on: 11/3/2013 16:10:00 - 16:30:00

Pyrolysis chemistry of lignin model compounds; β-O-4 and α-O-4 linkages

Fast pyrolysis is considered a promising thermochemical technology for converting solid lignocellulosic biomass into a liquid bio-oil. Unlike the relatively homogeneous structure of cellulose, lignin is an amorphous polymer of irregularly bonded hydroxyl- and methoxy- substituted phenylpropane units. These phenylpropane units are linked primarily by β-O-4 and α-O-4 bonds which occupy 45-60% and 6-15% of the total linkages, respectively, in most lignins. While significant effort has been spent to understand the pyrolysis chemistry of cellulose with key success, the complex structure of lignin has been a bottleneck in understanding lignin pyrolysis. To overcome this issue, researchers have used lignin model compounds. However, most experimental setups obscure primary reactions in fast pyrolysis, since long vapor residence times can lead to secondary reaction. Herein, we focused on the primary reactions with a micropyrolyzer-GC-MS/FID system allowing short vapor residence time of 15-20 milliseconds and high heating rate of 1000 °C/s. Phenolic dimers connected by the most common linkages in lignin, β-O-4 and α-O-4, were selected and pyrolyzed at 500 °C. The simplest β-O-4 compound is 2-phenoxyphenylethanol (PPE: $PhCHOHCH_2OPh$). To characterize the effects of different substituents, $-CH_2OH$, -OH, $-OCH_3$ groups, on the pyrolysis chemistry, three other compounds with the same base structure of PPE were used. For exploring pyrolysis chemistry of α-O-4 compounds, benzylphenyl ether (BPE: $PhCH_2OPh$) and BPE with $-CH_3$ group at C_α were pyrolyzed. Performing mole balances from GC-FID and initial pathway screening using the Gibbs free energies of the reactants and products provides insight into the primary reaction pathways during pyrolysis thereby leading to improved understanding of lignin pyrolysis chemistry.

(24d) Aromatic Hydrocarbons Production From Catalysis of Douglas Fir Sawdust Pellets Pyrolysis Vapor Over Zeolite Catalyst

Wang, L., Washington State University
Lei, H., Washington State University
Bu, Q., Washington State University
Zhu, L., Washington State University
Originally presented on: 11/3/2013 16:30:00 - 16:50:00

Direct catalytic cracking of biomass pyrolysis vapor into aromatics derived from Douglas fir sawdust pellets was tested using an upfront microwave pyrolysis process coupled with a packed-bed catalysis process using a ZSM-5 Zeolite catalyst. A central composite experimental design (CCD) was used to optimize the bio-oil and syngas yields. The effects of temperature and inverse weight hourly space velocity $(WHSV)^{-1}$ on the bio-oil composition were determined. GC/MS analysis results showed that the bio-oil contained aromatic hydrocarbons and no oxygen containing compounds. Aromatic hydrocarbons were enriched and become the most abundant compounds accounting for about 15-92.6% in upgraded bio-oils depending on the catalytic pyrolysis conditions. The aromatic hydrocarbons were mainly composed of benzene, toluene, xylene, naphthalene, and there derivatives. When the effect of reaction temperature on bio-oil chemical compositions was analyzed with a fixed $(WHSV)^{-1}$ (0.048h) we found that the aromatic hydrocarbons were increased from 0.72% in raw bio-oil (no catalyst added) to around 92.6% when the catalysis temperature was 500°C clearly demonstrating that high temperature with the ZSM-5 Zeolite catalyst favored the production of aromatic hydrocarbons. At a lower temperature fixed at 375°C, the aromatic hydrocarbon content was increased from 0.72% to 78.1% with $(WHSV)^{-1}$ increase from 0 to 0.075.

(24e) Catalytic Fast Pyrolysis of Aspen Lignin Via Py-GC/MS

Zhang, M., South Dakota State University
Resende, F., University of Washington
Moutsoglou, A., South Dakota State University
Originally presented on: 11/3/2013 16:50:00 - 17:10:00

Aspen lignin extracted by an organosolv process was pyrolyzed in the presence of catalysts and analyzed using Py-GC/MS. The pyrolysis products detected are mostly aromatic hydrocarbons and phenolic compounds. The lignin and catalysts were either mixed or arranged in layers. Two different micro-porous zeolite catalysts, HZSM-5 and HY, were compared in the study. The effects of the catalysts on the production of aromatic hydrocarbons and phenolic compounds were assessed quantitatively. The HZSM-5 catalyst was found to produce from 2.5 to 40 times more aromatic hydrocarbons than the HY catalyst, by converting phenolic compounds into aromatic hydrocarbons. For the HY catalyst, the effect of the catalyst placement is dependent on the catalyst/lignin ratio.For the HZSM-5 catalyst, the effects of placement of catalyst were not as pronounced as were for HY. The data suggest that for any catalyst type and placement there exists an optimum catalyst-to-lignin ratio that maximizes the total production of aromatic hydrocarbons and phenolic compounds. For HZSM-5 in the mixture arrangement, a catalyst-to-lignin ratio of 3:1 was found to maximize both the aromatic hydrocarbons and the aggregate sum of aromatic hydrocarbons and phenolic compounds. At this ratio, 23 wt % yield of aromatic hydrocarbons and 28 wt % yield of the aggregate sum were obtained. Also at these conditions, the products detected have a total oxygen content of about 5 wt % and Higher Heating Value of 46 MJ/kg, roughly the same as gasoline and diesel. Toluene and p-xylene were the two most abundant hydrocarbons formed in the presence of catalysts. Up to 6.6 wt % and 6.3 wt % of the original lignin were converted into toluene and p-xylene, respectively. We estimate that, if all the lignin produced in pulp and paper mills was catalytically pyrolyzed, the amount of toluene produced would be more than 30 wt % of the current annual production of toluene from fossil sources worldwide.

(24f) Fast Catalytic Pyrolysis of Biomass and Relevant Model Compounds Studied in a Spouted Bed Reactor: Effect of Catalyst Type and Loading

Du, S., University of Connecticut

Fleming, N., University of Connecticut

Valla, J., University of Connecticut

Bollas, G. M., University of Connecticut

Originally presented on: 11/3/2013 17:10:00 - 17:30:00

Fast Catalytic Pyrolysis of Biomass and Relevant Model Compounds Studied in a Spouted Bed Reactor: Effect of Catalyst Type and Loading

Shoucheng Du, Nicholas Fleming, Julia Valla, George M. Bollas

Department of Chemical and Biomolecular Engineering, University of Connecticut, Storrs, CT

Thermal decomposition of lignocellulosic biomass produces low-quality biooils and high char yields. Introduction of catalyst in biomass fast pyrolysis can enable in-situ catalytic upgrading, thus providing an effective way to directly convert biomass into relatively high-quality product. As is shown in Figure 1, increasing catalyst amount can improve the bio-oil quality by increasing the carbon yields of non-oxygenated compounds via deoxygenation reactions. On the other hand, besides the catalyst loading, different types of the catalyst have also been studied in order to improve the bio-oil quality (aromatic yield).

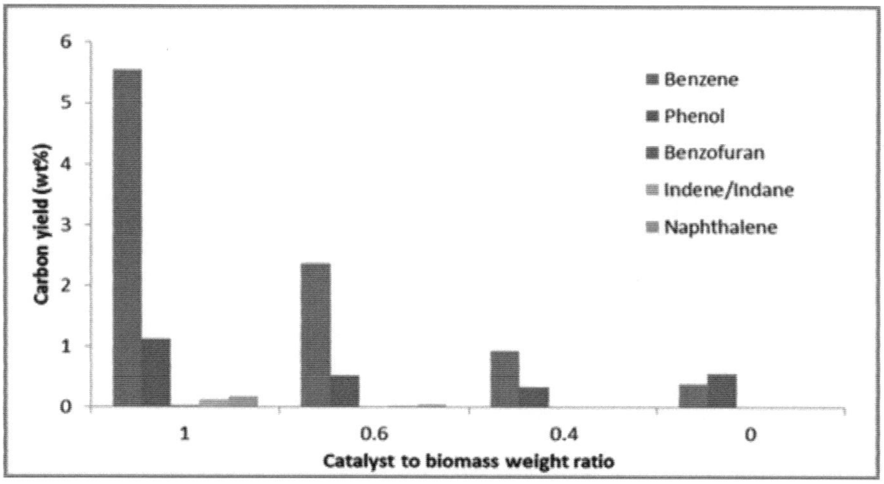

Figure 1 Carbon yields (calculated out of the total carbon in the biomass feedstock) of different compound groups as a function of catalyst to biomass weight ratio. Benzene (including Benzene; Benzene, 1-ethyl, 2-methyl; Benzene, ethyl; Toluene; Xylene; Styrene), Phenol (including Phenol; Phenol, 2-methyl; Phenol, 3-methyl; Phenol, 2,3-dimethyl), Benzofuran (including Benzofuran; Benzofuran, 7-methyl; benzofuran, 2-methyl), Indene/Indane (including Indene; Indane; 1-H-indene, methyl), Naphthalene (including Naphthalene; Naphthalene, 1,2-dihydro; Naphthalene, 2-methyl; Naphthalene, 1-methyl)

Carlson et al. [1,2] studied glucose fast catalytic pyrolysis in a micro-pyroprobe reactor and they showed that the coke oxygenates yield (mostly furan compounds) increase and the aromatic yield decreases as the catalyst-to-glucose ratio decreases (from 19 to 1.5). However, the aromatic selectivity is not a strong function of catalyst-to-glucose ratio. They also studied the effects of different catalyst types (ZSM-5, Silicalite, Beta-zeolite, Silica-alumina, and Y-zeolite) on the coke, oxygenate, aromatic yields and concluded that ZSM-5 was the best catalyst with respect to the aromatic yield due to its Brønsted acidity , pore size and structure. For the aromatic selectivity, they showed that Y-zeolite, beta-zeolite and SiO_2-Al_2O_3 selectively produce smaller aromatics, such as benzene and toluene, while ZSM-5 and silicalite, which have the same pore structure, are selective to larger aromatics, including naphthalene and indane. These observations show that the aromatic yield is affected by both the catalyst type and loading, while the aromatic selectivity is mostly affected by the catalyst type (pore structure and active sites). Moreover, Jae et al. [4] studied the effects of different structures (different pore size but all in the range of micropores) of zeolite catalysts on the aromatic yield in fast catalytic pyrolysis of glucose with a pyroprobe reactor. They concluded that ZSM-5 has optimal zeolite structure (pore size) for biomass conversion to aromatics. Furthermore, Foster et al. [5] performed fast catalytic pyrolysis of wood and glucose with modified ZSM-5 catalysts (with different silica to alumina ratios 23, 30, 50, 80 and mesopore structures) in a pyroprobe reactor. They found that the aromatic yield from glucose pyrolysis reached a maximum with zeolites of silica to alumina ratio of 30. They also showed that mesoporous ZSM-5 favored the production of larger alkylaromatics in both glucose and wood pyrolysis due to the relaxation of shape-selectivity, but did not affect the total aromatic yield significantly.

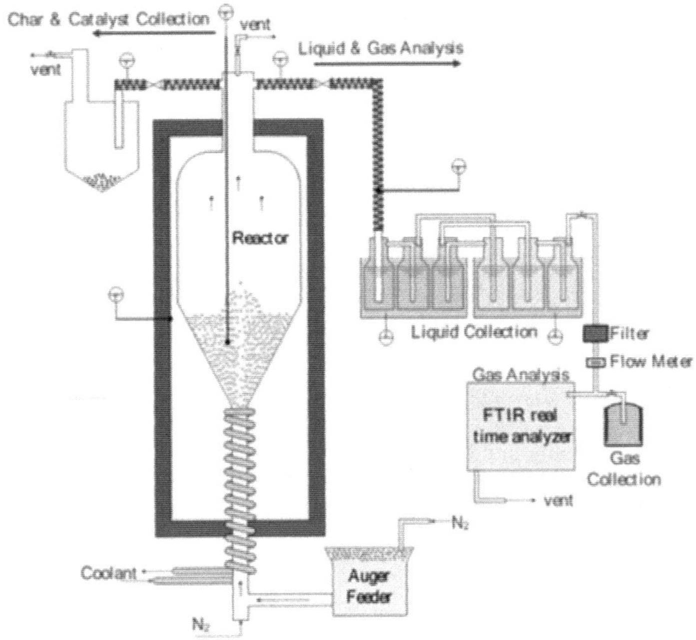

Figure 2 Schematic of the existing spouted bed reactor setup for biomass (catalytic) pyrolysis

However, most of the studies performed to study the catalyst effect on the quality of the resulting pyrolysis bio-oil are in micro-pyroprobe reactors. This arouses an interest for a similar investigation performed in a real fluidized bed reactor. Thus, in this study, investigations on the effect of catalyst types and loading on the bio-oil quality are performed in a spouted bed reactor, shown in Figure 2. Different biomass and relevant model compounds, such as pine sawdust, glucose and cellulose are used as the feedstocks. Comparison of the results regarding aromatic yield, aromatic selectivity between different feedstocks, catalyst types and catalyst loadings will be presented.

References:

[1] T.R. Carlson, G. A. Tompsett, W.C. Conner, G.W. Huber, Topics in Catalysis 52 (2009) 241.

[2] T.R. Carlson, J. Jae, Y.-C. Lin, G. A. Tompsett, G.W. Huber, Journal of Catalysis 270 (2010) 110.

[3] H. Zhang, R. Xiao, B. Jin, D. Shen, R. Chen, G. Xiao, Bioresource Technology 137C (2013) 82.

[4] J. Jae, G. A. Tompsett, A.J. Foster, K.D. Hammond, S.M. Auerbach, R.F. Lobo, G.W. Huber, Journal of Catalysis 279 (2011) 257.

[5] A.J. Foster, J. Jae, Y.-T. Cheng, G.W. Huber, R.F. Lobo, Applied Catalysis A: General 423-424 (2012) 154.

(24g) Fast Hydropyrolysis and Catalytic Hydrodeoxygenation of Cellulose in a Micro-Scale Batch Reactor

Choudhari, H. J.,
Delgass, W. N., Purdue University
Ribeiro, F. H., Purdue University
Agrawal, R., Purdue University
Originally presented on: 11/3/2013 17:30:00 - 17:50:00

Development of high energy density fuel for the transportation sector in a sustainable manner poses a significant challenge. The H_2Bioil[1] process proposed high pressure fast hydropyrolysis of biomass followed by catalytic vapor phase hydrodeoxygenation (HDO) and subsequent quenching to generate hydrocarbon-like fuel. Fast hydropyrolysis of biomass generates a complex mixture of oxygenated compounds which requires upgrading to reduce the oxygen content of the condensed bio-oil. Understanding the mechanism of fast pyrolysis of biomass and biomass model compounds is the key for tailoring the pyrolysis product distribution, thereby making it suitable for final downstream catalytic upgrading.

A micro-scale batch reactor, a modified commercial pyrolyzer, was used to study the pyrolysis of cellulose. A novel design allowed online sampling of pyrolysis vapor product composition under high pressure hydrogen (up to 350 psi) directly into the gas chromatograph and mass spectrometer (GC-MS) for analysis. An inline fixed-bed catalytic reactor was used to study the catalytic hydrodeoxygenation (HDO) of the pyrolysis products. Quantification of the pyrolysis products using the GC-MS accounted for 80-85% of the starting mass of the cellulose. Pyrolysis of cellulose was conducted in a temperature range 350°C-700°C. The major product observed was levoglucosan (~40%). The average molecular weight of the pyrolysis vapors decreased with an increase in the temperature of pyrolysis. Catalytic hydrodeoxygenation is required for upgrading pyrolysis vapors since fast hydropyrolysis alone did not lead to significant deoxygenation. Cellulose pyrolysis products were upgraded here over 2%Pt /ZrO_2 and 2%Ru/ZrO_2 catalyst at 300°C and 350 psi partial pressure of hydrogen. Complete hydrodeoxygenation of the pyrolysis products of biomass was observed under conditions of high catalyst to feed ratio (~20:1). The product distribution obtained over the two catalysts showed a lower degree of C-C bond hydrogenolysis over the 2%Pt/ZrO_2 catalyst versus 2%Ru/ZrO_2, with C4 and higher molecular weight hydrocarbons accounting for 40% of the carbon from the feed. The implications of these findings on the mechanism of cellulose pyrolysis and catalyst development for maximizing carbon conversion to fuel range molecules will be discussed.

1) R. Agrawal, N. R. Singh, Synergistic Routes to Liquid Fuel for a Petroleum Deprived Future, AIChE Journal,55,7,1898-1905, 2009.

(48a) Dual Fluidized Bed Reforming Pilot Test Results: Tar Conversion and the Fate of Sulfur

Mayer, K., Vienna University of Technology
Apanel, G., Rentech, Inc.
Proll, T., Vienna University of Technology
Hofbauer, H., Vienna University of Technology
Originally presented on: 11/4/2013 8:30:00 - 8:45:00

Synthesis routes for future generation of clean liquid fuels from gaseous or solid fuel feedstock (Coal to liquid, Gas to Liquid) are currently investigated. Reforming of hydrocarbons is widely used to provide synthesis gas for these processes. Even though there have been ideas of using dual fluidized bed technology for hydrocarbon reforming in the early 1940-ies, methane steam reforming and autothermal reforming are carried out mainly in fixed bed catalytic reactors. The dual fluidized bed system is an alternative to the conventional reforming systems. The reactor system consists of two reactors, interconnected with loop seals. The reforming catalyst circulates between these two reaction zones. One reactor is operated as the reformer and the second reactor as the regenerator, which is operated with air and additional fuel to provide the necessary heat for the reformer.

Using syngas derived from biomass gasification as carbon source for 2nd generation biofuel production two components present in the gas stream can cause problems. These two components are tar and sulphur containing species like H2S. Since CO and H2 are the valuable components in the synthesis of biofuels, the presence of tar causes a loss of carbon and hydrogen and lowers the efficiency of the process. Tar also causes problems during the gas cleaning process since it condenses. Using standard reforming technology (fixed bed and nickel based catalysts) to convert the tar into CO and H2, the H2S content in the gas causes catalyst deactivation and lowers the system performance or even prohibits stable operation.

In this study the above mentioned setup of a dual circulating fluidized bed reformer is used. A gas mixture of carbon dioxide, methane, carbon monoxide and hydrogen is used as model syngas in the similar composition to the gas derived from biomass steam gasification. To investigate the influence of tar and H2S on a nickel-based catalyst, 1-methylnaphthalein is used as model tar. The sulphur tolerance is tested with H2S. The tar loading in the reformer feed is 27 g/Nm^3 at a total gas feed of 15 Nm^3/h. This gas feed corresponds to a thermal fuel power of about 45 kW. The H2S concentration in the feed gas is 260 ppmv.

In the reformer off gas the H2S concentration and in the regenerator the SO2 concentration is measured. Since the formation of SO2 in the reformer is very unlikely due to the highly reducing conditions and the presence of H2S in the oxidizing regenerator atmosphere should not be possible, the Sulphur balance can be closed around the whole system. XRF analysis of the used catalyst is done to identify a possible Sulphur accumulation on the particles.

The results of this study show stable operation of the reforming catalyst in the dual fluidized bed reformer under severe conditions over several hours. Full conversion of tar is always achieved. The sulphur fed to the reformer is transported to the regenerator. Only 2% of the Sulphur passed through the reformer, the rest is released as SO2 in the regenerator. Methane conversion is nearly complete under all conditions and shows only a slight decrease at the presence of Sulphur.

(48b) Electricity Storage in Biofuels: Selective Electro-Catalytic Reduction of Levulinic Acid to Valeric Acid Or γ-Valerolactone

Xin, L., Michigan Technological University
Qi, J., Michigan Technological University
Chadderdon, D., Michigan Technological University
Qiu, Y., Michigan Technological University
Li, W., Michigan Technological University
Originally presented on: 11/4/2013 8:45:00 - 9:00:00

For a foreseeable future, the most desirable energy-carriers for transportations remain liquid hydrocarbon fuels, because of their ease, speed and safety of fueling, their efficient distribution network, and their very high energy densities. Direct storage of electrical energy in high energy-density liquid transportation fuels (electro-biofuels) using electro-catalytic (flow) cell devices shows unique characteristics as compared to other electricity storage devices and provides a promising and sustainable way to store renewable electricity. In this presentation, we reports an effective approach to electricity storage in biofuels via selective electro-catalytic reduction of levulinic acid (LA) to high energy-density valeric acid (VA) or γ-valerolactone (gVL) on a non-precious Pb electrode (99.9%) in a single polymer electrolyte membrane electro-catalytic (flow) cell device, which has demonstrated very high yield of VA (>90%), high Faradaic efficiency (>86%), promising electricity storage efficiency (70.8%) and low electricity consumption (1.5 kWh $L^{-1}{}_{VA}$). Applied potential and electrolyte pH were found to accurately control the reduction products: lower over-potentials favor the production of gVL, and higher over-potentials facilitate the formation of VA. The effect of molecular structure on electro-catalytic reduction of ketone and aldehyde groups of biomass-derived compounds and our current research on integrated electro-catalytic conversion of LA and formic acid to valeric acid over nanostructured metallic catalysts will also be presented. This ?electro-biofuel? technique will have potential to alleviate our long-time addiction to using non-renewable petroleum sources for production of transportation fuels.

(48c) Catalytic Deoxygenation of Octanoic Acid Over Supported Pd

Sun, K.,
Lamb, H. H., North Carolina State University
Originally presented on: 11/4/2013 9:00:00 - 9:15:00

Catalytic deoxygenation of octanoic acid (OA) to *n*-heptane was investigated over a series of supported Pd catalysts at 260°C and 1 atm using a fixed-bed micro-reactor with on-line gas chromatography. Under equivalent experimental conditions, a commercial 5 wt.% Pd/C catalyst was found to be more active and exhibit greater CO2 selectivity than Pd/SiO2, indicating a strong support effect for this reaction [1]. Particle size effects were investigated using Pd/SiO2 catalysts prepared by two different methods: a 2 wt.% Pd/SiO2 with an average Pd particle size of 7.5 nm and a broad size distribution prepared by incipient wetness, and a 0.6 wt.% Pd/SiO2 catalyst containing 1.5-nm particles with a narrow size distribution prepared by ion exchange. The catalysts were characterized by CO chemisorption, diffuse reflectance infrared Fourier transform spectroscopy (DRIFTS), and high-resolution transmission electron microscopy (HRTEM). The larger supported nanoparticles exhibited well-defined surface facets, as evidenced by HRTEM, and adsorbed CO preferentially at 2-fold bridging sites, as evidenced by DRIFTS. The smaller supported nanoparticles exhibited strong linear and bridging nCO bands. The more highly dispersed Pd/SiO2 catalyst deactivated rapidly and irreversibly in OA deoxygenation; whereas, the 2 wt.% Pd/SiO2 catalyst deactivated less rapidly with time-on-stream and could be regenerated by heating in H2 to remove OA residues. The catalyst deactivation mechanism is inferred to be self-poisoning by OA and CO. Addition of Au to Pd/SiO2 was found to improve catalyst stability without significantly reducing catalytic activity. A PdAu/SiO2 catalyst containing 0.47 wt.% Pd and 1.2 wt.% Au was prepared using a procedure previously reported to produce supported PdAu alloy nanoparticles [2]. The geometric and electronic effects of alloying Pd with Au reduce the CO adsorption energy and mitigate self poisoning by OA (and related species).

References:

[1] J. P. Ford, J. G. Immer, and H. H. Lamb, *Top. Catal.* **55**, 175-184 (2012).

[2] S. N. Reifsnyder, H. H. Lamb, *J. Phys. Chem. B* **103**, 321-329 (1999).

(48d) Aldol Condensation and Esterification of Oxygenates On Bifunctional Metal-TiO2 Catalysts

Wang, S.,
Goulas, K., University of California at Berkeley
Iglesia, E., University of California at Berkeley
Originally presented on: 11/4/2013 9:15:00 - 9:30:00

Aldol condensation of oxygenates via sequential C-C bond coupling and oxygen removal provides a promising way to convert biomass-derived feedstocks into useful fuels and chemicals [1,2]. In this presentation, we elucidate parallel condensation and esterification pathways of C_2-C_5 oxygenates on acid-base site pairs of TiO_2 with the aid of hydrogenation-dehydrogenation sites on metal catalysts (Cu, Pt), and the consequences of reactant structure and strength of acid-base site pairs in such cascade reactions.

Aldol condensation of alkanals and alkanones catalyzed by TiO_2 is equilibrium-limited and suffers rapid deactivation, because active sites on TiO_2 are readily blocked by unreactive residues derived from α,β-unsaturated carbonyl products. Hydrogenation of these unsaturated products to less reactive alkanals and alkanones on Cu or Pt co-catalysts drives the condensation forward and leads to stable condensation rates. In addition, alkanols can be directly feed as reactants and equilibrated with their carbonyl analogs on Cu surfaces (but not on Pt), which are then regarded as a reactant pool.

Kinetic studies showed that pool conversion rates (sum of condensation and esterification) are proportional to the amount of TiO_2 in the physical mixtures and to alkanal/alkanone pressures, whereas esterification/condensation rate ratios depend on alkanol/alkanal ratios in the equilibrated reactants. These data, taken together with a normal kinetic isotope effect (2.4, ethanol-D_6, 523 K) and the identity of the condensation products formed, indicate that enolates serve as common intermediates in condensation and esterification reactions, and their formations via α-H abstraction of alkanals/alkanones are kinetically relevant. Density functional theory (DFT) simulation on model TiO_2 oligomers supported that the enolate formations have higher activation free energy barriers than subsequent bimolecular C-C and C-O formation steps, in which calculated activation free energy barriers for C_2-C_5 oxygenates and the corresponding kinetic isotope effect were consistent with the experiments. The optimized structure of the enolate transition state and the independence of oxygenate reactivity with deprotonation energies for the α-carbon in reactants revealed the concerted nature of the α-C-H bond activation on Ti-O site pairs with O-atoms acting as the Lewis base and Ti centers as the Lewis acid, where the transition state is neither too early nor late.

The formed enolates then react via either nucleophilic attack with another alkanal/alkanone (to form α, β-unsaturated carbonyls) or with alkanols (to form unstable hemiacetals), followed by fast hydrogenations to saturated carbonyls or dehydrogenations to esters in the presence of a metal catalyst, respectively.

Esterification/condensation rate ratios increase with the size of reactant and do not show primary kinetic isotope effects (1.2, ethanol-D_6, 523 K), which also agreed well with the DFT simulations. Consistent with the kinetic branching pathways of the enolate intermediates, the esterification turnovers on the bifunctional metal-TiO_2 catalysts were decreased apparently by using Pt as admixture instead of Cu, because of the low reactivity of Pt in hydrogenation of C=O bonds, although the condensation rates were still maintained by the hydrogenation of C=C bonds on Pt surfaces.

Catalytic activity of TiO_2 in such oxygenate conversions is determined by the strength of acid and base sites on the surface. Acetone condensation rate on anatase TiO_2 was approximately ten times higher per surface Ti^{4+} cation than on rutile TiO_2. DFT calculations of model TiO_2 oligomers showed that basic sites on rutile were much stronger than anatase, which causes desorption of condensation products to be much harder from rutile than from anatase and is reflected by lower rates on rutile when compared to anatase. Analogous mechanisms to mediate oxygenate reactions by acid-base site pairs were also found on monoclinic ZrO_2, tetragonal ZrO_2 and CeO_2 surfaces. The relative reactivity of these oxide surfaces and the consistent theoretical predictions unveil the presence of site pairs of intermediate acid-base strength [3], which stabilize the transition states involving concerted interactions of the C and H atoms in C-H bonds with the Ti and O atoms of the Ti-O pairs on TiO_2 surface, respectively.

The authors acknowledge the financial support of BP through the XC^2 program and useful discussions with Drs. Glenn Sunley and John Shabaker (BP), Dr. Elif Gurbuz and Mr. Stanley Herrmann (UC-Berkeley), and Prof. Matthew Neurock (University of Virginia).

References

1. Sad, M., Neurock, M., Iglesia, E. *J. Am. Chem. Soc.* 133, 20384 (2011).
2. Rekoske, J. E., Barteau, M. A. *Ind. Eng. Chem. Res.* 50, 41 (2011).
3. Tanabe, K. in "Acidity and Basicity of Solids" (Fraissard, J. and Petrakis, L. Eds.) p. 353. Kluwer Academic Publishers, Dordrecht, 1994.

(48e) Catalytic Hydrodeoxygenation of Lignin Derived Compounds: Synergic Effect Between Pd and Fe

Hong, Y.,
Zhang, H., Wasington State University
Sun, J., Washington State University
Liu, C., Pacific Northwest National Lab
Karim, A., Pacific Northwest National Lab
Gu, M.,
Wang, C., Pacific Northwest National Laboratory
Hensley, A., Washington State University
Zhang, R., Washington State University
McEwen, J. S., Washington State University
Wang, Y., Pacific Northwest National Lab
Originally presented on: 11/4/2013 9:30:00 - 9:45:00

The shortage of energy feedstock is now becoming one of the major obstacles for the sustainable development of our hydrocarbon driven society. Among the various solutions proposed, conversion of renewable resources, especially biomass, to produce hydrocarbon fuels is considered to be the most promising approach. However, it remains challenging to remove oxygen from oxygen-rich biomass derived pyrolysis oil (O% = 10-30 wt%) for production of fuels with low oxygen content (O% < 2 wt%). Catalytic hydrodeoxygenation (HDO) is an effective and atom-economical approach for the oxygen removal of bio-oil. Among the various biomass feedstocks, lignin is the most featured one because of its vast availability, non-edible nature and relatively lower oxygen content. Therefore, the catalytic HDO of lignin derived phenolic compounds (guaiacol, cresol, etc.) to produce aromatics (benzene, toluene, xylene, etc.) became index reactions in the HDO catalysis research. Traditional sulfided Co-Mo and Ni-Mo catalysts employed in hydrodesulfurization and hydrodenitrogenation in petroleum industry require high H_2 pressure and usually deactivate along with sulfur stripping. Recently, supported base metal catalysts, such like Fe/SiO_2 and Fe/C, were found to be very selective for aromatics production in HDO of phenolics, albeit the low activity. On the other hand, noble metals (e.g. Pt and Pd) were demonstrated to be good additives to enhance the hydrogenation activity of metallic catalyst due to their superior activity in activation of H_2. Our previous report shows that the Pd additive in Fe/C catalyst notably promoted its activity in HDO of guaiacol to produce benzene, suggesting a synergic effect between Pd and Fe.

In this work, a series of Pd-Fe catalysts (Pd% = 0.1-5 wt%) were synthesized by incipient wetness impregnation of Pd precursor ($Pd(NH_3)_4(NO_3)_2$) onto as-prepared Fe_2O_3 nanoparticles to exclude the potential effect of carbon support and ensure direct interactions between Pd and Fe. The reduced samples featured with Pd entities sitting on the top of reduce Fe particles captured by unreduced iron oxides. TPR results showed that the addition of Pd, even at trace amount (0.1 wt%), will significantly change the redox property of Fe_2O_3, by facilitating the reduction of surface iron oxide to iron metal. Activity test for gas phase HDO of *m*-cresol showed that the addition of Pd remarkably promoted the activity of Fe catalyst, while the product distributions resembled with that of Fe catalyst, showing high selectivity towards production of toluene. Kinetic data obtained at low conversion level (< 10%) also suggested the synergic effect between Pd and Fe, showing the intrinsic activity of both metals in the bimetallic catalysts almost doubled, compared with their monometallic counterparts. In addition, the fully reduced (reduced at 450 °C, confirmed with XRD) catalysts suffered a deactivation with formation of iron carbide, while the partially reduced (reduced at 300 °C, confirmed with XRD) catalysts show distinct induction period, their activity increased along with time-on-stream, indicating the *in-situ* reduction of remaining iron oxide.

The interaction between Pd and Fe was further confirmed in XPS, STEM, XAS and theoretical calculations. A possible mechanism, including Pd facilitated H_2 dissociation and Fe dominated *m*-cresol activation, is proposed to be responsible for the higher activity and lower ring saturation selectivity in Pd-on-Fe catalysts. The Synergic catalysis derived from Pd-Fe interaction was found in this work, providing a promising strategy for designation of highly active and selective HDO catalysts, and an ideal model system for bimetallic catalysis research.

(48f) Generation of Novel Diesel Additives Via Furan Condensation and Partial Hydrogenation of Biomass-Derived Molecules

Sacia, E. R., University of California, Berkeley

Madhesan, B., University of California, Berkeley

Bell, A. T., University of California, Berkeley

Originally presented on: 11/4/2013 9:45:00 - 10:00:00

Generation of Novel Diesel Additives via Furan Condensation and Partial Hydrogenation of Biomass-Derived Molecules

Eric R. Sacia, Madhesan Balakrishnan, and Alexis T. Bell

Department of Chemical and Biomolecular Engineering,

University of California-Berkeley, Berkeley, California 94720

Production of alternative fuels from biomass sources remains one of the greatest contemporary challenges for catalysis. Due to its many advantages for growth and production, lignocellulosic biomass has been identified as a beneficial source of chemical energy for biofuels.[1] However, in order to synthesize fuels of the desired size and properties for blending into the diesel fuel supply, catalytic C-C bond formation is required. Towards this aim, 5-hydroxymethylfurfural (HMF) and furfural, dehydration products from the glucose and xylose sugars generated by cellulose and hemicellulose hydrolysis, have been broadly identified as useful platform molecules.[2-3]

While several pathways exist to produce diesel-range fuel additives through C-C bond formation, including aldol condensation and acid-catalyzed alkene oligomerization, the condensation of 2-methylfuran with biomass-derived aldehydes provides an innovative, high-yield pathway to fuel precursors.[4-5] The products of furan condensation have thus far been subjected to extensive hydrodeoxygenation to form branched 6-alkyl undecanes. In order to improve the feasibility of these reactions, we have elucidated methods for solvent-free acid-catalyzed furanylation at low temperatures followed by partial hydrogenation. We have demonstrated that this scheme generates products that display attractive fuel characteristics while requiring 38% less H_2 input than full hydrogenation to alkanes due to the beneficial effect of ethers on cetane number.[6]

The overall scheme for furfural condensation with 2-methylfuran is provided in Figure 1. By performing the furan condensation with furfural and a 10% stoichiometric excess of 2-methylfuran, 93% yields to the targeted product, **1**, could be achieved within 2 hours at 65°C with homogeneous catalysts under solvent-free conditions. In order to better facilitate acid catalyst recovery, heterogeneous Brønsted and Lewis acids were also screened. Brønsted acidic ionic liquids (IL) of tunable hydrophilicity were synthesized in order to create a liquid/liquid bi-phasic catalyst system in which the hydrophobic reactants and products would separate from the hydrophilic catalyst phase. Through modification of the hydrophilicity of the IL's anion, the acidic IL phase could catalyze the furan condensation with activity equal to that of homogeneous sulfonic acids.

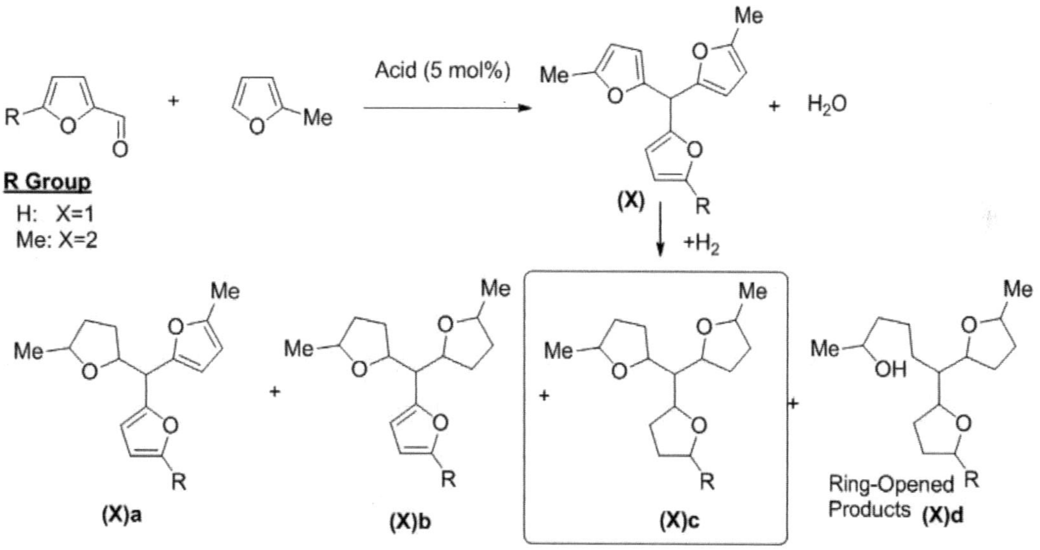

Figure 1. Scheme for furan condensation followed by partial hydrogenation to targeted products

Heterogeneous catalysts, such as sulfated zirconia and Amberlyst-15, were also tested for activity in furan condensation reactions but were demonstrated to display lower activity than the tested homogeneous counterparts. In order to more efficiently catalyze the furan condensation, several alkyl ?linkers? were used to tether sulfonic acid groups to high surface area silica. The resulting catalysts achieved complete conversion within 30 minutes while maintaining high selectivity and recyclability. Furan condensation reactions were also tested under these optimized conditions with 5-methylfurfural (97% yield), HMF (78% yield), and 5-methylfurfuryl alcohol (89% yield).

It was determined that condensation product **1**had a cetane number of 25, significantly below the target of 40, and also had too high of a viscosity to readily blend into diesel by itself. Therefore, we sought to exploit the high cetane number of ethers compared to aromatics to generate previously unreported diesel fuel candidates, **1c** and **2c**. Palladium supported on traditional supports proved unselective for the partial hydrogenation of **1** and **2** to their tetrahydrofuranyl analogs, often producing ring-opened products before complete ring saturation. However, we found that Pd supported on a surface-tethered ionic liquid on SiO_2 gave high selectivity to **1c** and **2c** (92 and 85% respectively), even in solvent-free conditions. Further, the products of partial hydrogenation were tested and found to have significantly improved cetane numbers and viscosities compared to the starting materials. The unique selectivity of these catalysts and their mechanism of production of high-value additives will be further discussed. The role of surface hydrophobicity and dispersion on the selectivity of this hydrogenation will also be explored.

References

1. *Somerville, C.; Youngs, H.; Taylor, C.; Davis, S. C.; Long, S. P., Feedstocks for Lignocellulosic Biofuels. Science 2010, 329 (5993), 790-792.*

2. *Huber, G. W.; Iborra, S.; Corma, A., Synthesis of transportation fuels from biomass: Chemistry, catalysts, and engineering. Chem Rev 2006, 106 (9), 4044-4098.*

3. *Alonso, D. M.; Bond, J. Q.; Dumesic, J. A., Catalytic conversion of biomass to biofuels. Green Chemistry 2010, 12 (9), 1493-1513.*

4. *Corma, A.; de la Torre, O.; Renz, M.; Villandier, N., Production of High-Quality Diesel from Biomass Waste Products. Angewandte Chemie International Edition 2011, 50 (10), 2375-2378.*

5. *Corma, A.; de la Torre, O.; Renz, M., High-Quality Diesel from Hexose- and Pentose-Derived Biomass Platform Molecules. ChemSusChem 2011, 4 (11), 1574-1577.*

6. *Murphy, M. J.; Taylor, J. E.; McCormick, R. L., Compendium of Experimental Cetane Number Data. NREL, Golden, Colo., 2004. ADDIN EN.REFLIST*

(48g) Catalyst Deactivation in Hydrothermal Decarboxylation

Yeh, T., University of Michigan
Savage, P. E., University of Michigan
Linic, S., University of Michigan
Originally presented on: 11/4/2013 10:00:00 - 10:15:00

Algae are a type of lipid rich aquatic biomass that has recently shown promise for the production of biofuels. It's been shown that the hydrothermal liquefaction of microalgae results in a bio-crude oil that can be potentially further upgraded to liquid hydrocarbon fuel. The major problem with the bio-crude is the high oxygen content, and much of the oxygen is contained within the fatty acids. The fatty acids, which result from the lipid content, are ideal candidates for upgrading because the removal of oxygen results in a long chain hydrocarbon. Pt based catalysts have been shown to be effective decarboxylation catalysts in hydrothermal environments for upgrading saturated fatty acids such as palmitic (C16:0) and stearic (C18:0) acids, but the longevity of the catalysts have not been well studied. This work focuses on the removal of oxygen from butyric acid using $PtSn_x$ catalysts at 350 C and 3000 psig. In both cases, Pt and PtSn catalysts were shown to deactivate over a 48 hour period. The addition of Sn decreased the rate of deactivation of the catalyst and also had an effect of the selectivity for the decarboxylation product of propane from butyric acid. The deactivation was shown to be due to a molecular poison as opposed to coking of the catalyst.

(48h) Selectively Activating the C=O Bond of Furfural Using Metal Carbide and Bimetallic Surfaces

Xiong, K.,
Chen, J. G., Columbia University
Originally presented on: 11/4/2013 10:15:00 - 10:30:00

Selectively activating the C=O bond of furfural is crucial for converting this important biomass-derived molecule to value-added fuels and chemicals such as 2-methylfuran and furfuryl alcohol. In this work, molybdenum carbide (Mo_2C) and Pt-based bimetallic surfaces were investigated in terms of their ability to selectively activate the C=O bond of furfural using surface science techniques.

The Mo_2C surface was found to selectively cleave the C=O bond of furfural to produce 2-methylfuran by the temperature programmed desorption (TPD) experiments. Results from surface vibrational spectroscopy showed strong interaction between the carbonyl group of furfural and the Mo_2C surface. These results indicate the potential application of Mo_2C as a selective catalyst for the hydrodeoxygenation of biomass-derived oxygenates.

On the other hand, Ni/Pt(111) bimetallic surfaces were studied in terms of their ability to hydrogenate furfural to make furfuryl alcohol. However, the TPD experiments showed that these surfaces mainly decomposed furfural to produce CO and H_2. Surface vibrational spectroscopy revealed that although furfural preferentially adsorbed onto Ni/Pt(111) surfaces through the carbonyl group, the binding energies of furfural on these surfaces were too strong for hydrogenation to occur. Instead furfural undergoes the reforming pathway to produce CO, $H_2$and atomic carbon.

The combined results on carbide and bimetallic surfaces demonstrate the possibility of selecting different catalytic surfaces for different routes of biomass conversion.

(48i) Modeling of Catalytic Cracking and Separation Units for Upgrading Heavy Oil and Residues

Riazi, M. R., Kuwait University

Originally presented on: 11/4/2013 10:30:00 - 11:00:00

"Modeling of Catalytic Cracking and Separation Units for Upgrading Heavy Oil and Residues" (Invited)

M R Riazi, Ph.D., AIChE Fellow

Kuwait University
Department of Chemical Engineering
College of Engineering and Petroleum
P. O. Box 1084 , Surra, Kuwait 45711
Tel: (+965) 24817662 / 24985772 Fax: (+965) 24811772 / 24839498

While world demand on lighter fuels such as naphthas and middle distillates of crude oils is on the rise, quality of world crude is decreasing and demand on heavy and unconventional oils is on the rise. Most refineries are moving to the direction of conversion of all residues to lighter oils through various separation and cracking processes. Safe and economical design and operation of these units mainly depend on accurate knowledge of feedstocks and products composition and properties.

In this presentation we demonstrate how a probability density function derived from a distribution model developed specifically for heavy oils can be used to predict the amount of conversions and product as well as products composition and properties. Proposed models are evaluated with data obtained from distillation of a crude oil sample and cracking of a heavy gas oil sample using data measured under various operating conditions. The proposed models can be used to optimize operating conditions in various separation and cracking units. The models may also be useful for online analysis of products produced from such plants.

(65a) Templated Ordered Mesoporou Materials for Supercapacitors and Lithium Ion Batteries

Lee, J., Pohang University of Science and Engineering

Originally presented on: 11/4/2013 8:30:00 - 9:00:00

For high performance supercapacitor and Li-ion battery electrodes, various mesoporous metal oxide materials have been synthesized by employing lab-synthesized block copolymers as a structure-directing agent or ordered mesoporous silica materials as hard templates. An ordered mesoporous WO_{3-X} material with high electrical conductivity was employed for use as a supercapacitor electrode in 2.0 M H_2SO_4. The electrode made of mesoporous WO_{3-X} exhibited a high rate capability and an excellent capacitance (366 Fcm^{-2}, 639 Fcm^{-3}), which were attributed to the large ordered mesopores, high electrical conductivity, and high material density. Fe_3O_4 nanocrystals confined in mesocellular carbon foam (MSU-F-C) are synthesized by 'Host-Guest' approach and tested as anode for lithium-ion batteries (LIBs). 13-27 nm magnetite Fe3O4 nanocrystals are formed inside ~30 nm pores of mesocellular carbon matrix. Fe_3O_4/MSU-F-C (45 wt% Fe_3O_4) exhibited 1007 mA h g^{-1} of first charge (Li de-insertion) capacity at 0.1 A g^{-1} and 111% of capacity retention at 150th cycle, and retains 37% capacity at 7 A g^{-1} (~7 C rate). For the first time, we synthesized mesostructured LTO/carbon composite (Lithium titanate/carbon) with large (> 15 nm) and uniform pores for high performance LIB anodes via a block copolymer assembly. Sn-mesostructured carbon nanocomposites have been synthesized by simple self-assembly of block copolymers with Tin and carbon precursors.

(65b) Reduced Graphene Oxide Wrapped FeS Nanocomposite for Lithium-Ion Battery Anode With Improved Performance

Fei, L., New Mexico State University
Luo, H., New Mexico State University
Originally presented on: 11/4/2013 9:00:00 - 9:15:00

The fast growing business of portable electronics (e.g. computers, mobile phones, cameras) has stimulated the rapid development of lithium-ion batteries (LIBs) due to their high energy density, flexible and lightweight design, and longer lifespan than conventional battery technologies. Graphite is the most popular commercial anode. However, its low theoretical capacity (372 mAh/g) renders it hard to satisfy the increasing market demands. Recently, metal sulfides with high theoretical capacity are widely studied as another promising class of anode materials with excellent electrochemical performance.

Here, we report FeS based nanocomposite as anode for lithium-ion batteries. FeS nanoparticles wrapped in reduced graphene oxide (RGO) are produced via a facile direct-precipitation approach. The resulting nanocomposite FeS@RGO structure has better lithium ion storage properties exceeding those of FeS prepared without RGO sheets. The nanocomposite delivered an initial discharge capacity of 1357 mAh/g and a charge capacity of 1116 mAh/g, giving the first cycle coulombic efficiency of 82 %. The composite shows a very stable cyclic performance with its capacity remaining as high as 978 mAh/g even after 40 cycles at a current density of 100 mA/g. This value is almost three times the capacity of commercial graphite anode, and is comparable to those reported in literature. At current density of 300 mA/g, the composite delivered discharge capacity of 927, 690, and 618 mAh/g in the first, 10th, and 30th discharge cycles, respectively; while FeS suffered rapid capacity decline, only delivering 885, 518, and 382 mAh/g in the first, 10th, and 30th discharge cycles. Electrochemical impedance demonstrate that FeS@RGO has a lower electrochemical resistance than that of FeS. The enhanced electrochemical performance is attributed to the robust sheet-wrapped structure and the synergetic effects between FeS and RGO sheets, such as increased conductivity, shortened lithium ion diffusion path, and the effective prevention of polysulfide dissolution.

(65c) Alkanethiol-Passivated Germanium Nanowires As High-Performance Anode Materials for Lithium-Ion Batteries

Tuan, H. Y., National Tsing Hua University
Yuan, F. W., National Tsing Hua University
Originally presented on: 11/4/2013 9:15:00 - 9:30:00

Rechargeable lithium-ion batteries have been widely used in portable electronic devices such as cell phones, laptop computers, and digital cameras, and have been considered as an electrochemical energy storage system for powering hybrid electric vehicles (HEV) and zero-emission vehicles (ZEV). It is necessary to develop lithium-ion batteries having a higher energy density and a higher power density to meet the increasing energy demand. Germanium (Ge) has emerged as a promising anode materials due to its high theoretical capacity of 1384 mAh/g - around four times greater than commercial graphite anode (372 mAh/g). However, the practical usage of Ge as an anode material is hindered by dramatic volume changes (~300%) during the insertion/extraction process of lithium ions, causing crack and pulverization, and loss of electrode contact.

Ge nanowires can effectively accommodate the volume changes, tolerate relaxed mechanical strain, provide channels for efficient electron transport, and alleviate pulverization. Although direct incorporation of Ge nanowires into slurry coating process shows improved electrochemical performance relative to microparticles, the anode performance decayed rapidly within a few tens of cycles. In this talk, we demonstrate that dodecanethiol monolayer passivation on Ge nanowire surface can significantly enhance their anode performance in lithium-ion batteries. The dodecanethiol-passivated Ge nanowires exhibit an excellent electrochemical performance with a reversible specific capacity of 1130 mAh/g at 0.1 C rate over 100 cycles, which is above 80% of the Ge theoretical capacity. The functionalized Ge nanowires show high-rate capability having charge and discharge capacities of ~555 mAh/g at rates as high as 11 C. An aluminum pouch type lithium cell using a $LiFePO_4$ cathode was also assembled to provide larger current (~30 mA) for uses on light-emitting-diodes (LEDs) and audio devices. Organic monolayer-coated Ge nanowires represent promising Ge-C anodes with very low carbon content (~2-3 wt %) for high capacity, high-rate lithium-ion batteries and are readily compatible with commercial slurry-coating process for cell fabrication.

(65d) A Facile Microwave-Assisted Route to Co(OH)2 and Co3O4 Nanosheet for Li-Ion Battery

Chen, G., New Mexico State University
Luo, H., New Mexico State University
Originally presented on: 11/4/2013 9:30:00 - 9:45:00

Cobalt hydroxide ($Co(OH)_2$), as one of the most important functional inorganic materials, can be used in the field of catalysis and electrochemical electrodes. The previous report revealed that $Co(OH)_2$ is a promising alternative conversion-based anode materials for lithium-ion batteries (LIBs) due to its high capacity. Unfortunately, like other conversion-based anode materials, the implementation in LIBs is greatly hampered by its poor cycling performance. One of the most promising strategies to tackle this obstacle is to make $Co(OH)_2$ into sheet-like structure to improve the structural stability in the electrochemical reactions. Compared with the bulk $Co(OH)_2$, $Co(OH)_2$nanosheet displays obvious advantages due to large surface area, highly oriented growth direction, tunable interlayer space for the de/insertion of lithium ions.

In this work, a facile microwave-assisted synthetic route has been successfully developed for preparing hexagonal $Co(OH)_2$ nanosheets with average width of 2 μm and thickness of 100 nm. $Co(OH)_2$ can further convert to porous Co_3O_4 nanosheets *via* thermal decomposition. Their phases, structures and morphologies were investigated by X-ray diffraction (XRD), atomic force microscopy (AFM), and transmission electron microscopy (TEM). Both $Co(OH)_2$ and Co_3O_4 nanosheets can serve as potential candidates for anodes of Li-ion battery. The electrochemical study revealed that $Co(OH)_2$ and Co_3O_4 nanosheets delivered a reversible capacity of 600 and 700 mAh g^{-1} after 40 cycles, respectively. Cyclic voltammetry (CV) curves also confirmed the relative stability of the as-synthesized electrode materials. This effective microwave-assisted route may be a promising approach for preparing other transition metal hydroxide/oxides for energy applications.

(65e) Li Adsorption and Intercalation in Graphene, Graphite and Carbon Onions By Reaxff Reactive Force Field

Raju, M., The Pennsylvania State University
Ganesh, P., Oak Ridge National Laboratory
Kent, P. R. C., Oak Ridge National Laboratory
van Duin, A. C. T., Pennsylvania State University
Originally presented on: 11/4/2013 9:45:00 - 10:00:00

A ReaxFF reactive force field has been developed to describe Li/C systems. The ReaxFF force-field parameters for Li-C systems were optimized against a density-functional theory (DFT) (with van der Waals corrections) based training set, containing a collection of data (energies, geometries, charges etc.) related to adsorption and diffusion of Li on graphene and in graphite. ReaxFF reproduces accurately the adsorption energies for Li on hollow, bridge and top sites on pristine graphene, formation energy of stage I compounds in graphite and intralayer Li migration barriers in graphite at separations of 3.35 Å and 3.7 Å. To validate this force field we have compared it against DFT calculations for Li diffusion in the direction perpendicular to the graphene sheet for graphene with single and double-vacancy. We performed grand canonical Monte Carlo simulations to obtain ground states for Li intercalation in graphite and find that ReaxFF describes the in-plane Li ordering and interlayer separations for stage I and II compounds in good agreement with previous DFT studies and experiment. The voltage profile for the Li-graphite system as a function of Li concentration obtained by ReaxFF matches experiment and DFT data. We study Li intercalation in presence of vacancy defects in graphite and find that the presence of vacancy defects increase the ratio of Li/C and shifts the voltage profile downwards, both in proportion to the number of vacancy defects. Li intercalation and diffusion kinetics in Carbon Onions was investigated for potential applications in batteries and electrochemical capacitors and we report a different loading behavior/voltage profile for these Carbon Onions as compared to graphitic systems. The Voltage vs Li concentration for Carbon Onions has a 'saw tooth' shape whereas the voltage profile for graphitic systems consists of plateaus.

(65f) Silicon-Rich Carbon Nanofibers From Water-Based Spinning to Control Combination of Silicon and Carbon for Li-Ion Battery

Joo, Y. L., Cornell University
Kim, Y. S., Cornell University
Kim, K. W., Cornell University
Cho, D.,
Hansen, N. S., Cornell University
Originally presented on: 11/4/2013 10:00:00 - 10:15:00

Silicon-rich carbon nanofibers from water-based spinning to control combination of silicon and carbon for Li-ion battery

Yong Seok Kim, Kyoung Woo Kim, Daehwan Cho, Nathaniel S. Hansen, and Yong Lak Joo*

School of Chemical and Biomolecular Engineering, Cornell University, Ithaca 14853, USA

ABSTRACT: Carbon (C) nanofibers with high silicon (Si) loading (> 65 wt%) were fabricated from water–based electrospinning of aqueous polyvinyl alcohol/Si nanoparticle solution for Li–ion battery anode applications. Our Si-rich C (SRC) nanofibers showed much facilitated charge transport features and increased activities because of continuous one–dimensional (1–D) carbon backbone structure, leading to superior battery performance to that of bare silicon nanoparticles. The presence of carbon as 1–D nanofibers could not only mitigate volume expansion of silicon but also avoid formation unstable solid electrolyte interface onto the surface of silicon. Our study on the optimum combination of carbon and silicon in the nanofibers for their electrochemical properties and battery performance revealed that the SRC nanofibers with 72.8 wt% Si content (27.2 wt% C) exhibits an adequate balance between high energy capacity of Si nanoparticles and dimensional stability with effective charge transport by carbon nanofibers, leading to an outstanding cycle life to maintain 1076 mAh/g normalized by total electrode mass and high Coulombic efficiency of about 99% for fifty cycles. Such scalable Si-rich C nanofibers made from water-based electrospinning using polyvinyl alcohol should be greatly contributed to a cost–effective development of high performance negative battery electrodes.

(65g) Electrochemical Properties of Carbon-Coated SnO2-Graphene Nanocomposites for Rechargeable Lithium-Ion Batteries

Ara, M., Wayne State University
Salley, D. S., Wayne State University
Ng, S. K. Y., Wayne State University
Originally presented on: 11/4/2013 10:15:00 - 10:30:00

SnO_2 nanoparticles were carbon-coated using two different methods and then dispersed on the reduced graphene oxide nanosheets prepared from modified Hummer's method. Carbon-coating was performed by a solvothermal method from glucose precursor material or by carbonization in Polyacrylonitrile (PAN) based carbon fibers.The later was prepared by dispersing SnO_2 nanoparticles in N-Methyl-2-pyrrolidone and then the dried powder was stabilized in air at 270°C for 30 minutes followed by the heat treatment at 900°C for 2 hours. The coated nanocomposites were characterized by X-ray diffraction, scanning electron microscopy (SEM), transmission electron microscopy (TEM), and Raman spectroscopy. From TEM observations, it was found that the SnO_2 nanoparticles were 2-6 nm in diameter and were uniformly distributed on the reduced graphene oxide nanosheets. Electrochemical properties were evaluated by galvanostatic charge-discharge cycling. Carbon-coated SnO_2-reduced graphene oxide nanocomposites showed higher capacity and more stable electrochemical performances than non-coated SnO_2-reduced graphene oxide nanocomposites. Carbon coating enhances the conductivity of the electrode and it reduces the stress developed by the volume changes of the SnO_2 nanoparticles during charge/discharge cycles. The synergistic effect of the amorphous carbon, SnO_2 nanoparticles and reduced graphene oxide nanosheets is also responsible for the superior specific capacity of carbon-coated anode material. Excellent rate capability and cyclic performances were also demonstrated with higher current densities of 500 and 1000 mA/g.

(65h) Multilayer Hybrid Electrodes of TiO2/Mwnt for Electrochemical Applications

Hyder, M. N., Massachusetts Institute of Technology
Kavian, R.,
Shao-Horn, Y., Massachusetts Institute of Technology
Hammond, P. T., Massachusetts Institute of Technology
Originally presented on: 11/4/2013 10:30:00 - 10:45:00

Rational design of next-generation electrochemical devices requires novel electrode materials and architectures to achieve high-energy, high-power density to bridge the gap between battery and capacitors. Conventional electrode fabrication processes such as doctor blading allows limited control over the electrode thickness and structure at the nanoscale, in addition to the incorporation of insulating binder materials that limits device performance. To address these challenges, we have engineered a route for the synthesis of highly stable, sub-8nm TiO_2 nanoparticles that were incorporated with acid-functionalized multi-walled carbon nanotubes (MWNT). Using electrostatic layer-by-layer nanofabrication, we assembled binder-free hybrid electrode of TiO_2/MWNT to obtain a synergistic effect from high electronic conductivity and excellent charge storage capacity. These thin film electrodes with highly controllable thicknesses shows well developed mesopores with finely dispersed, non-agglomerated TiO2 nanoparticles on MWNTs. Electrochemical measurements show high charge storage capacity (>150 mAh/$g_{electrode}$ at 0.1 A/$g_{electrode}$) with excellent cycling stability during charging and discharging up to 200 cycles indicating their promise as thin-film negative electrodes for future Li storage applications.https://aiche.confex.com/aiche/2013/t5/papers/index.cgi?username=341047&password=343273

(65i) Facile Preparation of Mesoporous α-Fe2O3 With Unique Cocoon-Like Hollow Nanostructure for Reversible Lithium Ion Storage

Zhu, J., Wayne State University
Ng, K. Y. S., Wayne State University
Deng, D., Wayne State University
Originally presented on: 11/4/2013 10:45:00 - 11:00:00

After 20 years of development since the first commercialization of rechargeable lithium ion batteries by Sony in 1991, carbon is still exclusively used as anode materials for lithium ion batteries. However, the theoretical capacity (372mAh g^{-1}) of graphite is almost achieved now and it becomes a bottleneck to further increase lithium-ion storage capacity using carbon. Carbon-alternative materials with higher capacity are currently being extensively explored to meet the increasing demand for energy in mobile electronic devices, electric vehicles and energy storage in grid. Transition metal oxides receive increasing attention as carbon-alternatives because of their much higher theoretical capacity. For example, α-Fe_2O_3 with a theoretical capacity of 1007 mAh g^{-1} is a promising candidate. Compared to other transition metal oxides, α-Fe_2O_3 is outstanding in terms of low cost, abundance, ease of fabrication, and environmental benignity. However, the poor cyclability caused by volume change during the insertion /extraction of Li ions and poor conductivity is the main challenge. Our strategy is to design and tailor α-Fe_2O_3 at nanoscale with structure control to improve their electrochemical performance. In this presentation, the strategy to facilely prepare mesoporous α-Fe2O3 with unique cocoon-like hollow nanostructure as anode materials will be reported. The formation mechanism is revealed by extensive characterization by FESEM, TEM and XRD for samples prepared under different controlled conditions. The improved electrochemical performance in reversible lithium ion storage will be discussed.

(114a) Catalytic Lignin Valorization By Liquid Phase Reforming

Jongerius, A. L., Utrecht University

Zakzeski, J., Utrecht University

Copeland, J. R., Georgia Institute of Technology

Foo, G. S., Georgia Institute of Technology

Sievers, C., Georgia Institute of Technology

Bruijnincx, P. C. A., Utrecht University

Weckhuysen, B. M., Utrecht University

Originally presented on: 11/4/2013 12:30:00 - 12:50:00

Catalytic Lignin Valorization by Liquid Phase Reforming

Anna L. Jongerius,[1] Joseph Zakzeski,[1] John R. Copeland,[2] Guo Shiou Foo,[2] Carsten Sievers,[2] Pieter C.A. Bruijnincx,[1] and Bert M. Weckhuysen[1*]

[1] Inorganic Chemistry & Catalysis, Debye Institute for Nanomaterials Science, Utrecht University, Utrecht, Netherlands

[2] School of Chemical & Biomolecular Engineering, Georgia Institute of Technology, Atlanta, GA

Introduction

Lignocellulosic biomass is becoming an increasingly important alternative resource for biofuels production. Lignocellulosic biorefinery operations must produce chemicals as well as energy vectors, however, in order to be competitive, making use of *all* the major components of lignocellulose including lignin. Indeed, the aromatic lignin polymer holds great potential for the sustainable production of renewable aromatics and phenolics, to be used as chemical building blocks or as fuel additives.[1] The catalytic valorization of lignin is highly challenging, however, as a result of its recalcitrance and highly heterogeneous nature. It is important to note in this respect that the structure of lignin depends both on the plant species as well as on the biomass pretreatment process. A generic method for catalytic lignin valorization for the production of value-added aromatics or phenolics needs to address issues with lignin solubility and agglomeration.

We previously reported on the aqueous phase reforming (APR) of lignin,[2] analogous to the APR of oxygenates such as glycerol and sorbitol, for the co-production of hydrogen and aromatics. Lignin APR was hampered by limited solubility of various lignins in water and extensive solid formation by recondensation under the hydrothermal conditions applied. Further improvements in the processing conditions of this lignin valorization process were therefore desired in order to improve yields and minimize recondensation. A simple switch to ethanol/water as solvent for the catalytic conversion of lignin, now referred to as liquid phase reforming (LPR), greatly improved solubilization, prevented agglomeration and resulted in an significant increase in aromatic monomer formation.[3]

Experimental details

Lignin solubilization and LPR reactions were studied using a semi-batch 200 mL autoclave equipped with quartz windows or a 40 mL Parr autoclave both equipped with a back-pressure regulator set at 58 bar. A typical treatment involved lignin in a 1:1 H_2O/ethanol mixture heated to 498 K with 1 wt% Pt/Al_2O_3 and a co-catalyst added for the LPR reactions. Catalyst stability was studied by XRD, ^{27}Al NMR, N_2 physisorption, TEM and TGA analysis.

Results and Discussion

Different types of lignin, i.e. kraft, organosolv and bagasse lignin, can be readily dissolved in ethanol/water mixtures at moderate temperatures without formation of agglomerates. Lignin solubilization is accompanied by a reduction of the molecular weight of the dissolved lignins due to cleavage of the lignin ether linkages, as shown by quantitative 2D HSQC NMR and GPC analysis (Figure 1). The enhanced dissolution of the various lignins, including kraft, organosolv, bagasse lignin, in ethanol/water mixtures at elevated temperatures, facilitates improved catalytic depolymerization with supported metal catalysts (Figure 2).

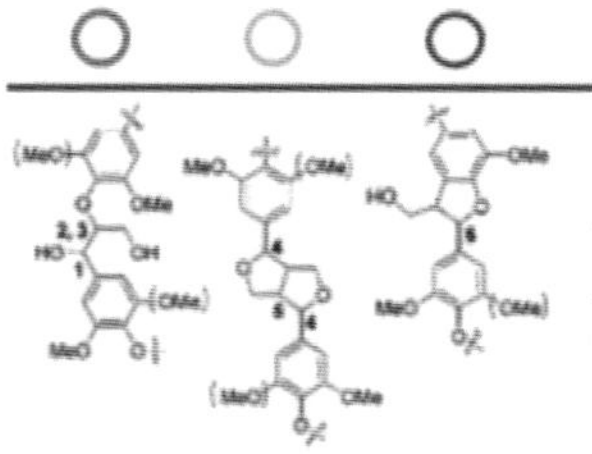

	Linkage	Decrease kraft lignin (%)	Decrease organosolv lignin (%)
1	β-O-4 C_{α}	53	63
2	β-O-4 $C_{\beta(\text{guaiacyl})}$	49	78
3	β-O-4 $C_{\beta(\text{syringyl})}$	51	79
4	β-β C_{α}	22	22
5	β-β C_{γ}	19	15
6	Phenylcoumaran	24	36
Mw	N/A	28	25

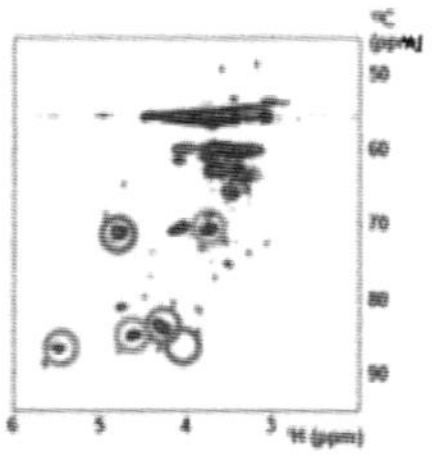

Figure 1. Influence of solubilization of kraft and organosolv in ethanol/water on M_w and extent of cleavage of selected linkages as determined by 2D HSQC NMR.

In the presence of a 1 wt% Pt/Al_2O_3 and either an acid or a base as co-catalyst, yields of up to 17% of isolated monomeric aromatics can be obtained, which is an order of magnitude higher than previously obtained with lignin APR. The yield and type of mono-aromatics is found to depend on the type of lignin, process conditions and catalyst used. The stability of the Pt/Al_2O_3 catalyst in ethanol/water under the typical liquid phase reforming conditions is studied in the presence of various aromatics and lignin itself. Boehmite formation and Pt particle sintering (typically observed under hydrothermal conditions [4]) was observed in ethanol/water, but was found to be greatly suppressed by the presence of lignin (Figure 2). [5]

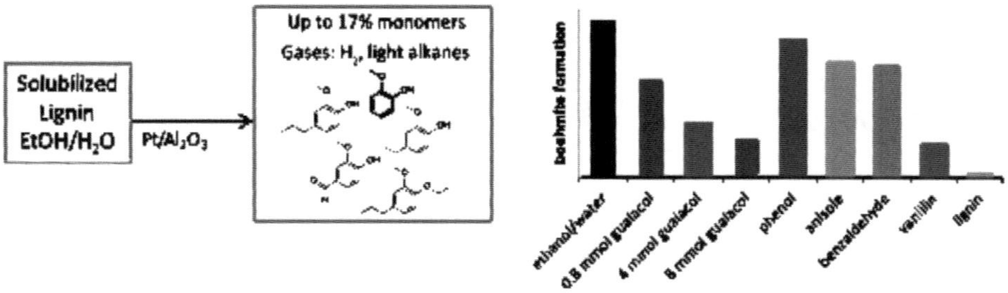

Figure 2. left: LPR results in improved monomer yield; right: influence of aromatics and lignin on catalyst stability (expressed as extent of boehmite formation).

Conclusions

The generic lignin solubilization holds potential for catalytic depolymerization by heterogeneous catalysts, as illustrated by the liquid phase reforming process. The explored route can thus contribute to the better integration of lignin in future biorefinery operations. The products obtained by liquid phase processing can, for instance, be subjected to a subsequent hydrodeoxygenation step to lower the oxygen content and reduce the complexity of the product mixture.

References

1. J. Zakzeski, P.C.A. Bruijnincx, A.L. Jongerius, B.M. Weckhuysen, *Chem. Rev.* **2010**, 110, 3552.

2. J. Zakzeski, B.M. Weckhuysen, *ChemSusChem* **2011**, 4, 369.

3. J. Zakzeski, A.L. Jongerius, P.C.A. Bruijnincx, B.M. Weckhuysen, *ChemSusChem* **2012**, 5, 1602.

4. R.M. Ravenelle, J.R. Copeland, W.G. Kim, J.C. Crittenden, C. Sievers, *ACS Catal.* **2011**, 1, 552.

5. A.L. Jongerius, J.R. Copeland, G. S. Foo, J.P. Hofmann, P.C.A. Bruijnincx, C. Sievers, B.M. Weckhuysen, *ACS Catal.* **2013**, 3, 464.

(114b) Highly Selective Fe/Mo Hydrodeoxygenation Catalysts

Hicks, J. C., University of Notre Dame
Rensel, D. J., University of Notre Dame
Originally presented on: 11/4/2013 12:50:00 - 13:10:00

Bio-oils and lignin contain many types of phenolic species that require removal in order to be used as transportation fuels.[1,2] Many catalysts have been used deoxygenate bio-oils and liquefy lignin including sulfides, carbides, oxides, and phosphides.[3] However, in many cases, these catalysts are not selective to hydrodeoxygenation and often hydrogenate aromatic components as well. Therefore, the development of new catalysts designed to selectively target the cleavage of C-O bonds in lignin is an important task. Here, we discuss the synthesis and characterization of a new bimetallic $Fe_xMo_yP_z$ catalyst that is highly selective for C-O bond cleaving and hydrodeoxygenation reactions.[4] Specifically, these catalysts are capable of selectively deoxygenating phenol to form benzene rather than completely hydrogenating the aromatic ring to produce cyclohexane (~96 % benzene). Various model compounds have been studied with this catalyst, including anisole, catechol, and phenethyl phenyl ether. Additionally, mechanistic D_2 studies were performed to determine how these catalysts deoxygenate various model compounds. These catalysts have been characterized in detail with scanning electron microscopy (SEM), transmission electron microscopy (TEM), x-ray diffraction (XRD), x-ray photoelectron spectroscopy (XPS), Raman, temperature programmed reduction (H_2-TPR), and FT-IR spectroscopy.

[1] Huber, G. W.; Iborra, S.; Corma, A., *Chem. Rev.,* **2006**, 106, 4044-4098.

[2] Zakzeski, J.; Bruijnincx, P. C. A.; Jongerius, A. L.; Weckhuysen, B. M., *Chem. Rev.,* **2010**, 110, 3552-3599.

[3] Hicks, J.C.; *J. Phys. Chem. Lett.,* **2011**, 2, 2280-2287.

[4] Rensel, D. J.; Rouvimov, S.; Gin, M.;, Hicks, J.C.; *J. Catal.,* 2013, in revision.

(114c) Mechanism of Ketonization of Acetic Acid Over Reducible Oxides and Zeolites

Pham, T. N., University of Oklahoma
Gumidyala, A., University of Oklahoma
Sooknoi, T., University of Oklahoma
Resasco, D. E., University of Oklahoma
Crossley, S., University of Oklahoma
Originally presented on: 11/4/2013 13:10:00 - 13:30:00

The conversion of acetic acid to acetone is a reaction of primary importance to the upgrading of bio oil and bio oil fractions. The conversion of corrosive acetic acid improves stability of the bio oil, while acetone can be utilized as a chemical building block for subsequent aldol condensation to liquid fuel range products. In this contribution, we discuss the mechanism of vapor phase ketonization over three different catalysts, TiO_2, Ru/TiO_2, and HZSM-5. A detailed Langmuir-Hinshelwood kinetic model was fit to the data over TiO_2 and Ru/TiO_2. Results indicate a second order rate dependence of acetic acid on the ketonization mechanism, as well as a strong inhibition of acetone on the reaction. Water and CO_2 also inhibit the ketonization, but to a lesser extent than acetone. For example, the adsorption constants of Water, CO_2, and Acetone at 280°C were found to be 0.029 and 0.016, and 0.403 $Torr^{-1}$, respectively. The activation entropies will be discussed as well to support the proposed reaction mechanism. The role of a β-keto-acid intermediate as well as the possible importance of α-hydrogen in an enolization step will be discussed.

Ketonization is also catalyzed by zeolites, such as HZSM-5, although the mechanism is not well understood partially due to the complications arising from subsequent aldol condensation reactions and excessive cracking at high temperatures. Conditions have been identified that favor high selectivity to ketones, and isotope labeling experiments have been conducted to propose the ketonization mechanism that occurs over zeolites and identify the importance of the α-hydrogen. Finally, the mechanism of ketonization over HZSM-5 will be contrasted with the mechanism over reduced TiO_2.

(114d) Investigation of Raney Nicu Catalysts for Hydrothermal Hydrodeoxygenation

Dickinson, J. G., University of Michigan, Ann Arbor
Savage, P. E., University of Michigan
Originally presented on: 11/4/2013 13:30:00 - 13:50:00

Hydrodeoxygenation (HDO) of bio-oils is an essential process for producing drop in replacement fuels for conventional petroleum because HDO increases the energy density and decreases the viscosity of bio-oils. HDO is complicated by the water content of bio-oils from their initial processing [1] (pyrolysis, hydrothermal liquefaction, etc.) and by the water produced in the reaction when oxygen is removed because water causes catalyst instability due to transformations of catalyst supports and oxidation of active metals [2]. Noble metal catalysts are active and stable for high temperature and pressure HDO reactions in the presence of water, but are undesirable due to their high cost.

In a previous study we developed a low cost hydrothermal HDO catalyst by adding ~7 wt.% Cu to a Raney Ni 2800® catalyst. The addition of Cu increased the product selectivity to liquid phase oxygenated and deoxygenated products, and away from gasification products in the hydrothermal HDO o-cresol. In this study, we examined the activity and selectivity of o-cresol HDO using Raney Ni doped with 1-40 wt.% Cu. Analysis of batch reactions indicated that at 380 °C the 1 and 2 % Raney NiCu catalysts gasified nearly all of the initial o-cresol to CH_4 and CO_2. Increasing the Cu content of the catalyst to 25 wt.% decreased the yield of gasification products to 12%, while maintaining a conversion of ~80%. We also examined the stability and of these Raney NiCu catalysts in a flow reactor. Catalyst characterization results from x-ray diffraction (XRD), x-ray photoelectron spectroscopy (XPS), and scanning electron microscopy (SEM) will also be presented.

REFERENCES

1. J. G. Dickinson, J. T. Poberezny, P. E. Savage, Appl. Catal. B: Environ. 123-124, (2012) 357-366.

2. T. M. Yeh, J. G. Dickinson, A. Franck, S. Linic, L. T. Thompson, P. E. Savage, J. Chem. Technol. Biotechnol. 88, (2013) 13-24.

(114e) Hydrocracking of a Heavy Feedstock Using Modified USY Zeolite-Based Catalysts

Agudelo V., J. L., Universidad Industrial de Santander
Giraldo D., S. A., Universidad Industrial de Santander
Hensen, E. J. M.,
Hoyos M., L. J., ECOPETROL S.A.
Originally presented on: 11/4/2013 13:50:00 - 14:10:00

Currently, refiners are facing several important challenges. They have to process lower quality feeds because of the depletion of conventional lighter crudes and, at the same time, to fulfill more stringent fuel quality specifications and satisfy the growing demand for middle distillates. In this scenario, hydrocracking technology has proven its usefulness because it is an extremely versatile refinery process. The most appealing example is that it can be used to produce high-quality diesel from even the most difficult-to-process feeds.

The catalyst plays a pivotal role in hydrocracking technology. The product distribution depends strongly on the nature of the acidic component. For production of middle distillates, amorphous silica-alumina and faujasite (Y) zeolite in its ultra stable form are the main acidic components in practical use. Comparatively, USY zeolite offers higher acidity and better stability, but it may impose diffusional limitations and lower middle distillates selectivity than amorphous silica-alumina. Nowadays, efforts are directed towards development of a new generation of zeolite Y-based catalysts with optimum number and strength of acid sites and suitable pore structure with high activity tailored to desired selectivity.

Despite many studies often using model feedstocks, there is still a lack of systematic studies using real feedstocks. Also, the role of extraframework aluminum species, the relation between Brønsted/Lewis sites and the effect of the method employed to create mesoporosity and/or to dealuminate the zeolite has not been systematically investigated. The present work reports on such parameters related to a USY zeolite for the hydrocracking conversion and yield to middle distillates of a heavy feedstock. Chemical treatments were performed with a chelating agent for selectively extracting extraframework aluminum species and with ammonium nitrate to create mesoporosity. The properties of the solids were characterized by means of ^{27}Al-MAS-NMR, ^{29}Si-MAS-NMR, XRD, XPS, TEM, N_2 adsorption, H_2-TPR, ammonia TPD, Pyr-IR, and elemental analysis. The catalysts were tested in a tubular reaction system at 350 °C, 1500 psig, H_2/feed: 1250 NL/L, LHSV: 1h^{-1} with a pretreated vacuum gas oil.

Overall activity and middle distillates yield was enhanced by the modifications performed to the USY zeolite. However, a progressive deactivation of the catalysts was observed, presumably due to coke formation. Although mesoporosity was significantly enhanced in all treatments, it will be shown that the activity most significantly depends on the density and strength of acid sites. ^{27}Al-MAS-NMR results evidenced that the treatment with the chelating agent caused partial reinsertion of some extraframework aluminum species, increasing in that way the density of Brønsted acid sites and favoring the cracking activity. According to the textural properties, ammonium nitrate treatments were effective to generate mesoporosity, but mass transfer limitations for molecules in the real feedstock seem to be still present.

(114f) Vapor Phase Catalytic Deoxygenation of Acetic Acid

Yung, M. M., National Renewable Energy Laboratory
Gomez, E., New Jersey Institute of Technology
Qayyum, E., University of South Florida
Kuhn, J. N., University of South Florida
Originally presented on: 11/4/2013 14:10:00 - 14:30:00

When biomass is converted to a liquid bio-oil through pyrolysis, it has a significantly higher oxygen content compared to petroleum fractions. In order to convert the pyrolysis products into infrastructure-compatible fuels, oxygen removal is required. Oxygen removal can be achieved by both hydrotreating (which requires the addition of hydrogen) and decarboxylation or decarbonylation, whereby oxygen is rejected as CO_2 and CO, respectively. Solid acid zeolite catalysts have been studied for bio-oil deoxygenation, but suffer from rapid deactivation due to coke formation which renders the pores and active sites inaccessible to reactants. In order to minimize wasted carbon and efficiently utilize biomass to create hydrocarbon fuels, selective deoxygenation is critical. In order to develop catalysts with an improved selectivity towards deoxygenation rather than coking, as compared to standard solid acid catalysts.

In the present contribution, Pt and Pd catalysts were tested for their activity and selectivity in deoxygenation of model biomass-derived oxygenated compounds (e.g., acetic acid, phenol). Pd catalysts on various supports have been studied, as well as various particles sizes, and Pd has been shown to be active for decarboxylation of long chain carboxylic acids. Comparison of catalytic activity of materials for different compounds, as well as material characterization results will be discussed.

(114g) Mechanistic Insights in Catalytic Conversion of Sugars to Platform Furan Derivatives

Choudhary, V., University of Delaware
Leon, M., University of Delaware
Mushrif, S. H., Nanyang Technological University
Ho, C., University of Delaware
Anderko, A., OLI Systems, Inc.
Nikolakis, V., University of Delaware
Marinkovic, N., University of Delaware
Frenkel, A. I., Yeshiva University
Sandler, S. I., University of Delaware
Vlachos, D. G., University of Delaware
Originally presented on: 11/4/2013 14:30:00 - 14:50:00

An approach to utilize biomass requires its conversion to various furan derivatives, which can potentially substitute petroleum precursors in producing chemicals, polymers and fuels. The objective of this work is to gain mechanistic understanding into the aldose conversion to platform furan derivatives using a Lewis acid with a Brønsted acid. For the first time, we employed kinetics experiments in conjunction with speciation modeling of the metal salts ($CrCl_3$ and $AlCl_3$) in aqueous media and revealed that among the various metal ions generated in aqueous media, partially hydrolyzed ions, such as $CrOH^{2+}$ or $AlOH^{2+}$, are the most active for the metal salt catalyzed aldose-to-ketose isomerization. Additionally, complex interactions between the two catalysts were explained: Brønsted acidity retards aldose-to-ketose isomerization by decreasing the equilibrium concentration of $[Cr(H_2O)_5OH]^{2+}$ or $[Al(H_2O)_5OH]^{2+}$. In contrast, Lewis acidity promotes side reactions during fructose dehydration and HMF rehydration reactions. Thus, an optimization between the Lewis and Brønsted acidities along with reaction conditions is required. Additionally, spectroscopic studies (NMR and EXAFS) revealed strong catalytic and mechanistic similarities among $CrCl_3$, $AlCl_3$ and Sn-beta catalyzed aldose-ketose isomerization.

(134a) Identifying Defect Sites in Lithium-Ion Battery Materials: Local Disorder in LiVPO4F and Its Influence On Bulk Properties

Messinger, R. J., CNRS
Menetrier, M., CNRS
Carlier, D.,
Ateba Mba, J. M., CNRS
Croguennec, L.,
Masquelier, C.,
Massiot, D.,
Deschamps, M.,
Originally presented on: 11/4/2013 12:30:00 - 12:45:00

The transition metal fluorophosphate $LiVPO_4F$ is a promising electrode material for lithium-ion batteries that is capable of operating both as a high-voltage positive electrode (4.25 V, $V^{III/IV}$) and a negative electrode (1.8 V, $V^{II/III}$).[1,2] While its Tavorite-like structure has been solved by a combination of X-ray and neutron diffraction, solid-state nuclear magnetic resonance (NMR) measurements reveal a significant extent of local disorder that has not been observed at longer length scales with scattering techniques. In particular, solid-state single-pulse ^7Li NMR spectra acquired under conditions of ultra-fast (64 kHz) magic-angle-spinning (MAS) reveal that $LiVPO_4F$ exhibits several additional ^7Li NMR signals of unknown origin that can account for up to 20% of the total lithium content. Electrochemical measurements in Swagelok-type cells with lithium metal as the negative electrode establish that materials with different populations of unknown lithium sites exhibit different cycling behavior, including differing extents of irreversible capacity and polarization. 2D $^7Li\{^7Li\}$ dipolar correlation NMR measurements establish the sub-nanometer proximities of the unknown lithium sites with respect to the $LiVPO_4F$ site, proving unambiguously that the unknown sites are local defects within the $LiVPO_4F$ structure and are not associated with separate impurity phases. In combination with magnetic measurements, variable-temperature NMR measurements reveal that the paramagnetic ^7Li NMR shifts of the different lithium sites vary linearly with the bulk magnetic susceptibility, further confirming that the sites are defects within the $LiVPO_4F$ structure. The results provide a general strategy for identifying lithium defect sites in lithium-ion battery materials and highlight the importance of understanding local disorder and its relationship to macroscopic material and device properties.

[1]J.-M. Ateba Mba, C. Masquelier, E. Suard, L. Croguennec, *Chem. Mater.* **2012**, *24*, 1223 – 1234.

[2]J.-M. Ateba Mba, L. Croguennec, N.I. Basir, J. Barker, C. Masquelier, *J. Electrochem. Soc.* **2012**, *159*, A1171-A1175.

(134b) Electrochemical Properties and Characterization of Li1.2Mn0.54Co0.13Ni0.13O2 Composite Cathode Powders Prepared by Ultrasonic Spray Pyrolysis

Langrock, A., University of Maryland
Ehrman, S. H., University of Maryland
Wang, C., University of Maryland
Originally presented on: 11/4/2013 12:45:00 - 13:00:00

Lithium-rich layered oxide $Li_{1.2}Mn_{0.54}Co_{0.13}Ni_{0.13}O_2$ (LMO) composite cathode powders are prepared by ultrasonic spray pyrolysis. The crystal structure and morphology of the prepared samples are characterized with X-ray diffraction and scanning electron microscope. The spherical shape and porous nanostructure of the as prepared powders are maintained after post-treatment at temperatures between 600 and 800°C. Grain growth in the powders increases with increasing post-treatment temperatures. The BET surface area of the composite cathode powders post-treated at 800°C is 12 m^2 g^{-1}. Discharge capacity of the cathode powders post-treated at 800°C decreases from 278 mAh g^{-1} to 208 mAh g^{-1} by the 50th cycle, the capacity retention being 75%. The post-treatment temperature strongly affected the electrochemical properties of the composite cathode powders. Composite cathode powders post-treated at high temperatures with high amounts of the Li_2MnO_3 phase had good cycle properties as well as large initial charge/discharge capacities.

(134c) Atomic Layer Deposition (ALD) On the Nanostructured Li-Mn-Rich Composite Li1.2Ni0.13Mn0.54Co0.13O2 Cathode Powder

Zhang, X., Institute of Nuclear and New Energy Technology, Tsinghua University
Belharouak, I.,
Originally presented on: 11/4/2013 13:00:00 - 13:15:00

Atomic Layer Deposition (ALD) on the Nanostructured Li-Mn-rich Composite $Li_{1.2}Ni_{0.13}Mn_{0.54}Co_{0.13}O_2$Cathode Powder

Xiaofeng Zhang,[a] Ilias Belharouak,[a,*]

[a] Chemical Sciences and Engineering Division, Argonne National Laboratory, 9700 S. Cass Ave, Argonne, IL 60439, USA, Email: belharouak@anl.gov(Corresponding author)

Li-Mn-rich (LMR) composite cathode materials $Li_{1+x}M_{1-x}O_2$ (M = Ni, Mn and Co) are of particular interest for plug-in hybrid electric vehicles (PHEVs), in which a high-energy battery is always desired for a higher mileage per charge.[1, 2] In the composite structure, the layered, the spinel and rocksalt structure are compatible at nano-scales, that many composites with different electrochemical properties can be synthesized by simply tuning the Li content in the materials. However, these materials would suffer from voltage decay and capacity fading with cycles, which significantly deteriorate their energy densities and life time for practical applications. In this type of materials, Li content is essential to the arrangement of the transition metal atoms in the composites and also to the structural stability against undesired reactions in a real electrochemical cell. It has been demonstrated that the occurrence of layered-spinel transformation induces the voltage decay. It is because the Li intercalation/deintercalation reactions may preferentially occurs at the tetrahedral site other than octahedral site, which facilitates the spinel formation.[1, 2] Further study by HR-TEM indicated that the composite particle surface could lose the layered feature after a few cycles, and the first-principle calculation suggested a defect-spinel structure forming on the surface due to the migration of transition metal ions into the Li layers.[3] Therefore, the surface chemistry and structure stability appear to be essential to limit the undesired reactions and phase transformation in the composite materials. Some achievements have been made *via* developing new synthesis processes, new chemistry or surface modifications (e.g., coating), which possibly altered the short-range and long-range ordering of transition metal atoms (Ni and Mn) near the surface, which lead to an improvement in the battery performance.[4-8]

Atomic layer deposition (ALD) is a well-established method to grow ultrathin films on flat surfaces and powders in atomic thickness using sequential, self-limiting surface reactions.[9-14] It has been demonstrated that the surface film can react as a surface barrier between the cathode active material and the electrolyte, which would reduce the loss of active materials in the electrode. [9-14] Consequently, less capacity loss is achieved for the coated cells as compared with the pristine materials. The well-established ALD process normally is using trimethylaluminium (TMA) and H_2O as precursors forming an Al_2O_3 surface film. Experimental results showed that the coated $LiCoO_2$ powders exhibited a capacity retention of 89% after 120 charge–discharge cycles in the 3.3–4.5 V vs. Li/Li+ range. [12, 15-17] The ALD surface coating can also be applied to battery electrodes at a low temperature <150 °C, which would not degrade the electrodes (e.g., melting of polymer binders) in this temperature range. It is reported that Al_2O_3-coated $LiCoO_2$ electrodes display superior electrochemical performance in comparison to the uncoated $LiCoO_2$ electrodes. However, Liu et al observed that the ALD surface coated Al_2O_3 could be lithiated forming passivation $LiAlO_2$ glass on the materials surface using in-situ TEM [18], which provided new insight into the electrochemical characteristic of the Al_2O_3surface film against Li.

In this paper, we are aiming for studying the importance of a conformal surface protection film using atomic layer deposition (ALD) technique on the high-capacity $Li_{1.2}Ni_{0.13}Mn_{0.54}Co_{0.13}O_2$ cathode materials in Li-ion batteries, which can be potentially significantly improved in their electrochemical performance. More precisely, we evaluated the effect of coating species, coating thickness and morphology on the rate performance, cycling life and high temperature performance of the high-capacity $Li_{1.2}Ni_{0.13}Mn_{0.54}Co_{0.13}O_2$ cathode materials in half-cell and full-cell configurations.[19] We found that an conformal Al_2O_3 thin film could be deposited on the powders using 10 ALD cycles of TMA and H_2O at 150 °C. TiO_2 coating was prepared using 20 ALD cycles of titanium isopropoxide (TTIP) and H_2O at 150 °C, which displayed a particulate morphology. Nevertheless, the surface coated TiO_2 film was electrochemically active, which could be lithiated forming Li_xTiO_2 as detected by X-ray photoelectron spectroscopy (XPS). The electrochemical reactivity of TiO_2 will cause degradation of the surface film, leading to capacity fading of the cell. Nevertheless, it was also found that the Al_2O_3 surface coating was not electrochemically active after repetitive cycles, which was contrary to the early report by Lui et al that Al_2O_3 can be lithiated into $LiAlO_2$. [18] To further explorate the phenomena, a systematic study on the ALD coating of Al_2O_3 and $LiAlO_2$was performed.

Refereances:

[1] C.S. Johnson, N. Li, J.T. Vaughey, S.A. Hackney, M.M. Thackeray, Electrochem. Commun., 7 (2005) 528-536.

[2] C.S. Johnson, N.C. Li, C. Lefief, J.T. Vaughey, M.M. Thackeray, Chem. Mater., 20 (2008) 6095-6106.

[3] B. Xu, C.R. Fell, M.F. Chi, Y.S. Meng, Energy & Environmental Science, 4 (2011) 2223-2233.

[4] J. Liu, B. Reeja-Jayan, A. Manthiram, Journal of Physical Chemistry C, 114 (2010) 9528-9533.

[5] S.H. Kang, M.M. Thackeray, J. Electrochem. Soc., 155 (2008) A269-A275.

[6] J.R. Croy, M. Balasubramaniam, D. Kim, S.-H. Kang, M.M. Thackeray, Chem. Mater., 23 (2012) 5415-5424.

[7] J.R. Croy, S.H. Kang, M. Balasubramanian, M.M. Thackeray, Electrochem. Commun., 13 (2011) 1063-1066.

[8] D. Kim, S.H. Kang, M. Balasubramanian, C.S. Johnson, Electrochem. Commun., 12 (2010) 1618-1621.

[9] M.Q. Snyder, S. Trebukhova, B. Ravdel, M.C. Wheeler, J. DiCarlo, C.P. Tripp, W.J. DeSisto, Mater Res Soc Symp P, 972 (2006) 331-338.

[10] M.Q. Snyder, S.A. Trebukhova, B. Ravdel, M.C. Wheeler, J. DiCarlo, C.P. Tripp, W.J. DeSisto, J. Power Sources, 165 (2007) 379-385.

[11] S.M. George, Chem. Rev., 110 (2010) 111-131.

[12] Y.S. Jung, A.S. Cavanagh, A.C. Dillon, M.D. Groner, S.M. George, S.H. Lee, J. Electrochem. Soc., 157 (2010) A75-A81.

[13] L.A. Riley, A.S. Cavanagh, S.M. George, S.H. Lee, A.C. Dillon, Electrochem. Solid-State Lett., 14 (2011) A29-A31.

[14] L.A. Riley, S. Van Ana, A.S. Cavanagh, Y.F. Yan, S.M. George, P. Liu, A.C. Dillon, S.H. Lee, J. Power Sources, 196 (2011) 3317-3324.

[15] Y.S. Jung, A.S. Cavanagh, L.A. Riley, S.H. Kang, A.C. Dillon, M.D. Groner, S.M. George, S.H. Lee, Advanced Materials, 22 (2010) 2172-2176.

[16] I.D. Scott, Y.S. Jung, A.S. Cavanagh, Y.F. An, A.C. Dillon, S.M. George, S.H. Lee, Nano Lett., 11 (2011) 414-418.

[17] Y.S. Jung, P. Lu, A.S. Cavanagh, C. Ban, G.H. Kim, S.H. Lee, S.M. George, S.J. Harris, A.C. Dillon, Advanced Energy Materials, 3 (2012) 213-219.

[18] Y. Liu, N.S. Hudak, D.L. Huber, S.J. Limmer, J.P. Sullivan, J.Y. Huang, Nano Lett., 11 (2011) 4188–4194.

[19] X. Zhang, I. Belharouak, L. Li, Y. Lei, J.W. Elam, A. Nie, X. Chen, R. Shahbazian-Yassar, R.L. Axelbaum, Advanced Energy Materials, (2013). 10.1002/aenm.201300269

(134d) Metal Fluoride/Graphene Composites for High-Performance Li-Ion Cathodes

Hayner, C. M., Northwestern University
Zhao, X., Northwestern University
Yu, Y. Y., Northwestern University
Kung, M. C., Northwestern University
Kung, H. H., Northwestern University
Originally presented on: 11/4/2013 13:15:00 - 13:30:00

Rechargeable batteries with improved energy and power density have many potential, high impact applications including advanced portable electronics and electric vehicles. However, most 'next-generation' materials capable of providing improved performance suffer from rapid capacity degradation or severe loss of capacity when rapidly discharged. Graphene, a unique two-dimensional material composed of atomically thin carbon sheets, has shown potential to address unsatisfactory rate capability, limited cycling performance and abrupt failure of these high capacity material. Therefore, we have been exploring graphene-based architectures, allowing us to exploit graphene's unique combination of high surface area, high in-plane electrical conductivity, excellent tensile modulus and mechanical durability.

Metal fluorides, including iron trifluoride (FeF_3), have the potential of both high energy delivery and high cyclability. We have recently developed a stable nanostructured iron fluoride (FeF_3)/graphene composite cathode. A facile route that combines co-assembly and photothermal reduction was developed to synthesize free-standing, flexible FeF_3/graphene papers. The papers contained a uniform dispersion of FeF_3 nanoparticles (< 40 nm) and open ion diffusion channels in the porous, conducting network of graphene sheets that resulted in a flexible paper cathode with high charge storage capacity, rate, and cycling performance, without the need for other carbon additives or binder. The composites demonstrate promising applications as cathodes in high-energy density Li-ion batteries. In a free-standing form, they showed a high storage capacity of >400 mAh/g and improved cycling performance compared to bare FeF_3 particles. In addition, we have investigated a novel mixed metal fluoride cathode system, containing tunable manganese and iron fluoride, for which characterization and electrochemical performance will be presented.

(134e) Nanostructured Li2S-Graphene Composites As Cathode for High Energy Density Li-S Batteries

Han, K., Northwestern University
Shen, J., Northwestern University
Hayner, C. M., Northwestern University
Yu, Y. Y., Northwestern University
Kung, M. C., Northwestern University
Kung, H. H., Northwestern University
Originally presented on: 11/4/2013 13:30:00 - 13:45:00

Lithium-ion batteries, although a promising technology for energy storage, is reaching their energy density limitation, making it difficult to fulfill the requirements of emerging applications such as electric vehicles and smart grid. Lithium-sulfur batteries are one of the most attractive systems for next generation high energy storage beyond Li-ion batteries. Conventional Li-S batteries mainly use elemental sulfur as the starting cathode material due to its high theoretical capacity of ~1675 mAh/g and high energy density of ~2500 Wh/kg based on the redox reaction $S_8 + 16Li \rightleftarrows 8Li_2S$. Sulfur is also low cost, abundant, and of low toxicity. However, significant hurdles such as rapid capacity fade and poor rate capability have to be overcome before practical application can be realized. Two important underlying reasons for the poor performance of this battery are the insulating nature of sulfur and the dissolution of the intermediate product polysulfide into the electrolyte resulting in redox shuttling. Although extensive research efforts have been expanded to solve these issues, most studies are limited to the simple cell configuration of sulfur as cathode and metallic lithium as anode. For safety and ease of operation, introducing Li into the cell in the form of Li_2S is more desirable than Li metal. Moreover, Li_2S, with a theoretical capacity of ~1166 mAh/g, could be paired with a lithium-metal-free anode such as Si or Sn.

Here, we report the use of a nanostructured Li_2S-graphene composite as cathode material. Li_2S nanoparticles (20~50nm) are imbedded in pocket-like structures in a three-dimensional graphene/reduced graphene oxide via a facile, wet chemical method followed by thermal treatment. Structural characterization by XRD and SEM provide clear evidence for the unique structure. The constructed 3-D graphene network could enable efficient use of Li_2S active material by providing excellent electrical contacts, offer long-term stability against loss of polysulfide and accommodate volume changes during charge and discharge processes. The as-prepared Li_2S-graphene nanocomposites exhibit high initial reversible capacity and good cycling performance. The effect of lithium polysulfide as additive of the electrolyte is also investigated, especially with respect to its contribution to the overall charge storage capacity. The structural and electrochemical characterization results of the nanocomposites will be presented.

(134f) Carbon/Sulfur Microspheres With Multi-Modal Pore Structures for Lithium-Sulfur Battery Cathodes

Zhao, C., University of Wisconsin-Milwaukee
Liu, L., University of Wisconsin-Milwaukee
Zhao, H., University of Wisconsin-Milwaukee
Li, Y., University of Wisconsin-Milwaukee
Originally presented on: 11/4/2013 13:45:00 - 14:00:00

Sulfur has received increasing attention as a cathode material for lithium-sulfur batteries due to its high theoretical specific capacity of 1675 mAh g-1, which is five times higher than that of conventional lithium-ion batteries. However, poor electronic conductivity of sulfur, dissolution of the lithium polysulfide intermediates and large volume expansion of sulfur during cycling limit the commercialization of Li-S batteries. In this work, we have demonstrated a cathode design with a novel structure made of multi-modal porous carbon microspheres. The tunable porous structure of sulfur-infiltrated carbon nanocomposites allows easy access to Li ion transport, inhibits loss of active sulfur material, and provides void space for sulfur expansion. An initial capacity of 1289 mAh g-1 at 1C and a Coulombic efficiency of 90% over 200 cycles are achieved.

(134g) Nanofiber-Based Cathode Electrocatalysts Supported On Carbon Substrates for Lithium-Air Rechargeable Batteries

Joo, Y. L., Cornell University
Yin, J., Cornell University
Kim, J.,
Originally presented on: 11/4/2013 14:00:00 - 14:15:00

Nanofiber-Based Cathode Electrocatalysts Supported on Carbon Substrates for Lithium-Air Rechargeable Batteries

*Jun Yin[a], Jangwoo Kim[a], and Yong Lak Joo[a]**

[a] School of Chemical and Biomolecular Engineering, Cornell University, Ithaca, New York 14853

In terms of usable energy density rivaling that of gasoline, rechargeable lithium-air batteries have attracted increasing interests presently. However, the practical capacity was highly affected by the effective catalyst materials in the air cathode in Li-air batteries. The demand for effective catalyst materials in the air cathode requires the design and synthesis multifunction catalysts for power density improvement. In this presentation, several types of metal oxide nanofibers (e.g., ZnO, Co_3O_4 etc) were synthesized by water-based electrospinning of highly loaded precursor solutions, followed by thermal treatment and their use as the cathode catalysts in single-cell measurements was described. In the meantime, the correlation between the various carbon-based substrates and the metal oxide fiber catalysts and how they synergistically influence the electrocatalytic activity will also be discussed. The metal oxide nanofibers are characterized using an array of techniques such as SEM, XRD, XPS, BET and Raman to determine their morphology and structural properties. Charging – discharging evaluation, cyclic voltammetric measurements in organic electrolytes were carried out to assess the mechanistic details as well.

(134h) Low-Temperature Dehydrogenation From Aminoborane Complexes Under Carbon Dioxide Atmospheres

Lee, J. W., The City College of New York
Xiong, R., CCNY
Zhang, J., The City College of New York
Originally presented on: 11/4/2013 14:15:00 - 14:30:00

Ammonia borane (AB) is a chemical hydride and one of attractive solid-state media for on-board hydrogen storage due to high hydrogen contents (19.6 wt. %) and low-temperature dehydrogenation compared to other metal hydrides. The hydrogen release from AB can be accelerated at the working temperature (~ 85 °C) of polymer electrolyte membrane fuel cell. However, this talk shows that CO_2-treatment can accelerate hydrogen release of AB polymer complex, polyaminoborane (PAB). Polyaminoborane (PAB) is the spent fuel of ammonia borane (AB, NH_3BH_3) thermolysis with around 1 mole hydrogen release per mol AB. Dehydrogenation of PAB by thermolysis generally commences at temperatures above 120 °C. However, we find gaseous CO_2 environments dramatically enhance the thermolytic dehydrogenation of polyaminoborane (PAB) at 85 - 90 °C. By exposing PAB to 2.07 MPa CO_2, 5.48 wt % hydrogen based on the mass of ammonia borane (AB) desorbs in less than 1 h, while 4 h is required to generate 5.04 wt % hydrogen by maintaining PAB at 120 °C without using CO_2. The total hydrogen desorbed from the initial AB is 11.38 wt % under the CO_2 atmosphere at 85 °C, including the hydrogen release during the transition from AB to PAB. The main contributing factors to the facile hydrogen release are the promoting effect of newly formed formic acid and the exothermic effect of the reaction between PAB and CO_2. We will explain the possible promoting mechanism by various spectroscopic and elemental analyses.

(134i) Intermittent Electrical Energy Storage By Liquid Organic Hydrogen Carriers

Arlt, W., University of Erlangen-Nuremberg
Muller, K., University of Erlangen-Nuremberg
Teichmann, D., University of Erlangen-Nuremberg
Originally presented on: 11/4/2013 14:30:00 - 14:45:00

The intermittent delivery of energy from sources like photovoltaics or wind demands a large and long-term storage of energy. Liquid Organic Hydrogen Carriers like "Carbazol" address this topic using the infrastructure of liquids like domestic fuel oil. In Germany, the situation is far more complicated, because the country decided to phase out nuclear energy in the year 2022. The lecture highlights the situation in the German state of Bavaria and the planning.

The lecture reports the progress of Liquid Organic Hydrogen Carriers (LOHC). A demo-project is under construction at a condo in the city of Erlangen/Bavaria. Important data are safety, efficiency of electrical energy storage and robust construction. The purpose of this demo is to investigate

1. The ratio between storage and heat demand of the building
2. The number of storage and delivery actions
3. The robustness of the process in a public area

The time frame is 2 years from beginning of 2014 on.

(134j) Synthesis of Nanoporous Palladium Powder With Controlled Pore and Particle Size for Hydrogen Storage Applications

Cappillino, P., Sandia National Laboratories
Jones, C.,
Hattar, K., Sandia National Laboratories
Clark, B.,
Hekmaty, M.,
Jacobs, B.,
Robinson, D. B., Sandia National Laboratories
Originally presented on: 11/4/2013 14:45:00 - 15:00:00

Synthesis of Nanoporous Palladium Powder with Controlled Pore and Particle Size for Hydrogen Storage Applications

by Patrick J. Cappillino, Christopher G. Jones, Khalid M. Hattar, Blythe G. Clark, Michelle A. Hekmaty, Benjamin W. Jacobs, and David B. Robinson

Materials capable of rapidly storing and delivering hydrogen are currently in demand for numerous energy applications. The high surface area exhibited by nanoporous palladium has the potential to greatly improve its kinetics of hydrogen charge and discharge *versus* nonporous metal while retaining favorable operating temperatures, pressures, and volumetric capacity. In addition, the void space imparted by porosity should accommodate the volumetric expansion of palladium upon hydriding, mitigating plastic deformation and improving cycle life.

By chemically reducing palladium salts in the presence of various surfactants, we are able to synthesize palladium having size-tunable mesopores from a few to tens of nm, in a scalable fashion. Our results suggest that Pd nanoparticles form and sinter around micelles present in the aqueous media. Changing the chemical composition of the surfactant affects the size of these micelles and ultimately determines the pore dimensions. We demonstrate that pore size affects not only surface area, but also pore thermal stability under vacuum, and in the presence of hydrogen gas.

In addition to pore geometry, particle size and shape are important factors in determining gas flow characteristics. Using nonporous copper particles as sacrificial reductants for palladium salts, in the presence of surfactants, we are able to synthesize micron-sized particles of uniform size and shape. This is in contrast to continuous aggregates and films that are obtained using soluble chemical reductants or planar electrodes.

We will present details of the synthesis of nanoporous palladium, demonstrating control of both pore and particle size. Bulk measurements of hydrogen storage properties and pore characteristics, as well as microscopic measurements, including *in situ* heated-stage TEM under vacuum and in the presence of hydrogen, will be included, as well as details on the kinetics of hydriding/dehydriding.

Sandia National Laboratories is a multi-program laboratory managed and operated by Sandia Corporation, a wholly owned subsidiary of Lockheed Martin Corporation, for the U.S. Department of Energy's National Nuclear Security Administration under contract DE-AC04-94AL85000.

(166a) Fossil and Bio Carbon Story in FCC Co-Refining

Thegarid, N.,
Fogassy, G.,
Meunier, F., Universite Lyon 1, CNRS, UMR 5256, IRCELYON, Institut de recherches sur la catalyse et l'environnement de Lyon
Schuurman, Y., Institut de Recherches sur la Catalyse et l'Environnement de Lyon (IRCELYON, UMR 5256, CNRS; Universite Claude Bernard Lyon 1
Mirodatos, C., Universite Lyon 1, CNRS, UMR 5256, IRCELYON, Institut de recherches sur la catalyse et l'environnement de Lyon
Originally presented on: 11/4/2013 15:15:00 - 15:35:00

Fossil and bio carbon story in FCC co-refining

Nicolas Thegarid, Gabriella Fogassy, Frederic Meunier, Yves Schuurman, Claude Mirodatos*

Ircelyon, CNRS-UCBL, 2 Avenue Albert Einstein, 69626 Villeurbanne, France.

claude.mirodatos@ircelyon.univ-lyon1.fr

Introduction. In order to meet the international binding renewable energy targets by 2020 (10% share in all forms of transportation fuels) [1], a realistic alternative to the first generation bio-fuels is to produce hybrid bio- and fossil fuels by co-refining biomass pyrolysis oil with crude oil fractions in a conventional oil refinery [2, 3]. However, co-refining leads to significant changes in products quality, such as higher gasoline aromaticity and coke deposits [4]. A detailed understanding is therefore necessary on how the oxygenated moieties affect the FCC mechanism and how both fossil and renewable carbon is dispatched among the FCC products (essentially in the gasoline and in the coke produced during the cracking and burned during the regeneration steps). In this study, up-graded pyrolysis bio-oils (hydrodeoxygenated, HDO or catalytic, CPO) are added to crude oil distillates (VGO or long residue) and the quality of the produced "hybrid" FCC fuels is analyzed and compared, focusing on coke generation and nature as a signature of the cracking process.

Experimentals. The various types of coke deposited during the cracking step were analysed by TPO, thermogravimetry under flowing air, DRIFT, TEM and ^{13}C-CP-MAS NMR. The coke occupancy in the porous volume was measured by BET/BJH nitrogen adsorption.

Results and conclusion. From pore volume changes upon co-refining and from coke analysis, the rate of coke deposition was found significantly higher (about two fold) in the presence of oxygenates in the feed, either from HDO or CPO oils, and still much higher when processing a pure HDO bio-oil. It was found in addition that the amount and nature of the deposited coke depends strongly on the composition of the reacting feed (either VGO alone, or a mixture VGO/upgraded bio oil, or the pure bio-oil). From TEM and BET-BJH analyses, the coke deposits from the VGO cracking are essentially graphitic and located inside or close to the Y zeolite microstructures, without affecting much their pore volume, while extra framework deposits, essentially amorphous, are formed in the meso/macroporous structure after bio-oil degradation/condensation. From NMR and DRIFT analysis, the coke formed from the oxygen containing bio-molecules is found to present more highly condensed aromatic rings than the coke formed from the feed hydrocarbons. From the radio carbon analysis for a typical co-processing experiment (ca. 9% of added bio-oils to VGO feed), it comes that the bio-carbon contained in the HDO-oil is concentrated mainly in the gas fraction (11%) and in the coke fraction (16%) while the targeted liquid product, gasoline, contains only around 7% of bio-carbon [5].

By acounting for the above results and others not presented here, a general mechanistic scheme is proposed where the lignin polymeric fragments of the added bio-oils are cracked into smaller methoxy phenols in the extra-framework space, leading also to a specific amorphous coke deposition. The latter tends to block the mesopores when fed at high concentration. In turn, the graphitic coke deposition arising from the hydrocarbons cracking essentially in the zeolite micropores is not significantly affected by the co-presence of oxygenated molecules.

These important and original findings can be considered as strong guide-lines to adapt the FCC catalysts design to future co-refining operating conditions.

References:

1. http://ec.europa.eu/energy/renewables/targets_eu.htm

2. F. de Miguel Mercader, J.A. Hogendoorn, S.R.A. Kersten, N.W.J. Way, C.J. Schaverien, M.J. Groeneveld, *Appl. Catal. A,* **96**, 57-66 (2010).

3. Y. Schuurman, G. Fogassy, C. Mirodatos. *Tomorrow's biofuels: hybrid bio-gasoline by co-processing in FCC units* in "The Role of Catalysis for the Sustainable Production of Bio-fuels and Bio-chemicals", Eds Triantafyllidis, Stöcker and Lappas, Elsevier, 2013, 321–349.

4. G. Fogassy, N. Thegarid, Y. Schuurman and C. Mirodatos. From biomass to bio-gasoline by FCC co-processing: effect of feed composition and catalyst structure on product quality. Energy & Environmental Sci. 4 (2011) 5068-5076.

5. G. Fogassy, N. Thegarid, Y. Schuurman and C. Mirodatos *The fate of bio-carbon in FCC co-processing products.* Green Chem., 14 (2012) 1367-1371

(166b) Control of Furfural Catalytic Hydrogenation Selectivity Using Alkanethiolate Self-Assembled Monolayers

Pang, S. H., University of Colorado Boulder
Schoenbaum, C. A., University of Colorado
Schwartz, D. K., University of Colorado Boulder
Medlin, J. W., University of Colorado Boulder
Originally presented on: 11/4/2013 15:35:00 - 15:55:00

Furfurals represent an important class of platform chemicals in the conversion of biomass to fuels and other valuable products. These highly reactive molecules present a challenge for *selectivity control* of heterogeneous catalytic hydrogenation due to the presence of multiple functional groups, all of which can interact with the catalyst surface. This multifunctionality results in a diverse range of products under hydrogenation conditions. In particular for group 10 metals, the furan ring adsorbs strongly to the surface, resulting in undesirable furfural decarbonylation and ring hydrogenation, rather than aldehyde hydrogenation and hydrodeoxygenation. This contribution focuses on experimental and computational investigations into the control and promotion of the hydrodeoxygenation pathway using alkanethiolate self-assembled monolayers (SAMs).

We have previously shown using a combination of surface science techniques on a Pd(111) single crystal that when furfural is allowed to adopt a flat-lying geometry, such as the case for low surface coverage of adsorbate, the decarbonylation pathway is strongly preferred. As coverage increases, some molecules adsorb through an oxygen lone pair with the ring extending into vacuum and the deoxygenation pathway becomes feasible. The degree of molecule-surface interaction can be tailored through the use of SAMs, which restricts the surface ensemble size by strongly bound sulfur head groups. We demonstrate that using a sparsely packed SAM results in a modest increase in selectivity to aldehyde hydrogenation and hydrodeoxygenation products. Furthermore, use of a dense SAM results in a large increase in selectivity to these products via the restriction of specific surface sites that are necessary for the undesired reactions. These results have implications for other reactions where ensemble size plays an important role in determining reaction selectivity.

(166c) Hydrodeoxygenation Or Reforming of Biomass - a DFT Study

Basaran, D., Technische Universitaet Muenchen
Chiu, C. C., Agency for Science, Technology and Research
Genest, A., Agency for Science, Technology and Research
Roesch, N., Technische Universitet Muenchen
Originally presented on: 11/4/2013 15:55:00 - 16:15:00

Cost effective conversion of plant and municipal solid waste material (biomass) into materials of higher energy density, to be utilized in the framework of the current carbon-based supply infrastructure, is one of the major challenges for a sustainable future. Aqueous-phase processes proceeding on supported metal catalysts may be a promising approach for the conversion of biomass to alkanes [1], but can also be used for the production of hydrogen [2], along with CO_2. These processes and their selectivities are studied for alcohols as model feedstocks to understand mechanisms involved in processing oxygen containing species.

We carried out electronic structure calculations at the DFT level on periodic slab models using VASP. We explored 1- and 2-propanol as models, adsorbed on flat and stepped surfaces of platinum [3-5].

We examined two scenarios generating either H_2 or alkanes. On Pt surfaces, hydrogen formation was calculated to be favored over alkane formation, in agreement with experiments [6]. First, we rationalized the experimentally found dissimilar behavior of primary and secondary alcohols in the formation of H_2 [3]. After dehydrogenating a primary alcohol in several steps, the C-C bond is cleaved, leading ultimately to carbon dioxide and ethane. Hereby the experimentally observed CO_2can be directly released or formed in a subsequent water-gas shift reaction after decarbonylation. Our calculations show that the propensity for C-C bond cleavage depends on the structure of the catalyst surface and on its capability to activate water [4].

The formation of alkanes, involving the cleavage of the alcohol C-O bond on the Pt surface, can proceed by various pathways. For 1-propanol, we present a dehydration-hydrogenation mechanism via propylene as intermediate and a hydrogenolysis mechanism. The former may be able to compete, under acidic conditions, with H_2formation [5].

References:

[1] J. N. Chheda, G. W. Huber, J. A. Dumesic, *Angew. Chem. Int. Ed.* **2007**, 46, 7164.

[2] R. D. Cortright, R. R. Davda, J. A. Dumesic, *Nature* **2002**, 418, 964.

[3] D. Basaran, A. Genest, N. Rösch: *J. Catal.* **2012**, 287, 210.

[4] D. Başaran, A. Genest, J. A. Lercher, N. Rösch, submitted.

[5] C.-c. Chiu, A. Genest, N. Rösch, *ChemCatChem* **2013**, DOI:10.1002/cctc.201300184.

[6] A. Wawrzetz, B. Peng, A. Hrabar, A. Jentys, A. A. Lemonidou, J. A. Lercher, *J. Catal.* **2010**, 269, 411.

(166d) Gas Phase Hydrodeoxygenation (HDO) of Fufural On Molybdenum Carbide Catalysts

Lee, W. S., Purdue University
Bhan, A., University of Minnesota
Originally presented on: 11/4/2013 16:15:00 - 16:35:00

The underlying challenge in converting biomass feedstock with a $(CH_2O)_n$ stoichiometry to fuels with $(CH_2)_n$ stoichiometry is the selective removal of oxygen without a concurrent decrease in the carbon chain length. For hydrodeoxygenation (HDO) reactions in which H_2 is usually in excess, it is important to utilize the H_2 for oxygen removal rather than unwanted hydrogenation reactions of C=C double bonds to avoid extra hydrogen consumption. In this work, we report the gas phase HDO of furfural over bulk Mo_2C in excess of hydrogen (H_2/furfural molar ratio ~390) at low temperatures (~150 °C) to selectively accomplish C=O scission to form 2-methylfuran with 60-65% selectivity; and the selectivity for the C-C bond scission product, furan, was less than 1%. The hydrogenation product of 2-methylfuran (2-methyltetrahydrofuran) is formed in negligible quantities suggesting that HDO is the dominant reaction pathway. Detailed kinetics studies including apparent activation energy and reaction orders, isotopic experiments, and in-situ titration studies were performed to elucidate the reaction mechanism and sites requirements for furfural HDO reactions and will be reported in this presentation.

(166e) Hydrodeoxygenation of Meta-Cresol On Pt/SiO2 and Pt/Hbeta

Nie, L., University of Oklahoma
Resasco, D. E., University of Oklahoma
Zhu, X., University of Oklahoma
Sooknoi, T., University of Oklahoma
Mallinson, R. G., University of Oklahoma
Originally presented on: 11/4/2013 16:35:00 - 16:55:00

Catalytic conversion of lignin derived meta-cresol (3-methylphenol) was investigated over Pt/SiO_2, HBeta and Pt/HBeta catalysts. Over Pt/SiO_2, 3-methyl-cyclohexanone, 3-methyl-cyclohexanol and toluene are obtained at 300°C. But at 400°C, only toluene was obtained. Acid site on HBeta catalyzes methyl transfer reactions (isomerization and transalkylation), yielding in cresol isomers, dimethylphenol (xylenol) isomers and phenol. Bifunctional catalyst Pt/HBeta catalyzes methyl transfer reactions and hydrodeoxygenation reactions in parallel. The turn over frequency for hydrodeoxygenation of meta-cresol is 3 times higher over Pt/HBeta than over Pt/SiO_2, highlighting the significant role of acid sites in hydrodeoxygenation of lignin derived phenolics. A reaction pathway has been proposed: m-cresol tautomerizes to an unstable ketone intermediate (3-methyl-3,5-cyclohexadienone). Hydrogenation of the C=C bonds of this intermediate produces 3-methyl-cyclohexanone and 3-methyl-cyclohexanol, which are only observed at 300°C and in the absence of acid sites. Over Pt/HBeta at 400°C, dehydration of 3-methyl-cyclohexanol followed by dehydrogenation results in toluene production. The interesting question is why toluene observed in the absence of acid sites and at low temperatures. The answer is that toluene can also derive from the ketone intermediate (3-methyl-3,5-cyclohexadienone) by hydrogenation of the carbonyl group that produces a very reactive unsaturated alcohol (3-methyl-3,5-cyclohexadienol), which can be easily dehydrated to toluene.

(166f) Surface Barrier Transport Limitations in MFI-Structured Zeolites

Teixeira, A. R., University of Massachusetts Amherst
Chang, C. C., University of Massachusetts Amherst
Ford, D., University of Massachusetts Amherst
Fan, W., University of Massachusetts Amherst
Dauenhauer, P., University of Massachusetts Amherst
Originally presented on: 11/4/2013 16:55:00 - 17:15:00

Zeolites are widely used to catalyze chemical reactions and improve selectivity to desired products through ordered, shape selective microporous channels [1]. As microporous materials are synthesized with length scales approaching that of a single lattice (pillared, nanosheets, membranes, etc.), the need to understand mass transfer limitations within small particles becomes paramount. Experimental evidence accumulated over the past several decades has led to the general phenomenon referred to as 'surface barriers,' which describes the hindrance to mass transfer at the surface of zeolite particles. While this timescale is often negligible when compared to the diffusion or reaction times, we show that it does contribute significantly to overall turnover in small particles. Chemical species exhibiting differing kinetic diameters and adsorption affinities were used to probe diffusion within MFI-structured zeolites to elucidate the role of guest molecule size and binding energy. Kinetic Monte Carlo simulations are used to mechanistically probe the effect of physical structures on overall transport timescales, including internal grain boundaries, regular surface restrictions and total pore blockages. The relative trends with particle size are compared to experimental measurements and potential mechanisms are proposed.

[1] C.Luke Williams, Chun-Chih Chang, Phuong Do, Nima Nikbin, Stavros Caratzoulas, Dionisios G. Vlachos, Raul F. Lobo, Wei Fan, Paul J. Dauenhauer, "Cycloaddition of Biomass-Derived Furans for Catalytic Production of p-Xylene," *ACS Catalysis* **2012**, 2(6), 935-939.

(166g) Production Of Renewable Diesel A NOVEL MICRO-Scale Based Reactor

Atadana, F., Oregon State University
Jovanovic, G. N., Oregon State University
Originally presented on: 11/4/2013 17:15:00 - 17:35:00

In this study a novel micro scale reactor was developed for conversion of palm olein into diesel range fuel with a thin layer of $NiMoP/Al_2O_3$ deposited on the walls of reactor. The study was performed with liquid feed concentrations of 5wt%, 25wt% and 50wt% palm olein in dodecane, temperature 275-350°C and pressure 10-35bar. Increasing temperature and pressure resulted in increased selectivity towards diesel range hydrocarbons. Decrease in hydrogen/ oil mole ratio with increasing palm olein feed concentration lowered conversion and selectivity towards diesel range products. The intensification of mass transfer afforded by the micro-scale reactor, achieved 100% conversion of palm olein at unconventionally high WHSV of 50-200h^{-1}.

(167a) Unravelling the Intial Decomposition Chemistry of Renewable Fuels

Van Geem, K. M., Ghent University
Marin, G. B., Ghent University
Simmie, J., National University of Ireland, Galway
Carstensen, H. H., Laboratory for Chemical Technology
De Bruycker, R.,
Originally presented on: 11/4/2013 15:15:00 - 15:35:00

In comparison to fossil derived hydrocarbons the chemistry of combustion and pyrolysis of oxygen containing renewable fuels is quite poorly understood[1]. The available kinetic models developed for combustion of mixtures of hydrocarbons and oxygenated compounds do a reasonable job in predicting auto-ignition, flame propagation parameters, and the formation of the main regulated pollutants. However they do quite poorly for predicting the formation of minor products (alkenes, dienes, aromatics, aldehydes) and soot, which have a deleterious impact on both the environment and on human health. For example most of the kinetic models simply do not include formation of aromatics, let alone soot. Moreover, the validation for toxic aldehydes such as formaldehyde is often impossible because these components are quite difficult to measure quantitatively.

In this contribution we will evaluate the decomposition chemistry of different candidates of renewable fuel (additives) ranging from longer alcohols (butanols, pentanols) to esters (saturated, unsaturated, cyclic) and furans both experimentally and theoretically.New experimental data have been acquired on a new setup that is both equipped with a plug flow reactor made out of quartz and Incoloy800HT. The latter allows to determine valuable kinetic data and at the same time evaluate the importance of surface reactions on the decomposition chemistry which could be important for real engines. The analysis section of the bench scale pyrolysis set-up has been discussed at length previously [2-6]. The effect of temperature on the product distribution was determined for two dilution regimes. During the currently discussed set of experiments the reactors are operated almost isothermally. The pressure drop over the reactor was found to be negligible. The analysis section of the pyrolysis set-up enables on-line qualification and quantification of the entire product stream, i.e. a wide boiling mixture containing permanent gasses (H_2, CO, CO_2, etc.), alcohols, aldehydes and ketones, esters, and hydrocarbons ranging from methane to polyaromatic hydrocarbons (PAH). As discussed previously[4], three different gas chromatographs are required: a refinery gas analyzer (RGA), a light oxygenates analyzer (LOA) and a GC×GC-FID/(TOF-MS). Response factors (RF_i) of all permanent gasses and light hydrocarbons (C_2-C_4) were determined with the aid of a gaseous calibration mixture provided by Air Liquide. The response factors of all C_{5+} hydrocarbons and methyl esters were determined using the effective carbon number concept, relative to methane.

The application Gaussian-09 [7] was used to carry out the computation of geometries, frequencies and energies of all the species. The model chemistries CBS-QB3, CBS-APNO and G3 were employed to determine the species enthalpy, *H*, and free energy, *G*, at 298.15 K giving a reasonable compromise between accuracy and computational expense [8-12]. The use of three distinctly different composite methods is crucial to an appreciation of the uncertainties with which energies can be determined[13, 14]. The three model chemistries differ substantially in their geometry optimization and vibrational frequency calculation procedures and in their corrections to the electronic energy. Visualization and manipulation of the Gaussian output was provided by Chemcraft[15]. The vibrational frequencies and moments of inertia of each molecule were computed at the B3LYP/6-311G(2d,d,p) level of theory (this is the default optimiser in the CBS-QB3 composite method) together with the keyword Freq=HinderedRotor. The latter choice forces an analysis of the vibrational modes and selects those that are more appropriately considered to be hindered rotors at the temperature at which the calculation is being carried out. The net effect is that primary statistical thermodynamic data alters as a vibration is re-classified as a hindered rotor. Choosing a different functional and basis set makes very little difference to the computed entropy; for example, using BMK/6-31+G(2df,p) with an entropy-specific scaling factor reduces the entropy of 5-methyl-2-acetylfuran by approximately 1% from the default B3LYP/6-311G(2d,d,p) values. In order to verify the above mentioned method the Thermo module of Multiwell [16] was used to calculate thermodynamic parameters under the rigid rotor harmonic oscillator approximation. In appropriate cases the barriers to internal rotations were computed by relaxed dihedral angle scans at the B3LYP/6-311G(2d,d,p) level and the results treated as 1-dimensional hindered rotors. The ab-initio calculations will be used to show the differences in how these different renewable components decompose under our conditions.

References

1. Metcalfe, W.K., J.M. Simmie, and H.J. Curran, *Ab Initio Chemical Kinetics of Methyl Formate Decomposition: The Simplest Model Biodiesel.* Journal of Physical Chemistry A, 2010. **114**(17): p. 5478-5484.

2. Harper, M.R., et al., *Comprehensive reaction mechanism for n-butanol pyrolysis and combustion.* Combustion and Flame, 2011. **158**(1): p. 16-41.

3. Van Geem, K.M., et al., *Accurate High-Temperature Reaction Networks for Alternative Fuels: Butanol Isomers.* Industrial & Engineering Chemistry Research, 2010. **49**(21): p. 10399-10420.

4. Pyl, S.P., et al., *Rapeseed oil methyl ester pyrolysis: On-line product analysis using comprehensive two-dimensional gas chromatography.* Journal of Chromatography A, 2011. **1218**(21): p. 3217-3223.

5. Chen, Q. and G.F. Froment, *Thermal-Cracking of Substituted Aromatic-Hydrocarbons .1. Kinetic-Study of the Thermal-Cracking of I-Propylbenzene.* Journal of Analytical and Applied Pyrolysis, 1991. **21**(1-2): p. 27-50.

6. Chen, Q. and G.F. Froment, *Thermal-Cracking of Substituted Aromatic-Hydrocarbons .2. Kinetic-Study of the Thermal-Cracking of Normal-Propylbenzene and Ethylbenzene.* Journal of Analytical and Applied Pyrolysis, 1991. **21**(1-2): p. 51-77.

7. M.J. Frisch, et al., *Gaussian 09*, 2009: Wallingford, CT.

8. Curtiss, L.A., et al., *Gaussian-2 Theory for Molecular-Energies of 1St-Row and 2Nd-Row Compounds.* Journal of Chemical Physics, 1991. **94**(11): p. 7221-7230.

9. Curtiss, L.A., et al., *Assessment of complete basis set methods for calculation of enthalpies of*

formation. Journal of Chemical Physics, 1998. **108**(2): p. 692-697.

10. Curtiss, L.A., et al., *Gaussian-3X (G3X) theory: Use of improved geometries, zero-point energies, and Hartree-Fock basis sets.* Journal of Chemical Physics, 2001. **114**(1): p. 108-117.

11. Ochterski, J.W., G.A. Petersson, and J.A. Montgomery, *A complete basis set model chemistry .5. Extensions to six or more heavy atoms.* Journal of Chemical Physics, 1996. **104**(7): p. 2598-2619.

12. Montgomery, J.A., et al., *A complete basis set model chemistry. VII. Use of the minimum population localization method.* Journal of Chemical Physics, 2000. **112**(15): p. 6532-6542.

13. Simmie, J.M., et al., *Enthalpies of Formation and Bond Dissociation Energies of Lower Alkyl Hydroperoxides and Related Hydroperoxy and Alkoxy Radicals.* The Journal of Physical Chemistry A, 2008. **112**(22): p. 5010-5016.

14. Simmie, J.M. and H.J. Curran, *Formation Enthalpies and Bond Dissociation Energies of Alkylfurans. The Strongest CÃ,XX Bonds Known?* The Journal of Physical Chemistry A, 2009. **113**(17): p. 5128-5137.

15. G.A. Zhurko and D.A. Zhurko, *chemcraft.*

16. Barker, J.R., *Multiple-well, multiple-path unimolecular reaction systems. I. MultiWell computer program suite.* International Journal of Chemical Kinetics, 2001. **33**(4): p. 232-245.

(167b) Neopentane Hydrogenolysis Over Supported Pd and Pt Catalysts and Alloys: A Kinetics Study of Particle Size and Alloy Effects

Childers, D., U. Illinois at Chicago
Schweitzer, N., The University of Michigan
Saha, A.,
Miller, J. T., Argonne National Laboratory
Rioux, R. M., The Pennsylvania State University
Meyer, R. J., University of Illinois - Chicago
Originally presented on: 11/4/2013 15:35:00 - 15:55:00

Neopentane Hydrogenolysis over Supported Pd and Pt Catalysts and Alloys: A Kinetics Study of Particle Size and Alloy Effects

David J. Childers[1], Neil Schweitzer[2], Arindom Saha[3], Robert M. Rioux[3], Jeffery T. Miller[2] and Randall J. Meyer[1]

[1]University of Illinois at Chicago, Chicago, Illinois 60607 (USA)

[2]Argonne National Laboratory, Lemont, Illinois 60439 (USA)

3Pennsylvania State University, University Park, PA 16802 (USA)Â

The demand for fossil fuels is continuing to grow despite the increasing emphasis on finding replacement fuel sources. Therefore, advances in reforming processes will be an important area for the foreseeable future. Selective isomerization of high n-alkanes to branched hydrocarbons is one method to improve to quality of these fuel stocks. There are competing pathways so selectivity is a very important consideration. Bimetallic catalysts have become a significant point in catalysis research in the past few decades [1]. Alloying has resulted in increased activity and selectivity for various reactions, as well as improving resistance to poisoning and sintering [1].

Both Pt and Pd monometallic catalysts had increased isomerization selectivity with increasing particle size. Pt and Pd catalysts displayed similar behavior for both metals. This trend for Pt has previously been observed by Foger et al [2]. This change in selectivity could be explained two different ways: ensemble effects and electronic effects. If a reaction requires more than one active site to proceed then a change in the number of ensemble sites could be the cause for the change in selectivity. Electronic effects on the other hand are the changes in the electronic properties of the metal. Metal atoms in different surface configurations and particle sizes have different energetics which can affect bond energies. Both these effects appear to be contributing to the change in isomerization selectivity. First we find that the isomerization selectivity appears to correlate well with CO adsorption strength: the weaker the adsorption of CO, the higher the selectivity to isomerization. In addition, alloying Pt and Pd together has shown an increase in selectivity compared to the single metal catalysts. In contrast, alloying Pd with Au has shown little change in the selectivity.

References

1. J.H. Sinfelt, Sci.Am. 253 (1985) 90-&

2. K. Foger, and J.R. Anderson, Journal of Catalysis 54 (1978) 318-335.

(167c) Mechanism of C-C Hydrogenolysis On Ir Catalysts

Hibbitts, D. D., University of California at Berkeley

Flaherty, D. W., University of California at Berkeley

Iglesia, E., University of California, Berkeley

Originally presented on: 11/4/2013 15:55:00 - 16:15:00

Mechanism of C-C Hydrogenolysis on Ir Catalysts

David D. Hibbitts[1], David W. Flaherty[1,2] and Enrique Iglesia[1*]

[1]Department of Chemical and Biomolecular Engineering, University of California at Berkeley, Berkeley, CA 94720, United States

[2]Department of Chemical and Biomolecular Engineering, University of Illinois at Urbana-Champaign, Urbana, IL 61801, United States

*iglesia@berkeley.edu

Catalytic hydrogenolysis of alkanes has been widely studied due to the necessity of selectively removing sulfur and oxygen through C-S and C-O hydrogenolysis as well as optimizing chain length of hydrocarbon mixtures through C-C hydrogenolysis. Hydrogenolysis is considered a structure-sensitive reaction in which the composition of the metal and the particle size have large effects on the turnover rates and activation energies of the reaction. Substituted alkanes contain distinguishable C-C bonds, which have distinct mechanisms that result in different dependencies on H_2 pressure and temperature, allowing for limited control over product selectivity. Kinetic analysis and density functional theory (DFT) of hydrogenolysis on well-defined catalyst particles are used to determine the mechanisms for C-C bond activation and how they differ with the substitution of the involved C-atoms. Ethane, the simplest alkane with a C-C bond, is investigated here to determine general trends in hydrogenolysis on Ir.

Ethane hydrogenolysis was examined on 3.0 wt% Ir/SiO_2 catalysts with mean particle diameters of 7.1 nm at 593 K at high H_2:C_2H_6 ratios (0.6-1.8 MPa H_2, 10-80 kPa C_2H_6). Atomic hydrogen (H*) is the most abundant surface intermediate (MASI) at these high H_2:C_2H_6 ratios and the rate of hydrogenolysis follows:

$$r = \alpha \frac{K_{C_2H_6}(C_2H_6)}{[K_{H_2}(H_2)]^{\lambda}} \tag{1}$$

where (H_2) and (C_2H_6) are partial pressures, α is the product of rate and equilibrium constants and λ was found to be 3.3 ± 0.2. Chemical interpretation of this rate expression suggests that the kinetically-relevant transition state is derived from C_2H_6 and λ represents rate-averaged number of H_2 molecules evolved from dehydrogenation of C_2H_6 and from desorption of H* to accommodate the kinetically-relevant transition state on the catalyst surface.

Previous studies have proposed that dehydrogenation of C_2H_6 is quasi-equilibrated on Ir surfaces and that the kinetically-relevant step for hydrogenolysis is the C-C bond cleavage, followed by quasi-equilibrated hydrogenation to form CH_4, Scheme 1.

$$H_2 + 2^* \rightleftharpoons 2H^* \tag{1.1}$$

$$C_2H_6 + {}^* \rightleftharpoons C_2H_6^* \tag{1.2}$$

$$C_2H_6^* + (6-x)^* \rightleftharpoons C_2H_{6-x}^* + xH^* \tag{1.3}$$

$$C_2H_x^* \longrightarrow CH_y^* + CH_z^* \tag{1.4}$$

$$CH_x^* + (4-x)H^* \rightleftharpoons CH_4 \tag{1.5}$$

For Scheme 1, a L-H expression can be derived which matches the functional form of Eq. 1:

$$r = \frac{\Sigma(k_j[{}^*C_2H_x{}^*])}{[K_{H_2}(H_2)]^2} = \frac{\Sigma\left(K_{C_2H_6}K_0 \ldots K_{j-1}k_{CC,j}(C_2H_6)\right)}{[K_{H_2}(H_2)]^{\frac{2+j}{2}}} \tag{2}$$

where $k_{CC,j}$ is the rate constant for the kinetically-relevant C-C bond cleavage reaction, K_j is the equilibrium constant for a quasi-equilibrated C-H activation and j represents the number of H atoms removed from the ethane. Taking into account isomers of C_2H_4, C_2H_3 and C_2H_2, there are ten possible dehydrogenated C_2 intermediates each with their own kinetically-relevant C-C bond activation, representing ten distinct possible mechanisms of the form shown in Scheme 1.

Periodic, plane-wave based DFT calculations were modeled on a four-layer Ir(111) slab using VASP with RPBE gradient corrections and PAW potentials. Reaction energetics confirmed that all dehydrogenation reactions were quasi-equilibrated from C_2H_6 to $^*C_2H_2^*$; further dehydrogenations to form *CHC* and *CC*, however, were not quasi-equilibrated. As such, there exists an equilibrium between C_2H_6 in the gas phase and eight $C_2H_x^*$ species, each of which can undergo a kinetically-relevant C-C bond activation with a rate exponentially proportional to the Gibbs free energy difference between the transition state with λ H_2 molecules in the gas phase and a H*-covered surface with C_2H_6 in the gas phase:

$$\Delta G^{\neq} = \Delta H^{\neq} + T\Delta S^{\neq} = G^{\neq} + \lambda G_{H_2} - 2G_{H^*} - G_{C_2H_6} \tag{3}$$

$DH^{\neq}$ and $DS^{\neq}$ were determined by measuring the dependence of the rate on temperature and are 213 ± 2 kJ mol^{-1} and 134 ± 5 J mol^{-1} K^{-1}, respectively at 20 kPa C_2H_6 and 1.8 MPa H_2.

In order to compute Gibbs free energy differences through DFT, vibrational frequency calculations were performed to determine zero-point vibrational energy corrections (ZPVE) as well as vibrational enthalpy and entropy of adsorbed species. For gas-phase species, additional rotational and translational enthalpy and entropy terms were computed from statistical mechanics. Results from theory suggest that C-C bond rupture occurs in *CHCH* with rates that would be 10^7 times greater than rates for any other potential intermediates. The predicted values of $DH^{\neq}$ and $DS^{\neq}$ are 218 kJ mol^{-1} and 126 J mol^{-1} K^{-1}, respectively, which are in excellent agreement with values from experiments.

These results on C_2H_6 hydrogenolysis with near-quantitative matching between experiment and theory establish a set of principles which will guide the investigation of larger n-alkanes and branched alkanes to determine how substitution across the C-C bond affects the mechanism for hydrogenolysis, ultimately affecting rates and selectivity.

(167d) A Four Step Catalytic Process for the Conversion of Aqueous Hemicellulose Stream to Hydrocarbons

Upadhye, A., University of Wisconsin
Olcay, H., Massachusetts Institute of Technology
Huber, G. W., University of Wisconsin
Originally presented on: 11/4/2013 16:15:00 - 16:35:00

A four step catalytic process for the conversion of aqueous hemicellulose stream to hydrocarbons

Aniruddha A. Upadhye,[1] Hakan Olcay[2], and George W. Huber[1, *]

1. Department of Chemical and Biological Engineering, University of Wisconsin, Madison, WI
2. Laboratory for Aviation and the Environment, Massachusetts Institute of Technology, Cambridge, MA

Depleting fossil fuel resources, increasing concerns over environmental effects arising from combustion of these fossil fuel resources, and growing global energy needs have driven today's research towards renewable energy technologies. Because of its abundance [1], biomass is considered as the most appropriate long-term alterative to fossil carbon. However, a large-scale transition to lignocellulosic "drop-in" fuels is constrained by the economic and technological challenges associated with biomass conversion processes.

Hemicellulose is one of the major components of biomass consisting of oligomers of C5 sugar. It can be easily separated in the form of aqueous stream from the lignocellulosic biomass using a pretreatment process. This aqueous hemicellulose stream is also a by-product of pulp and paper industry. The focus of this presentation is a four step catalytic process [2] for conversion of aqueous hemicellulose stream obtained from the pre-treatment processing. First, the oligomers of C5 sugars in hemicellulose stream (mainly xylose) are hydrolyzed and dehydrated to furfural in a continuous flow reactor with a homogeneous acid catalyst and a biphasic reaction scheme with tetrahydrofuran (THF) as co-solvent. Furfural yields up to 90% were obtained in this first step. The furfural in THF is then condensed with acetone to form a C13 precursor in a biphasic base catalyzed aldol condensation reaction. The C13 precursor is then hydrogenated over Ru/C catalyst at low temperatures to saturate the double bonds. This low temperature hydrogenation makes the C13 precursor more stable towards high temperature hydrodeoxygenation step. In this step, certain larger oligomers are also formed from partially hydrogenated product. The product of hydrogenation step is subsequently hydrodeoxygenated over Pt/SiO_2-Al_2O_3 at higher temperatures to obtain jet fuel range straight chain hydrocarbons. Using this process, we can produce hydrocarbons containing up to 31 carbon atoms. We have successfully demonstrated this process with aqueous hemicellulose stream obtained from real biomass. Here, we present the effects of different pretreatment processes on the downstream catalytic processing of the hemicellulose stream to hydrocarbons and provide directions for future research in catalytic conversion of biomass.

References:

[1] G. W. Huber, S Iborra, A. Corma, Chem. Rev., 2006, 106, 4044-4098.

[2] H. Olcay, A.V. Subrahmanyam, R. Xing, J. Lajoie, J.A. Dumesic, G.W. Huber, Energy Environ. Sci., 2013, 6, 205

(167e) Furfural Deoxygenation Over Carbon-Supported Noble Metal Catalysts

Thompson, S. T.,
Lamb, H. H., North Carolina State University
Originally presented on: 11/4/2013 16:35:00 - 16:55:00

Catalytic upgrading of fast-pyrolysis oils via hydrodeoxygenation (HDO) provides a viable approach for producing liquid transportation fuels from lignocellulosic biomass. Recent bio-oil upgrading research has demonstrated that carbon-supported noble metal catalysts, in particular Pd/C and Ru/C, provide higher yields of liquid products with lower oxygen contents than conventional hydrotreating catalysts, $NiMoS/Al_2O_3$ and $CoMoS/Al_2O_3$.[1] In this work, HDO of furfural, a hemicellulose pyrolysis model compound, was performed in a fixed-bed flow micro-reactor over Pd/C, Ru/C and PdRe/C catalysts at 200°C and 1 atm H_2. Two on-line gas chromatographs were employed for product analysis. Three main reaction pathways were observed: decarbonylation to furan, hydrogenation to furfuryl alcohol, and HDO to 2-methyl furan. Furan ring-opening products were not observed over any of the catalysts under these reaction conditions. Saturation of the furan ring was observed resulting in additional products: tetrahydrofuran (THF), tetrahydrofurfuryl alcohol, and 2-methyl THF. C_1 products included carbon monoxide, carbon dioxide, formaldehyde, methanol and methane. Pd/C was selective for furfural decarbonylation, yielding furan/THF and a stoichiometric quantity of CO. Furfural hydrogenation was also observed over Pd/C; furfural HDO was insignificant under these conditions. The observed catalytic performance of Pd/C is consistent with previous data for furfural HDO over Pd/SiO_2.[2] In contrast, PdRe/C produced nearly equal selectivity to decarbonylation and HDO. With regard to C_1 product selectivity, PdRe/C generated a small stoichiometric quantity of CO_2, indicative of water-gas shift activity. In comparison, Ru/C was selective to decarbonylation with markedly less ring saturation than either Pd/C or PdRe/C. The catalysts were characterized by CO chemisorption, temperature-programmed reduction, temperature-programmed hydride decomposition (TPHD), and extended x-ray absorption fine structure (EXAFS) spectroscopy. The bimetallic catalyst consisted of segregated Pd and Re nanoparticles as evidenced by EXAFS spectra at the Pd K- and Re L_{III}-edges and by TPHD, which indicated that a β-Pd hydride phase decomposes under 5% H_2at 55°C.

References:

1. Wildschut, J.; Mahfud, F. H.; Venderbosch, R. H.; Heeres, H. J. *Ind. Eng. Chem. Res.* **2009,** 48, (23), 10324-10334.

2. Sitthisa, S.; Resasco, D. E. *Catal. Lett.* **2011,** 141, (6), 784-791.

(167f) Effective Hydrodeoxygenation of Biomass-Derived Oxygenates By MoO3 Catalyst With Low H2 Pressures

Roman-Leshkov, Y., Massachusetts Institute of Technology
Prasomsri, T. A., University of Oklahoma
Nimmanwudipong, T., Massachusetts Institute of Technology
Originally presented on: 11/4/2013 16:55:00 - 17:15:00

Hydrodeoxygenation (HDO) is the most promising route to improve the quality of biomass-derived bio-oil. The key challenges faced by HDO processes is achieving a highdegree of oxygen removal, while simultaneously preserving the carbon backbone of the product, and minimizing hydrogen consumption. Several classes of catalysts have been explored for HDO including metal catalysts and non-precious metal catalysts. However, they are not selective toward carbon-oxygen bonds, and require high H2 pressures that result in the complete hydrogenation of all unsaturated bonds. The industrial HDO catalysts based on Co-Mo-Ni formulations give a superior performance in HDO, but undergo rapid deactivation due to coke formation and water poisoning. Metal oxides such as molybdenum oxide and tungsten oxide have been used for the oxidation of olefins to aldehydes and ketones due to their unique redox capability. Here, we found that biomass-derived oxygenate compounds are efficiently transformed into hydrocarbons by the use of MoO3 catalysts at low H2 pressures, presumably via reverse Mars-van Krevelen reactions. The model compounds used in this study were acetone, 2-hexanone, cyclohexanone, anisole, 2-methylfuran, and 2,5-dimethyfuran and all investigations were performed in packed-bed flow reactor. The effect of H2 pressure and water conten on catalyst performance is also reported.

We show that regardless of the type of oxygenate used, the HDO selectivity toward hydrocarbon production exceeded 97%. For instance, conversion, acetone was mainly converted to a mixture of propylene, hexenes. In contrast to acetone, no dimerized products were detected from the conversion of 2-hexanone. Instead, hexadienes, the dehydrogenated products of hexenes, were observed. The data show that the subsequent dehydrogenation of the deoxygenated product was more pronounced for cyclohexanone, compared to 2-hexanone. Interestingly, anisole was selectively converted to aromatic products. Specifically, at the reaction conditions investigated, the final products yielded benzene, toluene, xylenes, and alkylbenzenes, suggesting that this catalyst promotes deoxygenation and transalkylation reactions of the phenolic derivative. Finally, pentenes and hexenes were readily obtained from 2-methylfuran and 2,5-dimethylfuran, respectively. We hypothesize that a reverse Mars-van Krevelen cycle is responsible for the HDO behavior in which oxygenate compounds interact with oxygen vacancies to form an Mo–O bond, followed by cleavage of the C–O bond, and desorption of the resulting hydrocarbon products.

(167g) Aqueous-Phase Hydrodeoxygenation Activity of Pd/ZrO2 Catalysts

Liu, C., Pacific Northwest National Lab
Marin, O.,
Karim, A., Pacific Northwest National Lab
Brown, H., Pacific Northwest National Lab
Wang, Y., Pacific Northwest National Lab
Originally presented on: 11/4/2013 17:15:00 - 17:35:00

Aqueous-phase Hydrodeoxygation Activity of Pd/ZrO_2 Catalysts

Changjun Liu[†], Oscar Marin[†], Ayman M. Karim [*‡], Heather Brown[‡] and Yong Wang [*†,‡]

[†] The Gene & Linda Voiland School of Chemical Engineering and Bioengineering, Washington State University, Pullman Washington 99164, United States

[‡] Institute for Interfacial Catalysis, Pacific Northwest National Laboratory, Richland, Washington 99352, United States

Keywords: Aqueous-phase hydrodeoxygenation, bifunctional catalyst, biomass, zirconia

To meet the increase in worldwide energy demand, new processes that use a sustainable resource such as lignocellulosic biomass need to be developed [1]. A common step in all the processes that currently exist for the thermochemical conversion of biomass to liquid fuels/chemicals, (e.g. liquefaction fast pyrolysis/catalytic fast pyrolysis, and aqueous phase conversion of carbohydrates), is oxygen removal by hydrodeoxygenation (HDO). Our focus is on the aqueous phase HDO of sugar streams generated from the hydrolysis of biomass. Well-developed hydrolysis processes are capable of deconstructing the cellulose and hemicellulose in lignocellulose into their corresponding sugars (e.g., glucose, fructose, xylose, and et al.) and separate them from lignin[2]. Selective removal of oxygen from these biomass-derived compounds (monoses/polyols) is one of the key challenges in converting them into fuels and valuable chemicals[3]. Bifunctional catalysts with appropriate metallic and acidic sites were found capable of catalyzing the aqueous-phase HDO of polyols[4]. Wide ranges of Pt promoted solid acid catalysts have been tested[5]. However, in order to improve the selectivity to mono-oxygenates further understanding of the HDO reaction mechanism is required.

In this work, Pd/ZrO_2 catalysts promoted by various metal oxides (SnO_2, MoO_3, WO_3) were studied in the aqueous-phase hydrodeoxygenation of polyols with focus on its selectivity to C-O cleavage and C-C cleavage. Ethylene glycol and glycerol were used as model compounds of polyols to simplify the reaction routes and products. The interaction between Pd and WO_3-ZrO_2, the effects of palladium particle size, and the promotion effect of three metal oxides in aqueous-phase HDO of polyols were elucidated. The structures of the catalysts were characterized by CO FTIR and *in situ* XAFS. Addition of Sn was found to affect the Pd structure and accessibility due to PdSn alloy formation. The effect of the reaction conditions on the Pd and Sn structure and oxidation state will be discussed and correlated with the reaction pathways.

*Corresponding author

References

1. Huber, G. W.; Dale, B. E., Grassoline at the Pump. *Sci.Am.* **2009,** *301* (1), 52-56.

2. (a) Huber, G. W.; Iborra, S.; Corma, A., Synthesis of transportation fuels from biomass: chemistry, catalysts, and engineering. *Chemical Reviews* **2006,** *106* (9), 4044-4098; (b) Sun, Y.; Cheng, J., Hydrolysis of lignocellulosic materials for ethanol production: a review. *Bioresource technology* **2002,** *83* (1), 1-11.

3. Chen, L.; Zhu, Y.; Zheng, H.; Zhang, C.; Zhang, B.; Li, Y., Aqueous-phase hydrodeoxygenation of carboxylic acids to alcohols or alkanes over supported Ru catalysts. *Journal of Molecular Catalysis A: Chemical* **2011,** *351* (0), 217-227.

4. (a) Huber, G. W.; Cortright, R. D.; Dumesic, J. A., Renewable alkanes by aqueous-phase reforming of biomass-derived oxygenates. *Angewandte Chemie International Edition* **2004,** *43* (12), 1549-1551; (b) Li, N.; Huber, G. W., Aqueous-phase hydrodeoxygenation of sorbitol with Pt/SiO2-Al2O3: Identification of reaction intermediates. *Journal of Catalysis* **2010,** *270* (1), 48-59.

5. Li, N.; Tompsett, G. A.; Zhang, T.; Shi, J.; Wyman, C. E.; Huber, G. W., Renewable gasoline from aqueous phase hydrodeoxygenation of aqueous sugar solutions prepared by hydrolysis of maple wood. *Green Chemistry* **2011,** *13* (1), 91-101.

(167h) Utlilization of Char From Biomass Gasification for Catalytic Processing of Tars

Klinghoffer, N., Columbia University
Castaldi, M. J., City College of New York
Nzihou, A., Ecole des Mines d'Albi-Carmaux
Originally presented on: 11/4/2013 17:35:00 - 17:55:00

Char is a by-product of gasification and has the potential to be used in catalytic applications. It is a promising candidate for catalyzing the destruction of tar, which is another by-product of gasification. Tar is typically decomposed catalytically, however, tar reforming causes rapid catalyst deactivation. For this reason, it is expensive to use commercial metal catalysts for treatment of tars, since the catalysts will need to be replaced or regenerated frequently. Char, however, is produced on site (in the gasifier) and therefore can easily be replaced as it deactivates. Gasification conditions (co-reactant, temperature, residence time, etc.) can vary significantly, which will influence the physical and chemical properties of the char. This research focuses on demonstrating the catalytic activity of char from biomass gasification and understanding which char properties give rise to its catalytic activity.

The char studied in this work was produced from gasification of poplar wood chips. The biomass was gasified under CO_2 and steam at temperatures ranging from 550-920°C in a fluidized bed reactor. The char was recovered and its catalytic properties and performance were investigated. The char catalyzed the decomposition of methane, propane, and toluene, which is a major component in gasification tars. The catalytic performance of the char was compared to a commercial precious metal catalyst and the char had a lower light off temperature and higher extent of reaction than the commercial catalyst.

The char is primarily composed of carbon (~85%) and also contains oxygen (~7%) and trace amounts of many inorganic elements (ex. Ca, P, K, Na, Fe, Ni, Al, etc.). The influence of each of these components on the catalytic activity of the char was studied. The role of inorganics was studied by removing the inorganics from the char via an acid washing treatment (no other properties were modified from this treatment). It was found that removal of inorganic elements reduced the catalytic performance of the char, so the inorganics play a role in its catalytic activity. In addition, the inorganic elements were found to be highly dispersed, and lowering the dispersion decreased the catalytic activity. The presence of oxygen-based functional groups on the surface of the char was observed, and these functional groups are often considered to be important in the catalytic activity of carbon based catalysts (such as activated carbon). However, the oxygen functional groups desorbed at temperatures below 675°C which is lower than the temperature where hydrocarbon cracking reactions take place. Therefore, the catalytic performance was not influenced by oxygen functional groups for the reactions tested here. In order to understand the role of carbon, the carbon was removed (burned off), creating ash, and the catalytic activity of the ash was compared to the raw char. The activity of the ash was lower than the char. Therefore, the catalytic activity is improved by the presence of carbon, as well as the highly dispersed inorganic elements in the char.

Gasification conditions influenced the morphology of the char. The surface area and micropore volume were measured for char samples that were made in different gasification conditions. Surface area ranged from 429-687 m^2g^{-1}. Higher gasification temperature and longer residence times created chars with higher surface areas. In addition, the gasification environment influenced the porosity. Gasification in CO_2 produced char with more micropores (with diameter <1nm) compared to gasification with steam. As expected, the catalytic activity increased with increasing surface area. However, diffusion limitations were observed in the pores, leading to lower catalytic activity for chars with micropores, compared to non-microporous char with the same surface area. After the char was used as a catalyst the surface area was reduced by 20% and the pore volume was reduced by 30%, indicating that pore blocking is a mechanism of catalyst deactivation.

Calculations were done to determine if the quantity of char produced from a gasifier would be sufficient to reform all of the tar from the process. Based on the demonstrated catalytic activity of the char, and the deactivation profile of the char, the quantity of char is sufficient to reform all of the tar from a given gasifier. Therefore, utilization of char as a catalyst for tar reforming presents an economical and convenient solution for the reforming of tars, which remains one of the most significant issues with gasification processes. The activity of the char is attributed to the high surface area of the char, and the presence of carbon with dispersed inorganic elements.

(192a) Fabrication of Porous Carbon Nanofibers With Adjustable Pore Sizes As Electrodes for Supercapacitors

Tran, C., Drexel University
Kalra, V., Drexel University
Originally presented on: 11/4/2013 15:15:00 - 15:45:00

We report a facile method for obtaining extremely high surface area and uniformly porous carbon nanofibers for supercapacitors. Blends of polyacrylonitrile and a sacrificial polymer at different compositions have been electrospun into non-woven nanofiber mats with diameters in the range of 200–400 nm. Electrospun nanofiber mats are then subjected to carbonization to obtain porous carbon nanofibers (CNFs) as polyacrylonitrile converts to carbon and the sacrificial polymer decomposes out creating intra-fiber pores. Resultant porous CNFs exhibit specific surface area of up to 1600 $m^2\ g^{-1}$ with a large fraction of mesopores (2–4 nm). No additional chemical or physical activation process was used. We demonstrate the tunability of the pore sizes within CNFs by varying the composition. The non-woven fiber mats of porous CNFs are studied as free-standing electrode materials for supercapacitors eliminating the need for polymeric binding agents. Electrochemical measurements showed large specific gravimetric and volumetric capacitances of up to 210 F g^{-1} and 60 F cm^{-3} in 1 M H_2SO_4 at a high cyclic voltammetry scan rate of 100 mV s^{-1} due to the large fraction of mesopores. These materials retain 75% performance at a large current density of 20 A g^{-1} indicating excellent power handling capability.

(192b) Enhanced Energy Storage By Tunable Electrolyte Confinement in Structure-Directed CNT Arrays

Nicol, K. T., Mainstream Engineering Corporation
Zastrow, D., Mainstream Engineering Corporation
Hill, J. J., Mainstream Engineering Corporation
Originally presented on: 11/4/2013 15:45:00 - 16:00:00

The demand for power and energy dense storage devices is growing rapidly. While electrochemical double layer capacitors (EDLCs) offer high power density, many believe that high energy density is also obtainable through precise tuning of the electrolyte, electrode and structure of the overall cell. Multiwalled carbon nanotubes (MWNTs) are a promising electrode material for EDLCs because of their high electrical conductivity, chemical stability and exceptionally high surface area. Until single walled carbon nanotubes can be produced as solely conductive structures, dense and low-diameter MWNTs offer the best properties with regard to carbon-based EDLC electrode materials. When fabricated in anodic aluminum oxide templates, a MWNT is grown within each nanopore, resulting in an ultra-high density aligned array. This method produces a carbon electrode material with a significantly increased active surface area over commercially available materials. The energy density and capacitance of these materials can be further improved by tuning the size of the electrolyte domain around each MWNT. This is accomplished through partial and selective removal of the template, thereby confining the electrolyte domain. In typical EDLCs, the diffuse part of the double layer does not contribute to charge storage. However, we have observed an apparent charge storage enhancement in domain-confined MWNT electrodes. We will also discuss the tunable approach with regard to optimizing energy storage independent of the choice of REDOX couple or solvent.

(192c) Multifunctional Nitrogen-Rich "Brick-and-Mortar" Carbon As High Performance Supercapacitor Electrodes and Oxygen Reduction Electrocatalysts

Yu, D.,
Wei, L., Nanyang Technological University
Chen, Y., Nanyang Technological University
Originally presented on: 11/4/2013 16:00:00 - 16:15:00

Novel porous carbon materials with excellent electronic, chemical and structural properties and high nitrogen content are desirable for many applications. Here, we show the design and synthesis of a new multifunctional porous carbon material with unique “point to plane” architecture through a simple but effective activation-free procedure. Porous carbon derived from a new type of copolymer poly (vinylidene chloride-co-acrylonitrile) serves as the nitrogen-rich “mortar”. Reduced graphene oxide layers work as “bricks” with an aim to provide an open nanoscale scaffold and connect porous carbon, as well as modulate the ratio between mesopores and micropores. This new material has large surface area (957 m^2/g), high nitrogen content (6.6 at %), excellent conductivity (up to 5.1 S/cm), and favorable hierarchical meso- and microporosity. Benefiting from these intriguing features, this material shows ultrahigh specific capacitance at 361 F/g in aqueous electrolyte, far superior to that of many carbonaceous materials reported so far. The as-assembled asymmetric supercapacitors with the designed carbon material as negative electrodes and porous cobalt oxide nanorods as positive electrodes deliver a high-energy of 50.1 Wh/kg with a cell voltage of up to 1.6 V. Further, this material shows a superb electrocatalytic activity for oxygen reduction reaction in alkaline medium comparable to that of 20 wt% platinum-carbon electrodes, with much better methanol tolerant and long-term durability. We expect that this uniquely designed nitrogen-rich porous carbon material with its simple and scalable synthesis method will have great potentials for various applications in energy storage, energy conversion and catalysis.

(192d) Multiwalled Carbon Nanotubes With Tuned Surface Functionalities for Electrochemical Energy Storage

Wei, H., Lamar University
Gu, H., Lamar University
Guo, J., Lamar University
Wei, S., Lamar University
Guo, Z., Lamar University
Originally presented on: 11/4/2013 16:15:00 - 16:30:00

In the paper, surface functionalities effect on the electrochemical properties and energy storage properties of multiwalled nanotubes (MWNTs) were investigated via two different functionalized MWNTs, one modified with ether and carboxylic acid functional groups (denoted as MWNTs-EC), and the other with ester functional groups (denoted as MWNTs-E) as the supercapacitor materials. The functionalities onto the nanotubes were confirmed by Raman spectra and caused a minor enhancement in the band gap obtained from the Uv-vis spectra. Cyclic voltammetry results reveal that the capacitance of the MWNTs-EC, and MWNTs-E, was significantly increased by 41.5% and 123.2%, respectively, when compared to the as received MWNTs at lower scan rates. The energy density is also found to increase by 36.1 and 130.0% for the former and the latter, respectively, than that of the as received MWNTs-As in the galvanostatic charge/discharge measurement. The increases in both the capacitance and the energy density might arise from the enhanced surface wettability and larger pseudocapacitance contributions as confirmed by the electrochemical impedance spectroscopy. An equivalent circuit was proposed to disclose the nature of the capacitive behavior of the electrodes. The cyclic stability studies show an excellent long term cycling stability of the MWNTs, which retain 100.0% of the charge storage or discharge capacity even after 1000 charge-discharge cycles.

(192e) Three-Dimensional Core@Shell Nanostructured Array for Microscale Electrochemical Energy Storage

Mao, Y., Washington State University
Sun, X., Kansas State University
Originally presented on: 11/4/2013 16:30:00 - 16:45:00

Microelectromechanical systems (MEMS) based autonomous and biomedical devices play an ever increasing role in our society by monitoring the structural health of major civil works, the environmental presence of toxic or explosive compounds, and the physical health of human patients. Integration of robust and high power energy storage devices within these systems is important for reliable MEMS with necessary burst energy for data relay and dense components by reducing bulky external power sources. To overcome the shortcomings of conventional two-dimensional battery and supercapacitor configurations while developing more powerful energy storage devices, we focus on the development of three-dimensional architectures to take advantage of the vertical dimension-height. In this study, we designed a novel architecture composed of branched nanowire arrays, i.e. forest of nanotrees, to further boost electrochemical performance over one-dimensional nanowire arrays, inspired by branched trees with larger surface area to capture more sunlight. The prepared three-dimensional $ZnO@MnO_2$ core@shell branched nanowire arrays exhibit five times higher areal capacitance, better rate performance and smaller inner resistance than their nanowire array counterparts. These novel three-dimensional architectures offer promising designs for powering microelectronics and other autonomous devices on exceptionally small geometric scales.

(192f) Microfabricated Nickel Oxide Supercapacitors Based On High Aspect Ratio Concentric Cylindrical Electrodes

Armutlulu, A., Georgia Institute of Technology
Bidstrup Allen, S. A., Georgia Institute of Technology
Allen, M. G., Georgia Institute of Technology
Originally presented on: 11/4/2013 16:45:00 - 17:00:00

Supercapacitors are known as high-power energy storage devices, which narrow the application gap between the batteries and traditional electrostatic capacitors. As in the case of their electrostatic counterparts, they possess rapid charge and discharge capabilities with high degree of reversibility. Yet they can also store more energy per unit mass and/or volume which makes them superior to traditional capacitors and thus, yields a more attractive candidate for energy storage systems to be utilized in portable electronic devices. Supercapacitors can store energy electrostatically via separated charges in a Helmholtz double-layer that forms at the electrode/electrolyte interface, as well as electrochemically based on a phenomenon called pseudocapacitance that relies on reversible Faradaic charge transfer between the electrolyte and the surface of the electrode. In both energy storage mechanisms, electrode surface area plays a crucial role. In this study, our focus is on MEMS-enabled nickel oxide (NiO) based supercapacitors comprising high surface area 3D microelectrodes. The fabrication process consists of four major steps: formation of the array of high aspect ratio SU-8 pillars via backside exposure, deposition of alternating Cu and Ni layers on the SU-8 pillars by a sequential robotic electroplating process, mechanical and chemical etching of sacrificial Cu layers to obtain high surface area Ni structures in the form of concentric cylinders, and lastly conformal electrodeposition of the active material, i.e. nickel hydroxide ($Ni(OH)_2$), on the high surface area Ni backbone. Following the electrodeposition process of the $Ni(OH)_2$ layer, the structure is heated up to 300 °C in order to convert $Ni(OH)_2$ film to NiO, and the fabrication process is finalized. Performance tests for the resulting structures are carried out via cyclic voltammetry in 1 M KOH solution. Devices are cycled in excess of 500 times, and no significant capacitance degradation is observed. The areal and specific capacitance values of the supercapacitors are estimated to be 270 mF/cm^2 and 56 F/g, respectively.

(192g) Pretreatment Effects On Charge Storage of Early Transition-Metal Carbides and Nitrides

Djire, A., University of Michigan

Pande, P., University of Michigan

Sleightholme, A. E. S., University of Michigan

Deb, A., University of Michigan

Rasmussen, P. G., University of Michigan

Penner-Hahn, J., University of Michigan

Thompson, L. T., University of Michigan

Originally presented on: 11/4/2013 17:00:00 - 17:15:00

Pretreatment Effects on Charge Storage of Early Transition-Metal Carbides and Nitrides

Abdoulaye Djire[a], Priyanka Pande[a], Aniruddha Deb[b], Alice E. S. Sleightholme[a], Paul Rasmussen[ac], James Penner-Hahn[b], Levi T Thompson[ac]*

[a] Department of Chemical Engineering

[b] Department of Chemistry

[c] Hydrogen Energy Technology Laboratory

University of Michigan, Ann Arbor, MI_- 48109-2130

E-mail: ltt@umich.edu

Early transition-metal carbides and nitrides are promising candidates for use in supercapacitor electrodes due to their high electronic conductivities, high surface areas (can exceed 200 m^2/g), good electrochemical stabilities and high capacitance [1,2]. For example, the capacitance for VN has been reported to be as high as 1340 Fg^{-1} in aqueous KOH [3]. This high capacitance has been attributed to pseudocapacitive charge storage involving fast near-surface redox reactions. We observed that pretreatment of the surface to remove the passivation layer significantly improved the capacitances and electrochemical stabilities of early transition-metal carbides and nitrides in aqueous electrolytes. In this paper we present results from electrochemical characterization including cyclic voltammetry (CV), chronopotentiometry, and electrochemical impedance spectroscopy (EIS), and surface characterization using x-ray photoelectron spectroscopy (XPS) to explain changes in surface chemistry and electrochemical properties caused by pretreatment.

The early transition-metal carbides and nitrides of V, W and Mo were synthesized via temperature-programmed-reaction of their oxide precursors with 15% CH_4/H_2 or anhydrous NH_3 followed by passivation in 1% O_2/He at room temperature to form a oxygen-rich passivation layer preventing bulk oxidation on exposure to air [1]. Characterization of the structural properties was performed using nitrogen physisorption (BET surface area) and X-ray diffraction. Prior to the electrochemical characterization, the electrodes were pretreated in 0.3M NaOH aqueous solution for 1 min then rinsed with acetone, ethanol, and ultrapure water [4]. The CV was used to establish the stability windows and capacitances for these materials. The capacitance was deconvoluted into double-layer capacitance and pseudocapacitance by means of CV and EIS. Chronopotentiometry was used for further stability measurements and to determine changes in surface activities caused by pretreatment.

As shown in Figure 1, the specific capacitance for VN increased by 57% after pretreatment. This enhancement in performance suggests significant improvement in material electrochemical activities. We believe that pretreatment altered the charge storage for the carbides and nitrides. We also observed good electrochemical stability after successive cycling of the material for several thousand cycles. In general, the treated materials were more stable than the passivated materials.

Figure 2 shows the open circuit potential (OCP) or rest potentials for the passivated and treated VN in acidic medium. The OCP for the treated material is lower than that for the passivated material. We attributed this to changes in surface activities due to pretreatment. Using the Nernst equation, we found that the treated VN had a higher pseudocapacitance than the passivated VN. This pseudocapacitive charge-storage mechanism is being investigated using XPS and *in-situ* x-ray absorption spectroscopy.

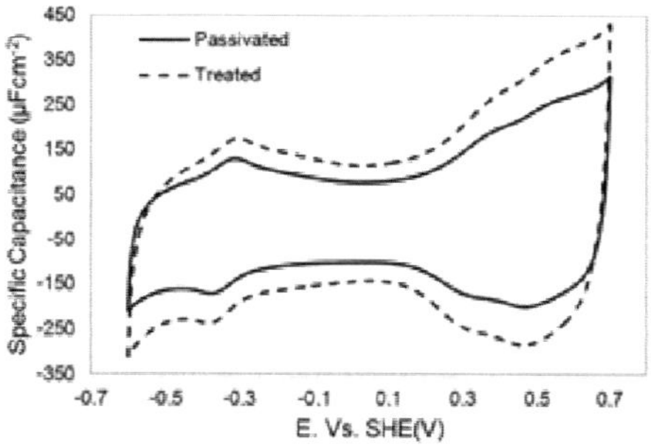

Figure 1: CV of passivated and treated VN in 0.1M H_2SO_4 at 50 mVs^{-1}.

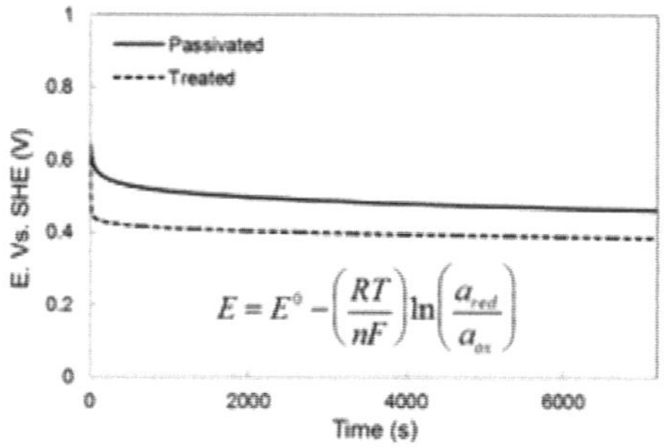

Figure 2: Rest potential of passivated and treated VN in 0.1M H_2SO_4

The physical and electrochemical properties will be described in terms of a proposed mechanism for psuedocapacitive charge storage.

References

(1) Cladridge, J. B.; York, A. P. E.; Brungs, A. J.; Green Malcolm L. H.; *Chem. Mater.* **2000**, 12, 132.

(2) Wixom, M.R.; Tarnowski, D. J.; Parker, J. M.; Lee , J.Q.; Chen, P. ?L.; Song, I.; Thompson, L. T.; *Mat. Res. Soc. Symp. Proc.* **1998,** 496, 643.

(3) Choi, D.; Kumta, P. N.; *Electrochem. Solid-State Lett.* **2005,** 8, 8, A418.

(4) Weidman, M.C.; Elposito, D.V.; Hsu, I. J.; Chen J. G.; J. Electrochem Soc.**2010**, 157, 12, F179

(192h) Doped Transition Metal Oxide Composite Electrodes for Supercapacitor Applications

Jampani Hanumantha, P., University of Pittsburgh
Kadakia, K., University of Pittsburgh
Hong, D. H., University of Pittsburgh
Poston, J., National Energy Technology Laboratory, U.S. Department of Energy
Ayyakkannu, M., National Energy Technology Laboratory
Kumta, P. N., University of Pittsburgh
Originally presented on: 11/4/2013 17:15:00 - 17:30:00

There is a dire need to provide a sustainable solution to the growing global demand for energy. It is this incessant demand that is the driving force fueling various current and emerging technologies including lithium ion batteries, supercapacitors and fuel cells. Supercapacitors occupy a unique place among these energy storage devices on account of their high power density and long cyclability. Supercapacitors are electrochemical charge storage devices capable of delivering large bursts of charge over a very short period of time. It is due to this unique ability that supercapacitors assume a very important role in the growing area of alternative energy storage systems. High surface area carbons, hydrated ruthenium oxide and MnO_2 are among the primary materials being considered for supercapacitor applications. Bulk oxide materials are traditionally hamstrung on account of their limited electrical conductivity for various electronic and electrochemical applications. In order to gain a fundamental understanding of the charge storage mechanisms in oxide materials, we deposited films of undoped transition metal oxides and doped transition metal oxides on current collectors and used various materials and electrochemical characterization techniques including X-ray photo-electron spectroscopy (XPS), cyclic voltammetry and electrochemical impedance spectroscopy to obtain a first-hand understanding of the underlying mechanisms. Effect of coating the various oxides on vertically aligned carbon nanotubes was also examined. Results of these studies will be presented and discussed.

(192i) Novel 3-D MnO2/Holey Graphene Nanostructure for Supercapacitor Applications With Enhanced Electrochemical Performances

Wang, L., Wayne State University
Meng, T., Wayne State University
Ara, M., Wayne State University
Deng, D., Wayne State University
Ng, S., Wayne State University
Originally presented on: 11/4/2013 17:30:00 - 17:45:00

As an excellent energy storage device with high-power density and long life cycle, supercapacitor has been intensely investigated in many aspects, such as electrode materials, electrode structure engineering, electrolyte compositions, separator improvements, as well as device design. Among them, one of the most urgent shortcuts is the around 10 times lower energy density compared to the traditional Li-ion batteries. This drawback has dramatically confined the applications of supercapacitors in electrical vehicles, intermittent energy storages and other occasions where both the power density and energy density are required. However, many researchers have been working on improving the energy density without significantly scarifying power density. These researches include increasing the voltage windows by employing organic electrolyte with larger voltage openings, applications of thinner and yet still mechanically strong separators, integrate materials with higher intrinsic capacitance, and so on. Despite all the effort on supercapacitor research, the current state-of-art on energy density still needs to be improved to meet the fast-growing application demanding in the emerging and potential applications. Therefore, we propose a mild method of preparing a novel 3-D MnO_2/holey graphene composite for supercapacitor electrodes. The nano MnO_2 nanorods in the composites serves as a capacitance booster due to its high intrinct capacitance and spacer to prevent 2-D graphene sheets from stacking back together. At the same time, graphene also serves both as active material to provide large surface area for charge storage and conductive agent to collect electrons generated by the surface redox reactions of semiconductive MnO_2. Holes on graphene sheets enhance the efficient accessibility of electrolyte to the surface of active materials. With this novel 3-D composite, an increasement of around 30% to 50% in capacitance is expected.

(192j) Electrochemical Energy Storage of Magnetic Carbon Nanocomposites: Role of Magnetocapacitance and Magnetohydrodynamics

Zhu, J., Lamar University
Wei, S., Lamar University
Guo, Z., Lamar University
Originally presented on: 11/4/2013 17:45:00 - 18:00:00

Magnetic field has been used in different chemical processes to improve the quality of electrodeposited films, yield of photoredox products, and oxidative reactivity of catalysts and *etc*. In this work, a small external magnetic field will be imposed on a supercapacitor unit trying to improve its electrochemical energy density. Magnetic microtubular carbon nanocomposite fabrics decorated with uniformly distributed metal oxides (iron, cobalt and nickel oxides) nanoparticles will be fabricated via both microwave-assisted and conventional annealing processes. These fabrics could serve as flexible electrodes for supercapacitors to deliver high energy density and exhibit excellent cycling stability. The effect of metal oxide species, annealing method and applied magnetic field on the capacitance of each nanocomposite material will be tested using cyclic voltammetry and galvanostatic charge/discharge measurements. The corresponding mechanism will be investigated using electrochemical impedance spectroscopy. Cycling stability of each composite with and without magnetic field will be also comparatively studied.

(192k) Synthesis and Characterization of Orange Peel As Electrodes in Li-Ion Capacitors

Arie, A. A., Parahyangan University

Originally presented on: 11/4/2013 18:00:00 - 18:15:00

Orange peel was utilized as the raw material for activated carbon-based electrodes in Li-ion capacitors. The carbon samples weres synthesized
using a chemical activation process followed by a surface modification method. The surface modification method showed no significant effect on the specific surface area, however it changed the surface property of the carbon samples. Experimental results demonstrated that the electrochemical characteristics of un-treated activated carbons had a better specific capacitance compared with those obtained by surface modification method. Based on our results, orange peel waste can potentially be used as a cheap raw material for the mass production of activated carbons as electrodes in Li-ion capacitors.

(237a) Biomasspyrolysisrefinery: Hydrodeoxygenation of Liquid Phase Pyrolysis Oil

Pucher, H., Graz University of Technology
Feiner, R., Graz University of Technology
Schwaiger, N., Graz University of Technology
Pucher, P., BDI BioEnergy International AG
Siebenhofer, M., Graz University of Technology
Originally presented on: 11/5/2013 8:30:00 - 8:50:00

BiomassPyrolysisRefinery:

Hydrodeoxygenation of Liquid Phase Pyrolysis Oil

H. Pucher, R. Feiner, N. Schwaiger, P. Pucher*, M. Siebenhofer

Graz University of Technology, Institute of Chemical Engineering and Environmental Technology, *BDI-BioEnergy International AG

Non-renewable energy resources such as oil and coal will not suffice the increasing demand of energy in the future. It is assumed that renewable resources such as biomass are able to fill this dawning gap because they have been considered as the most promising energy sources in the next century [1]. In addition the European Commission has released a directive to increase the volume of renewable biofuels for transportation to 10% by 2020 [2].

To fill the upcoming gap and meet the requirements of the directive a biomass liquefaction concept from renewable resources has been developed. The BiomassPyrolysisRefinery concept consists of two main process steps. In the first step lignocellulosic biomass like wood and straw, is converted in pyrolysis oil and pyrolysis char through liquid phase pyrolysis. In the second step these intermediate products are upgraded. Two promising upgrading technologies were investigated: hydrodeoxygenation [3] [4] of pyrolysis oil, and hydrogenation of liquid phase pyrolysis char. The main focus of this project was to investigate and optimize upgrading of liquid phase pyrolysis oil to maximize the output of the carbon rich fraction, raise stability and identify potential fields of application. Investigations were carried out in lab size batch mode. Through hydrodeoxygenation the polar, highly acidic, water and oxygen rich liquid phase pyrolysis oil is converted into an appropriate, carbon rich, oxygen poor and water poor product phase. The upgraded product is seemingly ready for further processing in conventional oil refining.

Depending on the field of application, the liquid phase pyrolysis oil can be dehydrated and the water content can be reduced by 90%.

The operation parameters for hydrodeoxygenation, like hydrogen pressure, temperature, reaction time and catalysts, were determined. Table 1 shows representative results.

Table 1: Comparison of Liquid Phase Pyrolysis Oil, Pyrolysis Oil dehydrated and Pyrolysis Oil upgraded.

		Liquid Phase Pyrolysis Oil	Pyrolysis Oil dehydrated	Pyrolysis Oil upgraded
Water Content	[%]	60	7	4
Lower Calorific Value	[kJ/kg]	7600	20500	30160
Carbon Content	[%]	22	52	69
Hydrogen Content	[%]	10	7	9
Oxygen Content	[%]	68	41	21
Nitrogen Content	[%]	<1	<1	<1

70 percent of the organic carbon of the liquid phase pyrolysis oil was transferred to a carbon rich product phase (upgraded pyrolysis oil). The upgraded pyrolysis oil has an oxygen content of about 21% and a hydrogen content of about 9%.

In conclusion trough dehydration as well as hydrodeoxygenation a dehydrated/upgraded pyrolysis oil with lower oxygen and water content but higher energy density is produced. Composition and characterization of the products were determined by SEC, GC-MS, elemental analysis, Karl Fischer Titration, TAN, NDIR continuous gas analysis and by a solvent-solvent extraction according to the Oasmaa guideline [6].

[1] Wang, G. et al., "The direct liquefaction of sawdust in tetralin," *Energy Sources, Part A,* pp. 1221-1231, 2007.

[2] Richtlinie_2009/28/EG, "EU Richtlinie zur Förderung der Nutzung von Energie aus erneuerbaren Quellen und zur Änderung und anschließenden Aufhebung der Richtlinien.".

[3] Mercader, F.M. et al., "Hydrodeoxygenation of pyrolysis oil fractions: Process understanding and quality assessment through co-processing in refinery units," *Energy & Environmental Science ,* pp. 985-997, 2011.

[4] Wildschut, J. et al., "Insights in the hydrotreatment of fast pyrolysis oil using a ruthenium on carbon catalyst," *Energy Environ. Sci.,* pp. 962-970, 2010.

[5] Mortensen, P.M. et al., "A Review of catalytic upgrading of bio-oil to engine fuels," *Applied Catalysis A: General,* pp. 1-19, 2011.

[6] Anja, O., Fuel Oil Quality Properties of wood-based Pyrolysis Liquids, 2003.

(237b) Production of Transportation Fuels Range Hydrocarbons From Pyrolysis Oil Via A Low Hydrogen Catalytic Deoxygenation

Tanneru, S., Mississippi State University
Steele, P., Mississippi State University
Originally presented on: 11/5/2013 8:50:00 - 9:10:00

Biomass derived fast pyrolysis oil is a potential alternative replacement for conventional fuels. But the negative properties such as lower energy density, higher water content and acidity prevent the direct use of pyrolysis oil as a fuel. Conversion of pyrolysis oil to hydrocarbons has been successfully performed by deoxygenation with hydroprocessing catalysis with high heat and pressure. However, rapid deactivation of the catalysts and consumption of a large amount of expensive hydrogen have remained a problem for this technology and have prevented commercialization. Therefore, development of an efficient and reduced hydrogen deoxygenation method would be desirable. In this study, we have developed recently a bio-oil pretreatment method patent pending by oxidation to produce an acidified product. This pretreatment of raw bio-oil process allowed to performance of a partial deoxygenation step with a low amount of hydrogen during the hydroprocessing of bio-oil. This partially deoxygenated product was then fully deoxygenated with pure hydrogen to produce mixed hydrocarbons. Properties of the resultant hydrocarbon fuels were analyzed by ASTM standards for transportation fuels. The hydrocarbon mixture obtained by our process was analyzed by gas chromatography mass spectroscopy (GC-MS) to identify all compounds in the fuel. Detailed hydrocarbon analysis (DHA) was performed to provide the carbon numbers associated with the hydrocarbons. Simulated distillation was performed to analyze the gasoline, jet fuel and diesel fractions comprising the hydrocarbon mixture.

Keywords: Pyrolysis oil, oxidation, deoxygention, hydroprocessing, GC-MS and DHA.

(237c) Woody Biomass to Liquid Hydrocarbon Fuel

Zhu, T., United Technologies Research Center
Emerson, S. C., United Technologies Research Center
Vanderspurt, T. H., United Technologies Research Center
She, Y., United Technologies Research Center
Colket, M.,
Hautman, D., United Technologies Research Center
Davis, T., ClearEdge Power
Originally presented on: 11/5/2013 9:10:00 - 9:30:00

This work describes the production of jet fuel range hydrocarbons from woody biomass. Both untreated yellow poplar (Liriodendron tulipifera) and commercially available 100-mesh wood flour were used in experiments. A novel, low temperature process was developed in which the lignocellulosic feedstock, particularly the lignin fraction, is minimally broken down into processable fragments through liquid phase hydro-pyrolysis in the presence of a base and a solid catalyst. The resulting fuel precursors mainly consist of phenol, cyclohexanol, alkyl-phenols and alkyl-cyclohexanols, derived from the lignin fragments of the biomass. The hydropyrolysis process was carried out in a 0.5-L Inconel stirred autoclave. The precursor yield was studied at temperatures of 200 °C to 300 °C and at wood/base mass ratios of 0.27 to 3 in the presence of hydrogen. The results showed that low temperatures and a low wood/base ratio favored higher precursor yield. The maximum fuel precursor yield achieved was 55.1%. The hydropyrolysis process was also carried out in a 1.8 L autoclave to scale up the process. The results of the scale-up process will be presented.

The fuel precursors were separated from the unconverted biomass, mainly cellulose and organic acid byproducts, and were coupled into C_9-C_{16} hydrocarbons in the presence of a solid catalyst and hydrogen. Both precious metal and base metal catalysts were studied for the precursor coupling reaction. With a model compound mixture, such as phenol and cyclohexanol (2:1 by weight), a maximum C_{11}+ yield of 68% was achieved at 275 °C and the major product was 2-cyclohexylcyclohexanol. With increasing coupling temperatures, a more hydrogenated product, 1,1'-bicyclohexyl, was produced with a slight decrease in overall C_{11}+ yield. Water was also found to inhibit the yield of the coupling process. Cobalt-based catalysts showed more resistance to water inhibition than palladium-based catalysts. Using solvent-extracted (water-free) fuel precursors obtained from the hydro-pyrolysis of a hardwood flour mixture, a 30% yield of C_9-C_{12} hydrocarbons was obtained with a Pd-based catalyst. The produced fuel was composed of hydrocarbons in the C_8 to C_{16} range. The estimated density of the fuel sample was about 0.802 g/cm^3 and the average boiling point was approximately 159 °C.

In the presentation, the different steps of the proposed biomass to fuel process and associated experimental results will be discussed. Techno-economic analysis results based on a HYSYS material and energy balance process model for fuel production from wood will also be presented.

(237d) Comparison of In-Situ and Ex-Situ Catalytic Fast Pyrolysis of Pinewood and Phragmites Australis

Hammer, N. L., USDA
Garrido, R., Villanova University
Welch, S.,
Coe, C. G., Air Products and Chemicals, Inc.
Smith, M. A., Villanova University
Satrio, J. A., Villanova University
Originally presented on: 11/5/2013 9:30:00 - 9:50:00

One of the routes to provide sustainable chemicals and fuels from renewable sources that is receiving considerable attention is the catalytic pyrolysis of lignocellulosic biomass, such as pinewood and *Phragmites Australis*. Comparison of traditional *in-situ* catalysis, where the catalyst is directly in contact with the biomass at 500°C, to *ex-situ* catalysis where conditions are milder for catalytic upgrading of pyrolyzed vapors before condensation. These pyrolysis studies will be conducted in two reactors; CDS micropyrolyzer (µg scale) connected to a GC/MS and a bench scale tubular furnace (g scale). Both have an external fixed bed catalytic reactor downstream from the pyrolysis reactor. This study analyses the effect of various acidified amorphous silica based catalysts on the products formed from catalytic cracking during fast pyrolysis both *in-situ* and *ex-situ* for the two pyrolysis reactor systems. The preliminary results from the *in-situ* catalytic pyrolysis of pinewood at 500°C shows that mesoporous SBA-15 silica-based catalyst acidified with 75% monolayer coverage of sulfuric acid exhibits unusual behavior. The lignin derived phenolics are completely absent and the main condensable products are furfural and levoglucosenone (LGO). In contrast, under the same pyrolysis conditions and acid loading, an amorphous silica catalyst, having a larger porosity and a higher pore volume, gives a much broader distribution of products which includes alcohols and phenolics in addition to LGO and furfural. Based on these preliminary findings, additional work is underway to explore the structural and compositional features of silica based catalysts that are contributing to the high selectivity to LGO as well as looking at how the catalytic pyrolysis conditions and reactor configuration influence the product distribution. Resent results from these studies will be presented.

(237e) Upgrading Biomass-Derived Syngas to Mixed Hydrocarbons Over Three-Dimensionally Ordered Macroporous (3DOM) Iron Fischer-Tropsch Catalysts

Hu, J., Mississippi State University
Yu, F., Mississippi State University
Lu, Y., Mississippi State University
Zhou, P., Mississippi State University
Originally presented on: 11/5/2013 9:50:00 - 10:10:00

Nowadays, our society relies heavily on traditional fossil fuels. It was estimated that the global energy need will continue rising due to world population explosion [1]. However, traditional fuels will soon deplete in the near future because of very limited reservation [2]. The extensive burning of traditional fossil fuels can produce and emit large amounts of greenhouse gas, such as carbon dioxide, into atmosphere and cause some other severe environmental issues. Thus, it is necessary and critical to find alternative solutions to suffice our energy needs [3-5].

The potential route of renewable fuels production, such as gasoline, diesel and jet fuel, through Biomass to Liquid via Fischer-Tropsch Synthesis (BTL-FT) process has been gaining increasing interests from academia and industry because of its potential ability to produce carbon neutral and environmentally friendly clean fuels, such kind of fuels can help to meet the globally increasing energy demand and also to meet the stricter environmental regulations in the future [4, 6]. In the BTL-FT process biomass, such as woodchips, is firstly gasified with air, oxygen, and/or steam to produce raw bio-syngas. Then cleaning process is applied to the raw bio-syngas to remove contaminants like small char particles, ash, and tar. The cleaned bio-syngas is then conducted into catalytic reactor to perform FT synthesis to produce renewable liquid fuels [7, 8].

Compared with the bulk transition-metal oxide catalysts, 3DOM catalysts possess unique physicochemical properties because of their uniform structure, interconnected porous channel, higher surface areas, and good catalytic performance [9-12], which make them show promising potentials in the Fischer-Tropsch synthesis. In order to investigate this little-studied area, we prepared and characterized series of 3DOM Fischer-Tropsch catalysts and utilize them to catalytically convert biomass-derived syngas into renewable liquid transportation fuels, such as gasoline, diesel and jet fuel.

(1) Kaygusuz, K. *Renewable and Sustainable Energy Reviews,* **2012**, 16, 1116-1126.

(2) Veziroğlu, T.N.; Şahi˙n, S. *Energy Conversion and Management,* **2008**, 49, 1820-1831.

(3) Davis, S.J.; Caldeira, K. *Proceedings of the National Academy of Sciences,* **2010**, 107, 5687-5692.

(4) Street, J.; Yu, F. *Biofuels,* **2011**, 2, 677-691.

(5) Le Quere, C.; Raupach, M.R.; Canadell, J.G.; Marland, G. *Nature Geosci.* **2009**, 2, 831-836.

(6) Street, J.; Yu, F.; Wooten, J.; Columbus, E.; White, M.; Warnock, J. *Fuel.* **2012**, 96, 239–249.

(7) Demirbas, A. *Energy Education Science Technology,* **2006**, 17, 27–63.

(8) Tijmensen, M.J.A.; Faaij, A.P.C.; Hamelinck, C.N.; van Hardeveld, M.R.M. *Biomass and Bioenergy*, **2002**, 23, 129-152.

(9) Li, J.; Han, T. *Biosensors and Bioelectronics*, **2009**, 25(4), 773-777.

(10) Bosco, J. P.; Sasaki, K. *Chemistry of Materials*, **2009**, 22(3), 966-973.

(11) Lai, C.-Z.; Joyer, M. *Journal of Solid State Electrochemistry*. **2009**, 13(1), 123-128.

(12) Wei, N.; Xin, J. *Biosensors and Bioelectronics*, **2011**, 26(8), 3602-3607.

(237f) Pyrolytic Biodiesel From Vacuum Pyrolysis of Plant Oil Asphalt Catalyzed By Sodium and Potassium Compounds

Tang, Q., Tsinghua University
Zheng, Y., Tsinghua University
Li, J.,
Wang, J., Tsinghua University
Originally presented on: 11/5/2013 10:10:00 - 10:30:00

Plant oil asphalt (POA) is the residue biomass generated in biodiesel and fatty acids (FAs) industry. POA is named after its characteristics as high-density, high-viscosity black liquid. Recently, production capacity of POAs presents a rapidly increase owing to the booming development of biodiesel (FAMEs, fatty acids methyl esters) and FAs industry. At present, POA is mainly landfilled or used as boiler fuel, which cause adverse effects on economy and environment.

A POA sample from a biodiesel plant in China was used as raw material for pyrolysis. Our previous work found that the POA is a mixture of polymeric products of FAMEs and some natural components like waxes origin from plant oils. Non-catalytic vacuum pyrolysis of POA could obtain 71 wt% of pyrolytic oil. It contains 65wt% C6-C19 FAMEs/FAs and 33 wt% C7-C24 hydrocarbons. A pyrolytic biodiesel product was obtained via esterification of the pyrolytic oil. Compared with hydrocarbons fuel, the pyrolytic biodiesel had a few disadvantages stemming from the high content oxygen at 9.10 wt%, namely less than ideal storage stability, unfavorable cold flow properties and relatively low heating value. Therefore, it is objective to obtain pyrolytic biodiesel with more hydrocarbons content via enhancing deoxygenation reaction during pyrolysis.

In this work, vacuum pyrolysis of POA for hydrocarbons catalyzed by sodium and potassium compounds was investigated in a semi-batch reactor. The catalysts employed include NaOH, Na_2CO_3 and KOH, K_2CO_3. Both sodium and potassium compounds could enhance deoxygenation reactions, contributing to pyrolytic oil with higher oxygen content than non-catalytic. However, the catalysts improve the pyrolytic oil yields in the order of KOH (73 wt%) > K_2CO_3 (71 wt%) ~ no catalyst (71 wt%) > NaOH (55 wt%) > Na_2CO_3 (49 wt%). Sodium compounds increase yields of bio-gas and bio-char simultaneously. While for potassium compounds, increase of bio-gas and decrease of char offset each other. When 5 wt% KOH was added, the pyrolytic oil consists of 74 wt% C6-C24 hydrocarbons and 23 wt% C7-C19 FAMEs/FAs. Finally, the oxygen content of the pyrolytic biodiesel was decreased to 2.17 wt% compared with non-catalytic.

Besides, a pyrolysis pathway and kinetics for thermal decomposition of POA were explored.
We believe that POA can become a new valuable biomass resource and the results from this work lay a foundation for further study on pyrolysis of POA.

(237g) Aqueous-Phase Hydrogenation of Model Bio-Oil Compounds Under Mild Condition Over CNTs Supported Catalysts

Zhou, M., Southeast University
Zhu, H., Southeast University
Xiao, G., Southeast University
Xiao, R., Southeast University
Originally presented on: 11/5/2013 10:30:00 - 10:50:00

Abstract:

Bio-oil is an attractive bio-fuel in view of the development of renewable energy sources, which could offer sustainable resources and reduce our dependence on limited fossil fuels in the future. However, the quality of bio-oil should be improved by the partial or total elimination of the oxygenated functionalities. In general, bio-oil upgrading can be accomplished in three steps: water extraction, low temperature hydrogenation of the water-soluble phase, and aqueous-phase reformation (to prepare H_2) or dehydration/hydrogenation (to alkanes). In this work, hydrogenation of four model compounds of the bio-oil in aqueous phase, including furfural, hydroxyacetone, cyclopentanone and acetaldehyde, was carried out under mild reaction conditions ($T \leq 473K$) in a high-pressure autoclave. A series of nickel-based catalysts(La, Co,Mo, etc. as additives) with HNO_3-pretreated CNTs as support were synthesized using impregnation and calcination methods, which were successfully applied for the upgrading of model compound in bio-oil. The investigated compounds were accordingly converted to alcohol, such as cyclopentanol, 1,2-propanediol, ethanol and so on. Effects of nickel loading amount, reaction temperature, reaction time and hydrogen pressure on conversion rate as well as selectivity for desired products were investigated in detail, which could provide possible alternatives for the upgrading of bio-oil.

(262a) Effects of Composition and Compositional Distribution On the Optoelectronic Properties and Function of Semiconductor Ternary Quantum Dots

Han, X., University of Massachusetts Amherst
Pandey, S. C., University of Massachusetts
Maroudas, D., University of Massachusetts Amherst
Originally presented on: 11/5/2013 8:30:00 - 8:49:00

The function of nanometer-size quantum dots (QDs) of ternary compound semiconductors, such as $In_xGa_{1-x}As$, $ZnSe_{1-x}S_x$, and $ZnSe_{1-x}Te_x$, used in the fabrication of optoelectronic and photovoltaic devices can be optimized by precise tuning of their electronic band gap through control of the QD composition (x) and diameter. In this presentation, we report results on compositional distributions in ternary QDs and how they affect the QDs' electronic band gap. We follow a hierarchical modeling approach that combines first-principles density functional theory (DFT) calculations and classical Monte Carlo simulations with a continuum model of species transport in spherical nanocrystals. In certain cases, the model predicts the formation of a concentration boundary layer near the nanocrystal surface, i.e., the formation of core/shell-like structures with shell regions rich in the surface segregating species. A systematic parametric analysis generates a database of transport properties, which can be used to design post-growth thermal annealing processes that enable the development of thermodynamically stable QDs.

Based on the continuum transport model, we have analyzed species interdiffusion kinetics in ternary semiconductor nanocrystals during their thermal annealing and used the modeling results to interpret the evolution of near-surface species concentration according to X-ray photoelectron spectroscopy (XPS) measurements. We have also explored the impact of composition and compositional distribution on the ternary QDs' electronic band gaps based on first-principles DFT calculations and found that ternary QDs with thermodynamically stable compositional distributions allow for precise band-gap tuning. Our findings lead to a proposal for an efficient one-step ternary QD synthesis method followed by annealing to promote self assembly of the thermodynamically stable configuration for optimal optoelectronic function in devices.

(262b) Molecular Modeling of Nanoparticles and Conjugated Polymers During Synthesis of Photoactive Layers of Organic Photovoltaic Solar Cells

Mortuza, S.,
Cisneros, C.,
Dela Cruz Bartolo, M.,
Banerjee, S.,
Originally presented on: 11/5/2013 8:49:00 - 9:08:00

Molecular modeling of nanoparticles and conjugated polymers during synthesis of photoactive layers of organic photovoltaic solar cells

S.M. Mortuza (1), Corinna Cisneros (1), Mark Dela Cruz Bartolo (2) and Soumik Banerjee (1)

1. School of Mechanical and Materials Engineering, Washington State University, Pullman, Washington, 99164, USA

2. California State University San Marcos, San Marcos, CA 92096

Solar cells are widely considered as potential alternative energy devices that can meet the increasing global energy demand. In particular, organic photovoltaic (OPV) solar cells, which employ organic nanoparticles in conjugated polymers as photoactive layers for energy conversion, have generated significant scientific interest due to their ease of manufacturing, relative low costs and flexibility. However, the power conversion efficiency (PCE) of OPV solar cells is still limited to 10-12% in part due to recombination of excitons before they can reach a nanoparticle-polymer interface. Such losses can be minimized by synthesizing photoactive layers with specific morphologies that result in large contact surface area between polymer and nanoparticles in the photoactive layer achieved through solution-based synthesis process. The scientific literature lacks detailed study of the effect of synthesis parameters, such as choice of solvent, processing temperature and relative concentration of nanoparticles and polymers, on the morphology of the photoactive layer. In order to understand the effects of these synthesis parameters on the agglomeration of nanoparticle, we performed molecular dynamics (MD) simulations of binary systems that comprise [6,6]-phenyl-C61-butyric acid methyl ester (PCBM), a commonly used acceptor, and various solvents in our previous study. The study suggested that indane is a better choice of solvent than other commonly used solvents for fabricating the photoactive layer of OPV solar cells. A systematic study of the effect of temperature and concentration of PCBM on agglomerate formation of PCBM showed that the extent of agglomerate formation increases with increasing temperature and concentration.
In an effort to further determine the agglomeration of PCBM in the conjugated polymer, we simulated systems comprising PCBM in poly-hexylthiophene (P3HT) that mimic the structure of a realistic

photoactive layer after spin coating. We evaluated the agglomeration behavior of PCBM by determining sizes of the largest clusters of PCBM in P3HT. To obtain further insight into the agglomerate structure of PCBMs, we evaluated the radial distribution functions (RDFs), which describe the coordination of one group of atoms or molecules, such as PCBMs or P3HTs, with the same or other group of atoms/molecules in the agglomerate. Our results indicate that the solubility of PCBM is greater in P3HT than in indane. Our simulations also demonstrate that PCBMs have a tendency to agglomerate in P3HT and the extent of agglomerate formation increases with increasing concentration of PCBM. We also analyzed the orientation of P3HTs to determine the extent of crystallinity. The results show that P3HTs have greater likelihood of parallel stacking in the system comprising PCBM and P3HT in weight ratio 1:2 than in the system that comprise 1:1 weight ratio of PCBM and P3HT. Figures 1 (a) and (b) show a magnified snapshot of a subsection of the simulation domain that illustrates the molecular arrangement of PCBM and P3HT in systems comprising PCBM and P3HT in weight ratios 1:1 and 1:2 respectively at 310 K. The aggregation pattern of PCBM and P3HT in Figure 1(a) demonstrates a greater contact area of PCBM and polymer compared to that in Figure 1 (b). Overall, our results show that relatively low concentration of PCBM in P3HT leads to sparse networks for charge transfer. In contrast, greater concentration of PCBM in P3HT leads to formation of suitable networks for transporting holes to photoanode and electrons to photocathode respectively. Results presented in this study provide fundamental insight that can help in selecting favorable processing parameters during synthesis of photoactive layer for OPV applications.

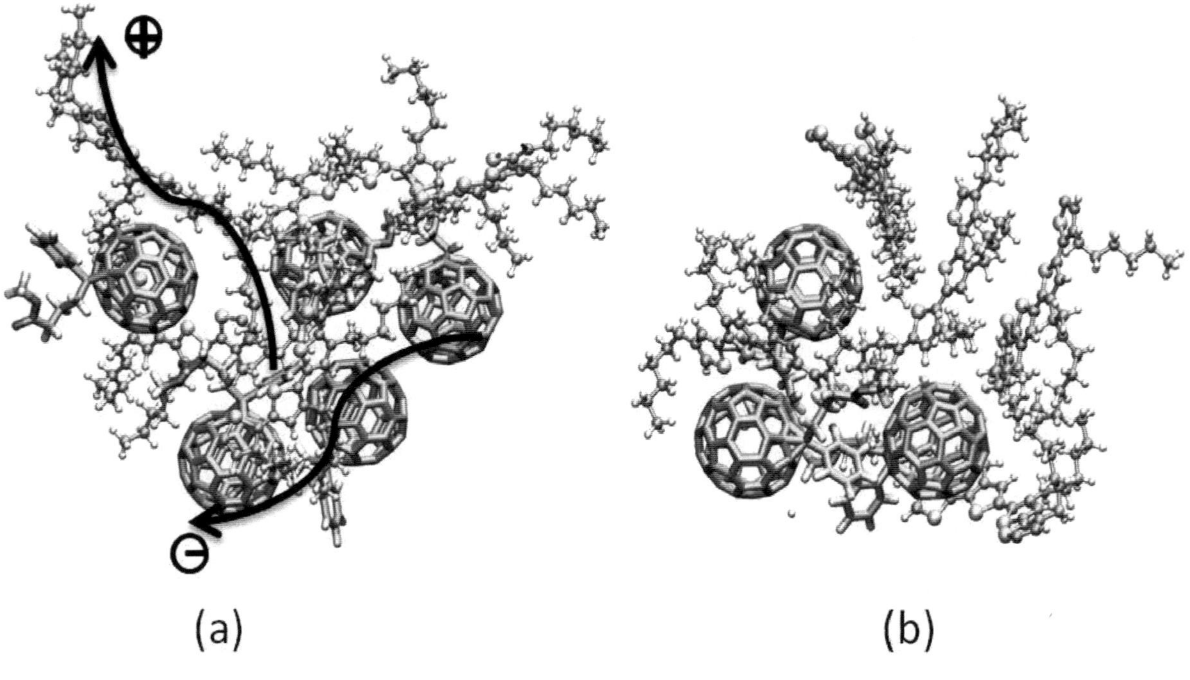

Figure 1: The molecular arrangement of PCBM and P3HT in systems of (a) 1:1 PCBM-P3HT and (b) 1:2 PCBM-P3HT at 300 K.

(262c) Integrating Photosystem I Proteins With Advanced Materials for Biologically Inspired Solar Energy Conversion

Jennings, G. K., Vanderbilt University
LeBlanc, G., Vanderbilt University
Gunther, D.,
Yang, S., Vanderbilt University
Cliffel, D., Vanderbilt University
Originally presented on: 11/5/2013 9:08:00 - 9:27:00

Photosystem I (PSI) is a ~10 nm protein complex that drives photosynthesis. Because of its nanoscale size and photodiode-like properties, PSI has drawn a surging interest for use in biologically inspired energy conversion devices upon extracting it from plants and assembling it as films on electrode surfaces. The research described in this presentation will focus on the molecular and electronic integration of spinach-extracted PSI at electrode surfaces, including metals, rectifying semiconductors, and graphene, to produce wet photoelectrochemical cells. Specifically, I will describe how monolayers of PSI can be deposited onto electrodes using self-assembly and Langmuir-Blodgett/Schaefer methods as well as how multilayers of PSI are deposited through a vacuum-assisted drop-casting approach. Thicker films of PSI absorb more light and yield much greater photocurrents as compared to monolayer films. The electrode that these films are deposited on is vitally important for the performance of the film. We have observed that micron-thick PSI films on p-doped silicon produce 200-times greater photocurrents than those obtained by deposition on gold. The position of the Fermi level combined with the band gap of silicon enables unidirectional electron transfer to PSI so that we avoid current cancellation from non-optimally aligned proteins. Adsorption of PSI on graphene provides flexibility in cell design due to the conductive, transparent nature of the atomically thin electrode. Furthermore, integrating a PSI film with reduced graphene oxide improves film conductivity and enhances observed photocurrents. These discoveries are leading to vastly improved performances for PSI-biohybrid solar cells.

(262d) Exciton Diffusion in Quantum-Dot Thin Films

Tisdale, W. A., MIT

Originally presented on: 11/5/2013 9:27:00 - 9:46:00

Colloidal quantum dots (QD), also known as semiconductor nanocrystals, are an intriguing material platform for solution-processable optoelectronic devices, such as solar cells, light-emitting diodes, thermoelectric modules, and flexible electronics. In QD thin films, the efficient movement of charge, heat, and excitonic energy is intimately linked to the QD surface chemistry and the amount of structural or energetic disorder in the material. In excitonic solar cells, the efficient diffusion of excitons to charge-separating interfaces is central to device operation. On the other hand, exciton diffusion to quenching interfaces in light-emitting diodes is a process that can limit luminescence efficiency. While the diffusion of singlet and triplet excitons in organic semiconductors has been studied extensively, exciton diffusion in colloidal QD thin films remains largely unexplored. In this talk, I will detail my group's combined experimental and computational efforts to obtain a deeper understanding of excitonic energy transport phenomena in colloidal QD materials. These efforts include spectrally-resolved transient photoluminescence spectroscopy, photoluminescence quenching, time-resolved optical imaging, and kinetic Monte Carlo simulation.

(262e) Photonic Curing of Nanocrystals for Photovoltaics

Stolle, C. J., The University of Texas at Austin
Harvey, T. B., The University of Texas at Austin
Pernik, D. R., The University of Texas at Austin
Du, J., The University of Texas at Austin
Rhee, D.,
Korgel, B. A., The University of Texas at Austin
Originally presented on: 11/5/2013 9:46:00 - 10:05:00

Nanocrystal inks are being explored as a route to low cost high efficiency photovoltaic (PV) devices. The efficiency of nanocrystal devices is limited by charge transport through the nanocrystal film and higher efficiencies can be achieved by sintering the nanocrystals into polycrystalline films. For copper-indium-selenide (CIS), this is done by thermal annealing under a selenium rich atmosphere above 500°C (called selenization). This processing step is relatively expensive and difficult to control.

Therefore, we have been exploring nanocrystal film processing strategies that might alleviate this step. Here, we present results on the use of rapid photonic curing using a rapid broadband pulsed light source to transiently heat the sample, which allows for sintering of CIS films on metallic contacts and helps prevent the outgassing of selenium from the film. We demonstrate a working solar cell fabricated from a photonically sintered CIS film. The choice of back contact material is critical for achieving working sintered devices, as the thermal conductivity and optical properties of the contact have a large impact on the film heating. We also demonstrate the removal of organic capping ligand from the film without inducing grain growth. In this case, nanocrystals come into direct contact with each other and charge transport is greatly increased. Devices made from nanocrystal films without capping ligands demonstrate high short circuit current and external quantum efficiency, although the overall device efficiency is limited due to a low open circuit voltage.

(262f) Chloride Surface Modified Cadmium Telluride Nanocrystals for Photovoltaics

Hellebusch, D. J., University of California, Berkeley
Alivisatos, A. P., Lawrence Berkeley National Laboratory

Originally presented on: 11/5/2013 10:05:00 - 10:24:00

Semiconductor nanocrystals (NCs) are of great scientific and commercial interest due to their optoelectronic properties. However, improvements in nanocrystals for solar cell applications, specifically cadmium telluride, require the ability to controllably manipulate their surface. Halides are ideal for nanocrystal surface functionalization as their small size provides advantages in passivation, coupling, and processing compared to the ubiquitous long-chained aliphatic surfactants. Here, we present the controlled surface modification of cadmium telluride nanocrystals with chloride while maintaining colloidal stability in organic solvents with a secondary labile ligand. After deposition of NCs from solution, the secondary ligand is easily removed with vacuum leaving behind an entirely organic-free NC film, offering a processing advantage for solution-based semiconductor thin films. X-ray Photoelectron Spectroscopy, Rutherford Backscatter Spectroscopy, and H-NMR supports successful substitution of the native ligand with the halide species while UV-Vis Absorption Spectroscopy and Transmission Electron Microscopy confirm maintenance of particle size, shape and solubility in organic solvents. Furthermore, we investigate the effect of the chloride on thermally treated, or sintered, NCs films. X-Ray Diffraction and Small Angle X-Ray Scattering reveal recrystallization and grain growth while Capacitance-Voltage, Field Effect Transistors, and Time-Resolved Photoluminescence probe enhancements in electrical and optical properties of sintered NC films. Our material system offers a unique method to deposit films from solution with easily removable secondary ligands that tend to contaminate films with electrically insulating materials. This study introduces a nanocrystal system with a quantitative handle on surface manipulations for a broad range of investigations and applications for photovoltaics.

(262g) Multistep Selenization of Copper Indium Gallium Selenide (CIGS) Nanocrystal Photovoltaics

Harvey, T. B., The University of Texas at Austin
Bogart, T., The University of Texas at Austin
Stolle, C. J., The University of Texas at Austin
Du, J., The University of Texas at Austin
Pernik, D. R., The University of Texas at Austin
Korgel, B. A., The University of Texas at Austin
Originally presented on: 11/5/2013 10:24:00 - 10:43:00

Copper Indium Gallium Selenide (CIGS) nanocrystal inks have been used to fabricate photovoltaic devices without post-deposition thermal treatment, but have been limited to relatively low efficiency (~3%) by poor electron transport. The insulating organic nanocrystal capping ligands are largely responsible for this. Annealing the nanocrystal films at high temperature under selenium atmosphere—i.e., selenization—can be used to improve device performance by removing ligands and increasing grain size in the film. The selenization process, however, is difficult to control due to non-uniform film sintering, the formation of insulating layers from residual carbon in the film, and film delamination with complete conversion of the Mo back contact to $MoSe_2$.

We have found that a one hour thermal anneal under argon at 525 °C before selenization of the nanocrystal film can significantly improve device efficiency. This pre-selenization anneal removes organic ligands, which appears to be important for uniform nanocrystal sintering. Lowering the anneal temperature from 525 °C to 425 °C drastically lowers device efficiency although the capping ligands are still removed. Sodium has been shown to improve CIGS grain growth in co-evaporated films and a similar connection is found in selenized nanocrystal film performance. X-ray photoelectron spectroscopy (XPS) measurements taken after the pre-selenization anneal show increased Na presence at the Mo surface when the annealing temperature is increased due to greater diffusion from the soda-lime glass substrate.

Additionally, this pre-selenization anneal accelerates grain growth in the film, shortening the necessary selenization time. Shortened selenization length reduces $MoSe_2$ formation at the back contact, enabling the use of repeated selenization treatments without film delamination. Devices with over 7% efficiency have been fabricated by depositing and selenizing successive CIGS nanocrystal layers.

(262h) Exciton Dissociation and Charge Carrier Generation At Core/Shell Heterojunction in Quantum Dots

Chakraborty, A., Syracuse University
Originally presented on: 11/5/2013 10:43:00 - 11:02:00

Electron-hole pairs (excitons) are generated by photoexcitation of electrons to excited electronic states. Exciton dissociation and generation of free charge carriers is central for light-harvesting applications of photoactive nanoparticles. This talk will focus on computational investigation of effect of heterojunctions on exciton dissociation in core/shell quantum dots. It has been shown in various experimental studies that presence of heterojunctions in nanocrystals can greatly facilitate exciton dissociation. This study will present the following four computed properties: (1) exciton binding energy, (2). electron-hole recombination probability (3). electron-hole separation distance and (4).electron-hole pair density as metrics for analyzing effect of core/shell structures in nanocrystals. Results for CdSe/ZnS core/shell quantum dots will be presented for a series of dots sizes. Effect of dot size and thickness of the shell of exciton dissociation will be presented and comparison with experimental results will be discussed. The electron-hole interaction energy is quantified by calculating the exciton binding energy and the spatial distribution of the charge carrier in the quantum dot is analyzed by electron-hole pair density functions. The local environment of the electron and hole quasiparticles in the quantum dot is probed using the average electron-hole separation distance and the electron-hole recombination probability. Exciton dissociation is a complex process and the four metrics mentioned above provide a multi-faceted description of this phenomenon.

(350a) On the Origin of Catalytic Activity of Zirconia in Aqueous-Phase Partial Oxidation of Cellulose to Levulinic Acid

Lin, H., University of Nevada, Reno
Yang, L., University of Nevada Reno
Liu, Y., University of Nevada Reno
Su, J., University of Nevada Reno
Originally presented on: 11/5/2013 15:15:00 - 15:35:00

The production of levulinic acid from cellulosic biomass has been known for more than 150 years through acid catalysis with mineral acids as the catalyst. A well-accepted reaction pathway of converting cellulose to levulinic acid is that cellulose is hydrolyzed to glucose which is dehydrated to 5-hydroxymethyl furfural (HMF), and finally HMF is rehydrated to equal molar levulinic acid and formic acid. However, the commercialization of this conventional acid catalyzed process is hindered by strong corrosivity, high separation cost, and significant waste disposal. Our group has reported that a high yield of levulinic acid was produced by directly converting cellulose over a ZrO_2 catalyst in a catalytic aqueous phase partial oxidation (APPO) process. We proposed an alternative pathway to synthesize levulinic acid from cellulose: cellulose is partially oxidized to gluconic acid which is then transformed into levulinic acid via consecutive decarboxylation, dehydration, and rehydration reactions.

Herein, the investigation of APPO reaction mechanism was performed via radical scavenger, probe reaction, and isotopic labeling studies, as well as electron paramagnetic resonance (EPR) spectrometry. We found that the *in-situ* generated superoxide radical anions may play an important role in the partial oxidation of cellulose. The influences of ZrO_2 morphology on the catalyst performance were also investigated. Electron paramagnetic resonance (EPR) and pyridine-FTIR results show that, relative to tetragonal zirconia (t-ZrO_2), monoclinic zirconia (m-ZrO_2) has a higher amount of Zr^{3+} centers and higher strength of acid sites, which leads to better catalytic performance: the yield of LA is more than 50% and lactic acid yield is 8% (on carbon molar basis) under optimized reaction conditions. Moreover, experimental results show that the reaction APPO of cellulose follow a radical mechanism and the Zr^{3+} centers are more active than F-centers in the formation of superoxide radical anions in subcritical water.

(350b) Mechanistic Requirements for the Removal of Oxygen Heteroatoms and Formation of Intra- and Inter-Molecular C=C Bond in Biomass Derived Oxygenates

Lin, F., University of Toronto

Chin, Y. H., University of Toronto

Originally presented on: 11/5/2013 15:35:00 - 15:55:00

Mechanistic Requirements for the Removal of Oxygen Heteroatoms and Formation of Intra- and Inter-Molecular C=C Bond in Biomass Derived Oxygenates

Fan Lin and Ya-Huei (Cathy) Chin*

Department of Chemical Engineering and Applied Chemistry, University of Toronto, Toronto, Canada

*cathy.chin@utoronto.ca

Small alkanals from the effluent of biomass pyrolysis are valuable precursors for the production of aliphatics and aromatics as drop-in liquid fuels. Such catalytic transformations may occur over acid [1], base [2], or acid-base bifunctional [3] sites protected within solid matrixes to remove oxygen heteroatoms and lengthen the carbon chain by creation of inter- and intra-molecular C=C bonds, leading to diverse alkenes and aromatics previously undetected in homogeneous reactions. The kinetic phenomena have been reported and widely investigated, but details of catalytic pathways, rate dependencies, site structures, and the factors that govern the reaction specificities have yet to be established.

We report the catalytic pathways and requirements for the initial C=C bond formation of alkanal (RCH_2CHO, $R=CH_3$, C_2H_5, C_3H_7) in two competing routes of uni-molecular deoxygenation and bi-molecular condensation reactions catalyzed by Brønsted acid sites confined in microporous or mesoporous solid structures to different extents (H-MFI, H-Y, H-MOR, WO_x/ZrO_2, SO_4/ZrO_2, and $H_4SiW_{12}O_{40}/SiO_2$) without the use of external hydrogen sources based on rate and isotopic assessments, transient studies, acid site titrations, and temperature programmed techniques. Alkanal reactions on Brønsted acid sites form intra- or inter-molecular C=C bonds via competitive unimolecular or bimolecular routes to evolve alkenes or larger unsaturated alkenals, respectively, as the primary products. These primary products undergo sequential reactions to produce diverse oxygenates, aromatics (C_6-C_{15}), and olefins (C_2-C_9). Here, we establish the rate and selectivity trends across the family of alkanal homologues and solid matrixes with different pore/cage dimensions used for containing the acid sites. Brønsted acid sites confined within medium pore microporous crystalline structures (H-MFI) catalyze the uni-molecular deoxygenation reaction much more effectively than sites confined within larger pore structures. In contrast, unconfined Brønsted acid sites (WO_x/ZrO_2, SO_4/ZrO_2, $H_4SiW_{12}O_{40}/SiO_2$) catalyze bi-molecular condensation reactions much more efficiently than the confined structures. Within a specific solid structure, larger alkanals prefer to undergo the uni-molecular deoxygenation reaction rather than bi-molecular condensation reaction and thus selectively form alkenes over larger alkenals and aromatics.

The reactivity trends across the various solid structures and reactant dimensions are interpreted using a thermochemical cycle formalism that connects the catalytic rates to the activation enthalpy and entropy required to evolve the transition states of the respective kinetically-relevant steps from the gas phase alkanal reactants. The inter-molecular C=C bond formation involves larger and late aldol-like transition state, and therefore the activation enthalpy reflects the combination of deprotonation enthalpy of the Brønsted acid sites, heat of alkanal condensation reaction, protonation enthalpy of the aldol intermediate, and ion-pair interaction energy between the transition state complex and the zeolitic pores. As the molecular dimension of alkanal increases from propanal to pentanal, the heat of condensation reaction remains essentially constant, while the protonation enthalpy of the aldol intermediate increases by 30 $kJ \cdot mol^{-1}$. These thermochemical data indicate that the decrease in the inter-molecular C=C bond formation rate with increasing alkanal size in the confined structures (e.g. H-MFI) must be caused by the increase in steric constraint within the pores, as required for the formation of the larger, bimolecular typed transition state, with increasing reactant sizes. In contrast, the uni-molecular deoxygenation reaction involves hydrogen transfer and the formation of smaller, cationic transition state. The smaller transition state ?senses? the confinements provided by the zeolitic pores to a lesser extent than the larger, bimolecular transition state and, as a result, the rates for intra-molecular C=C bond formation vary much less sensitively with increasing molecular dimension.

The effects of transition state confinement demonstrated here influence the relative rates and reaction specificity during alkanal deoxygenation reactions, and the connections between the extents of site and transition state confinements established here allow for the tuning of reaction pathways to attain specificity for bi-molecular condensation or uni-molecular deoxygenation reactions, required for the selective formation of aromatics and olefins via kinetic coupling of these reactions to the alkylation steps.

References

[1] T.Q. Hoang, X. Zhu, T. Sooknoi, D. E. Resasco, R. G. Mallinson, J of Catal. 271 (2010) 201?208.

[2] K. K. Rao, M. Gravelle, J. S. Valente, F. Figueras, J of Catal. 173 (1998) 115?121.

[3] M. J. Climent, A. Corma, H. Garcia, R. Cuil-Lopez, S. Iborra, V. Fornes, J of Catal. 197 (2001) 385?393.

(350c) Kinetic Studies of Succinic Acid Hydrogenolysis Using Multimetallic Catalysts in a Batch Slurry Reactor

Torres, A. A., University of Kansas
Subramaniam, B., University of Kansas
Chaudhari, R. V., University of Kansas
Originally presented on: 11/5/2013 15:55:00 - 16:15:00

Commercial production of succinic acid (SAC) using renewable sugar feedstocks is becoming a reality with recent announcement of bio-based succinic acid plants (BioAmber, Myriant, Succinity, and Reverdia). This opens an opportunity to produce useful high value added chemicals such as 1,4-butanediol (BDO) and tetrahydrofuran (THF) from renewable resources via the hydrogenation of bio-based succinic acid. However, high activity and selectivities to desired products under mild reaction conditions are some of the major challenges confronting the industrial deployment of bioSAC hydrogenolysis. To this end, the activity and selectivity of multimetallic catalysts containing combinations of metals such as Ru, Re, Pt, Co, and Sn were tested in an isothermal stirred batch slurry reactor using different solvents under a wide range of hydrogen pressures (1-10 MPa) and temperatures (373-523 K). Solvent effects using water, alcohols, and other inert organic compounds were also explored. Several reactions were conducted using SAC hydrogenolysis intermediates, such as γ-butyrolactone (GBL) and propionic acid (PA), as substrates to provide insight into the reaction pathways and relative importance of the different reactions steps. Finally, intrinsic kinetic parameters for SAC hydrogenolysis were evaluated using a subsystems approach, in which the kinetic parameters of the hydrogenolysis of the intermediates (GBL, PA, BDO, etc.) were evaluated and then used as initial guess values for the overall SAC hydrogenolysis system. This approach reduces uncertainty in the fitted parameters and increases the reliability and robustness of the models proposed. This presentation will describe optimal conditions (temperature, pressure, catalyst composition, and solvent) for maximum conversion of SAC and selectivity to desired products, address details of different Langmuir Hinshelwood models considered using rigorous and systematic discrimination criteria, and their utility in developing more complete and realistic rate models for SAC hydrogenolysis.

(350d) Synthesis, Charaterization of Bimetallic Ruthenium-Rhenium Catalysts By Strong Electrostatic Adsorption for Hydrogenation of Levulinic Acid to Gamma-Valerolactone

Cao, S.,
Regalbuto, J. R., University of South Carolina
Originally presented on: 11/5/2013 16:15:00 - 16:35:00

With the worldwide petroleum resources dwindling, it is urgent to look for renewable fuels to replace petroleum. As the one of major products of plant biomass, γ-valerolactone (GVL) is renewable, safe to store and could be used as a liquid fuel, food additive, green solvent and an intermediate in the synthesis of many value added chemicals. The synthesis of γ-valerolactone (GVL) is commonly employed by hydrogenation of levulinic acid (LA) using Ru based catalysts. In this work, we studied the catalytic activity of Ru C and Ru γ-Al_2O_3 catalysts prepared by strong electrostatic adsorption (SEA), in comparison to catalysts prepared by the conventional catalyst preparation method, dry impregnation (DI) for the LA hydrogenation into GVL. Ru C and Ru γ-Al_2O_3 catalysts prepared by SEA favored the formation of metallic Ru with the size of 1.3nm and 0.9nm (STEM), respectively. Due to the formation of more well dispersed particles, catalysts prepared by SEA show higher catalytic activities than those prepared at the same weight loading by DI. Furthermore, the presence of K promoter significantly enhanced the catalytic activity for this reaction. Bimetallic RuRe/C and RuRe/ γ-Al_2O_3 catalysts with different metal composition were also prepared by Co-SEA method those are more active than either K dopant catalyst or the monometallic Ru catalysts we mentioned above. The metal content, particle size, dispersion and reduction characteristics of supported Ru catalysts were characterized by ICP-AES, XRD, STEM, TPR-H_2 and XPS. After reaction, all samples were analyzed by XRD, the data indicated the sintering of metal particle occurs for all catalysts prepared by DI, which are not observed for the catalysts prepared by SEA.

(350e) Stabilization By Atomic Layer Deposition of Copper Catalysts for Liquid Phase Reactions

O'Neill, B. J., University of Wisconsin-Madison

Jackson, D., University of Wisconsin-Madison

Crisci, A. J., University of California, Santa Barbara

Farberow, C. A., University of Wisconsin - Madison

Lu, J., Argonne National Laboratory

Dietrich, P. J., Purdue University

Li, T.,

Ribeiro, F. H., Purdue University

Miller, J. T., Argonne National Laboratory

Winans, R., Argonne National Laboratory

Elam, J. W., Argonne National Laboratory

Greeley, J., Purdue University

Mavrikakis, M., University of Wisconsin - Madison

Scott, S. L., University of California, Santa Barbara

Kuech, T. F., University of Wisconsin

Dumesic, J. A., University of Wisconsin-Madison

Originally presented on: 11/5/2013 16:35:00 - 16:55:00

Precious metal catalysts are used extensively in the petrochemical industry and are anticipated to have a significant role in future biorefineries. While it would be desirable to replace precious metal catalysts (e.g., platinum) with base metals that are more abundant (e.g., copper), these base metal catalysts are susceptible to sintering and leaching under liquid-phase reaction conditions. Previously, we have reported a method to stabilize copper nanoparticles on g-Al_2O_3 by atomic layer deposition (ALD) of an alumina overcoat for the liquid phase hydrogenation of biomass-derived furfural. This overcoat provides resistance against sintering and leaching of the nanoparticles without significantly affecting the reaction kinetics. In order to broaden the impact of this result, a more fundamental understanding of the phenomena that lead to the pore opening of the overcoat and the stabilizing interaction between the overcoat and metal nanoparticle surface is required.

In order to achieve such a mechanistic understanding, traditional surface science and reaction kinetics studies are combined with *in situ* and *operando* measurements. For example, *in situ* and *operando* XAS measurements demonstrate that the copper nanoparticles simultaneously undergo sintering and leaching during reaction, but ALD overcoated catalysts remain stable during reaction. Additionally, hydrogen chemisorption, Fourier transform infrared spectroscopy, and scanning tunneling microscopy are used to demonstrate that a selective interaction between the overcoat and the least coordinated copper surface atoms causes the increased stability of the ALD overcoated catalysts. This fundamental understanding of the interfacial interaction is crucial to extending this result to other metals and overcoat materials.

An understanding of the pore opening process in the overcoat also has the potential to broaden the impact of the new overcoating technique by providing a degree of control over the final pore size. Solid state ^{27}Al magic angle spinning nuclear magnetic resonance demonstrates significant restructuring in the overcoat during heat treatment. Powder X-ray diffraction also shows that the overcoat has restructured after heat treatment, and *in situ* small angle X-ray scattering (SAXS) shows the evolution of the pores during the heat treatment process. Using SAXS, we are also able to probe the effect of different heat treatment parameters (time, temperature, ramp rate) on the resulting pore size, which gives us the potential to further refine the selectivity of the catalyst via size exclusion.

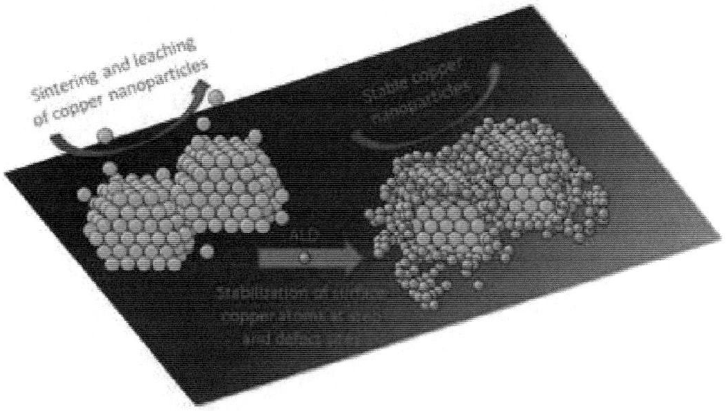

(350f) Understanding Solvation Effects On Biomass Derived Platform Chemicals: A Combined Spectroscopic and Theoretical Approach

Tsilomelekis, G., University of Delaware
Bagia, C., University of Delaware
Josephson, T. R., University of Delaware
Caratzoulas, S., University of Delaware
Nikolakis, V., University of Delaware
Vlachos, D. G., University of Delaware
Originally presented on: 11/5/2013 16:55:00 - 17:15:00

Understanding solvation effects on biomass derived platform chemicals: A combined spectroscopic and theoretical approach

G. Tsilomelekis, C. Bagia, T.R. Josephson, S. Caratzoulas, V. Nikolakis, D.G. Vlachos

Catalysis Center for Energy Innovation & Department of Chemical and Biomolecular Engineering, University of Delaware, Newark, DE 19716

The dehydration of carbohydrates, and especially fructose to 5- hydroxymethylfurfural (HMF), has attracted increasing interest since HMF is considered as a "top value-added chemical" because it can be transformed to non-petroleum based building blocks for chemicals and fuels [1,2]. Even though it is well established that the HMF selectivity of fructose dehydration reactions can be significantly affected by using organic co-solvents or by carrying out the reaction in organic solvents or in ionic liquids, the reasons behind these observations remain elusive[3,4]. Furthermore, studies correlating the effect of different solvents in HMF molecular structure in solution with HMF stability are lacking.

With the ultimate goal to unravel structure-activity relationships using insights from solvation, the present work focuses on the understanding the structural as well as the vibrational properties of dissolved HMF in DMSO-D_2O mixtures in a wide range of binary and ternary compositions by employing ATR/FTIR spectroscopy and *ab-initio* DFT calculations. Solvent effects, such as the formation of hydrogen bonds (HBs), are investigated by probing both HMF as well as DMSO and D_2O vibrational modes of specific functional groups under different conditions. The HMF conformational equilibrium is considered too, for the first time. HMF rehydration experiments are also carried out starting with HMF/DMSO/D_2O ternary mixtures of different compositions.

The analysis of the HMF ATR/IR spectra focused mainly on the change of the vibrational behavior of HMF functional groups, C=O and OH, located at ~1670 and ~3300-3450 cm^{-1} respectively. The position, the width as well as the shape of the bands are analysed by exploring all the plausible reasons such, vibrational coupling, formation of hydrogen bonds and –cis/-trans HMF conformational equilibrium. The latter is also studied for the first time using *ab-initio* calculations. Calculations indicate that the *trans*-HMF is the more stable conformer. The predicted IR spectra are in very good agreement with our experiments. The effect of solvent composition on the C=O and OH vibrational modes of HMF, reveals significant differences which are ascribed to intermolecular interactions between HMF and DMSO or/and water. The C=O stretching vibration appears clearly in higher frequency in the case of HMF/DMSO mixtures than in HMF/water. In contrary, the OH stretching vibration of HMF appears in lower wavenumbers in DMSO compared to water, indicating that the hydrogen bonds between the H of the HMF OH and the oxygen of DMSO to be stronger. The experimental observations are confirmed with our *ab-initio* calculations. In the case of HMF/DMSO/D_2O ternary mixtures, the addition of D_2O molecules around HMF seems to affect significantly the frequency of the HMF OH group while, the C=O stretching band is affected only when the D_2O molar fraction is higher than 0.3. More specifically, as the D_2O content increases a shift to lower wavenumbers is observed. This effect is ascribed to a perturbation in bond length (elongation) and strength (weakening) of the C=O band caused by the formation of C=O…D hydrogen bonds between HMF and D_2O molecules. The implications of spectroscopic analysis for solvent selection and improved biomass processes will also be discussed.

References

1. Chheda, J.N., Y. Roman-Leshkov, and J.A. Dumesic, Green Chemistry, 2007. **9**(4): p. 342
2. Rosatella, A.A., et al.. Green Chemistry, 2011. **13**(4): p. 754
3. Chidambaram, M. and A.T. Bell, Green Chemistry, 2010. **12**(7): p. 1253-1262
4. Lee, Y.Y. and K.C. Wu, Phys Chem Chem Phys, 2012. **14**(40): p. 13914-7

(350g) Structured Solvents for Catalytic Conversion of Cellulose to Bio-Chemicals

Zhang, Z. C., Dalian National Laboratory for Clean Energy
Zhou, J., Dalian National Laboratory for Clean Energy
Huang, T.,
Jia, S.,
Liu, K.,
Originally presented on: 11/5/2013 17:15:00 - 17:35:00

The conversion of solid cellulose as a biomass feedstock to bio-chemicals in high yield to a specific product involves multiple consecutive steps in a solvent, including cellulose dissolution, cellulose depolymerization, and direct or indirect glucose transformations to an end product. Because of the uniquely high solubility of certain ionic liquids for cellulosic biomass, ionic liquids have been widely studied as solvents, in which catalysts are added to enable a sequence of catalytic reactions starting from cellulose. However, similar to the conversion of carbohydrates in aqueous systems, the end products are typically recovered from the solvents by extraction with another solvent of low polarity. To minimize the end product degradation during the process of carbohydrate conversion, biphasic reaction systems have been used to separate and to protect the product in an organic solvent phase. In this work, structured solvent systems have been designed and used to efficiently convert cellulose to 5-hydroxymethylfurfural. Such a system takes the advantage known to the biphasic systems, but with largely increased interface areas.

(375a) Dye-Sensitized Solar Cells: Using Over 100 Natural Dyes As Sensitizers

Attanayake, I., University of Moratuwa

Originally presented on: 11/5/2013 15:15:00 - 15:35:00

DYE-SENSITIZED SOLAR CELLS: USING OVER 100 NATURAL DYES AS SENSITIZERS

C I.F. Attanayake [1], C. de Silva[2], B.A.J.K. Premachandra[1], A.A.P. de Alwis[1], G.K.R. Senadheera[3].*

1. *Department of Chemical & Process Engineering, University of Moratuwa, Moratuwa, Sri Lanka*

2. *Department of Chemical Engineering, City College of New York, New York, U.S.A.*

3. *Institute of Fundamental Studies, Hantana, Sri Lanka and Open University of Sri Lanka, Polgolla, Sri Lanka*

* *Corresponding Author, E-mail: ivancamilus2011@gmail.com*

ABSTRACT

Dye-sensitized Solar Cells (DSSC's) are one of the most promising new energy generation systems of photovoltaic technology to address the impending fossil fuel ? based power crisis utilizing renewable energy. These have emerged as a renewable energy source as a result of exploiting several new concepts and materials such as nanotechnology and molecular devices.

In the early 1990's Professor Michael Gratzel and his group developed a DSSC consisting of TiO_2 electrode sensitized with Ruthenium (II) Complex dye, organic liquid electrolyte with Iodine/Iodide redox couple and Platinum deposited counter electrode. In these devices a monolayer of the dye is directly attached to the semiconductor surface via carboxyl to the semiconductor surface via carboxyl group, which could realize an efficient injection of charge carriers from a photo-excited dye to the semiconductor. However, this sensitization of TiO_2 for solar applications requires not only efficient but also stable and inexpensive sensitizers. So far, several organic dyes and inorganic metal complexes have been employed to sensitize nanocrystalline TiO_2 semiconductors and one of the most efficient sensitizers is a transition metal coordination compound (Ruthenium polypiridyl complex). This complex has intense charge ? transfer (CT) absorption in the whole visible range, long excited life time and highly efficient metal-to-ligand charge transfer (MLCT).

Other organic dyes, such as phythalocyanine, cyanine dyes, xanthalene dyes, coumarin dyes etc usually perform poorly in DSSC's because of their weak binding energy with the TiO_2 film and low charge transfer absorption in the whole visible range, but these dyes are very cheap and can be prepared easily, compared to Ruthenium polypiridyl complexes. On the other hand, in nature, the fruit, flower, stem, bark, root and leaf of plants show various colours form red to purple and contain various natural dyes that can be extracted by simple procedures. Therefore, it has been emphasized by many researchers to obtain useful dyes as photosensitizers for DSSC's from natural products because of the simple preparation techniques, widely available sources and low cost. Due to these reasons, the importance of the development of low cost solar energy to electricity conversion units in principle in emphasized in this paper. Over 100 natural dyes extracted from fruits, leaves, flowers, stems, bark and roots of plants growing in Sri Lanka were used as sensitizers to fabricate dye-sensitized solar cells.

Preliminary investigation on the identification of nanocrystalline n-type TiO_2 were carried out. Fresh extracts of various fruits and vegetables were employed as sensitizers in thin layer sandwich type photo electrochemical dye-sensitized solar cells. After electrical and electronic analysis of several natural dyes of local plants, it was observed that dye extracts of Mangoostein fruit rind was found to be superior to those obtained from other dyes, and were Jsc=2.56 $mA.cm^{-2}$, Voc = 685.3 mV, ff = 60.02 % and h = 1.053%. Also Ekkiriya wood, Egg plant, Karawala kabilla and Banana flower yeilded Jsc, Voc, ff and h of 2.32 $mA.cm^{-2}$, 414.2 mV, 56.86%, 0.547% respectively ; 2.096 $mA.cm^{-2}$, 410.4mV, 56.42% and 0.485% respectively; 1.395$mA.cm^{-2}$, 443.5mV, 58.58% and 0.362% respectively; 0.763 $mA.cm^{-2}$, 414.0mV, 45.2% and 0.357% respectively. Whilst Fire fern leaf (not endemic to Sri Lanka) was found to display Jsc, Voc, ff and h of 4.128 $mA.cm^{-2}$, 405.1mV, 47.97% and 0.802% respectively.

(375b) Dye-Anchored Nanocatalysts for Improved Solar Energy Conversion Efficiency in Dye-Sensitized Solar Cells

Liu, G., University of Connecticut
Galoppini, E., Rutgers University
Agrios, A., University of Connecticut
Originally presented on: 11/5/2013 15:35:00 - 15:55:00

With the demand for sustainable and renewable energy, solar cell devices have attracted much attention which can directly convert solar energy into electrical power with the least environmental impact. Dye sensitized solar cells (DSSCs) are a very promising technology due to their attractiveness as cost-effective alternatives to conventional silicon-based photovoltaic devices. In the DSSCs, the dye is left oxidized after photo injecting an electron. A redox couple usually comprised of I^-/I_3^- in the electrolyte regenerates the oxidized dye to its neutral state and carrys the positive charge to the platinized counter-electrode. I^-/I_3^- has been commonly used as the redox couple due to the high ion mobility and slow recombination of electrons from TiO_2 to the oxidized species, triiodide (I_3^-). However, the regeneration of oxidized dye by I^- is not a simple one step reaction, which causes a large voltage loss of about 0.5–0.6 V due to the formation of diiodide radicals ($I_2^{-\bullet}$).

The talk will be focusing on reducing the overpotential of dye regeneration in DSSCs, including the preparation of Pt nanocatalyst, synthesis of new dye molecule, binding the dye to Pt nanoparticles and the assembling of TiO_2–dye–Pt system. In the earlier study, Pt nanoparticles have been prepared as the localized catalyst for the dye regeneration. A new dye molecule is synthesized with dithiolane groups for the binding to Pt nanoparticles. The aim of this talk is to demonstrate the proof-of-concept of using Pt nanoparticles to catalyze the regeneration of dye by I^- to reduce the overpotential hence improve the performance of DSSCs.

(375c) Bulk Heterojunction Polymer Solar Cells Based On Organic Semiconductor Nanomaterials

Jenekhe, S. A., University of Washington
Ren, G., University of Washington
Hwang, Y. J., University of Washington
Li, H.,
Subramaniyan, S., University of Washington
Xin, H., University of Washington
Originally presented on: 11/5/2013 15:55:00 - 16:15:00

Polymer solar cells have received increasing attention as a promising alternative for clean and low-cost solar-to-electrical energy conversion technologies. The bulk heterojunction (BHJ) photoactive layer of polymer solar cells requires rational tuning of the nanostructures towards a bicontinuous morphology to achieve maximal power conversion efficiency. The vast number of polymer semiconductors based on various donor- and acceptor-moieties has also enabled a variety of nanostructures. We have synthesized a series of new D-A copolymer semiconductors based on thiazolothiazole or benzobisthiazole building blocks and found them to be excellent p-type materials in polymer/fullerene bulk heterojunction (BHJ) solar cells with high power conversion efficiency (5-6%). Towards air-stable organic solar cells, we have also developed a series of new oligomers and polymers based on naphthalene diimide (NDI) moieties that function as efficient acceptor materials in BHJ polymer solar cells. BHJ solar cells based using NDI acceptors have showed one of the highest photovoltaic efficiencies achieved to date in *non-fullerene* polymer solar cells. Efficient all-nanowire BHJ solar cells were fabricated based on donor polymer nanowires and acceptor oligomer nanowires. Our results show that donor-acceptor architecture is a powerful means of tuning the molecular structures and nanostructures of polymer/oligomer semiconductors and their photovoltaic properties toward achievement of highly efficient polymer solar cells.

(375d) Engineering Energy Levels At the TiO2:P3HT Interface Using Atomic Layer Deposition

Dorman, J., University of California at Los Angeles
Weickert, J.,
Putnik, M.,
Schmidt-Mende, L.,
Originally presented on: 11/5/2013 16:15:00 - 16:35:00

It has been shown over the last two decades that conversion efficiencies up to 12% can be achieved using TiO_2 based solar cells, such as in liquid electrolyte dye sensitized solar cell. Recently, there has been a push to replace these hazardous electrolytes with organic materials to create environmentally friendly devices with extended lifetimes. However, one of the limitation of these hybrid solar cells is electron-hole interaction across the metal-oxide and organic hole transporter interface.

In this work, a core-shell nanostructured hybrid solar cells is synthesized with a single crystal core in order to increase electron mobility and light scattering without additional recombination effects. A combination of hydrothermal and atomic layer deposition (ALD) was used to fabricate TiO_2 nanowires (NWs) in order to take advantage of the directed electric field within the structures. During the growth process a Sn^{4+} dopant is introduced to create a doped Sn:TiO_2 nanostructured array with increased electron mobility. After the NW growth, an additional ALD step is used to deposit a TiO_2 layer, creating the core-shell structure. The dopant gradient within the core-shell structure causes the electrons to migrate toward the core of the nanowire due to the lower energy conduction band, potentially reducing electron-hole recombination at the TiO_2:P3HT interface. The conduction band engineering has a similar effect as that seen with dipole modification of the interface. Additionally, the surface can be further modified with dye molecules. This doped core-shell structure has resulted in a conversion efficiency of 2 % with a surface treatment of the squaraine dye SQ2. This efficiency is due to the contribution of the dye and the P3HT in the photon current generation.. The engineered energy levels and interfacial modifiers have a significant effect on the external quantum efficiency and internal resistances, as determined using various characterization methods.

(375e) Layered Double Hydroxides As An Effective Additive in Polymer Gelled Electrolyte Based Dye-Sensitized Solar Cells

Ho, H. W., National Tsing Hua University
Wei, T. C., National Tsing Hua University
Lu, S. Y., National Tsing-Hua University
Originally presented on: 11/5/2013 16:35:00 - 16:55:00

Dye-sensitized solar cells (DSSC) continue to draw a great deal of research attention as a promising alternative clean energy device, because of the simplicity and low cost involved in the manufacturing process and the reasonable power conversion efficiency (PCE). The boost in PCE to 12.3 % reported by Yella et al. [1] in 2012 makes DSSCs even more competitive. One of the key issues yet to be fully resolved is the volatility, leakage, and poor long term stability associated with the use of liquid electrolytes in DSSCs. Intensive and extensive research efforts have been devoted to the development of gel or solid electrolytes to tackle the issue. For solid electrolytes, the poor contact at the electrolyte/electrode interface often hampers the boost in PCE. Gel electrolytes, on the other hand, possess the advantages of low volatility and superior long term stability as compared with liquid electrolytes, and of good electrolyte/electrode contacts as compared with solid electrolytes. In the past few years, use of polymer gelled electrolytes containing performance boosting inorganic additives has been a promising approach. Particularly, clay materials, because of their high chemical stability, ion exchange ability, and natural abundance, have been a popular candidate as the inorganic additive [2]. Ho and co-workers added exfoliated montmorillonites to polymer gelled electrolytes to reduce the electrolyte/electrode interfacial impedances and to enhance the reduction rate of I_3^- at the cathode [3,4]. Geng et al. found that addition of montmorillonites to poly(ethylene oxide) (PEO) gelled electrolytes can inhibit the charge recombination at the anode [5]. Lai et al. used polyvonydiene fluoride-cohexafluoro propylene (PVDF-HFP) gelled electrolytes with mica nanoparticles as the additive to reduce the crystallinity of the polymer, leading to higher PCEs [6].

The above mentioned layer-structured materials, montmorillonites and mica, belong to the class of cationic clays, carrying negative charges on their hydroxide layers counter-balanced by positive inter-layer ions. The popular electrolyte redox couple for DSSCs, I^-/I_3^-, however, contains anions. The use of anionic clay materials, containing inter-layer cations, may prove advantageous for DSSCs [2]. In this work, layered double hydroxides (LDH), a class of anionic clay materials, were developed as the inorganic additive for PVDF-HFP gelled electrolytes. ZnAl-CO_3-LDH and ZnAl-Cl-LDH were prepared with a separated nucleation and aging method and low supersaturation co-precipitation method, respectively. The addition of the two LDHs, in terms of PCEs, significantly improved over the plain PVDF-HFP gelled electrolyte and competed favorably with the liquid electrolyte, 8.13% for the liquid electrolyte, 7.48% for the PVDF-HFP gelled electrolyte, 8.11% for the ZnAl-CO_3-LDH/PVDF-HFP gelled electrolyte, and 8.23% for the ZnAl-Cl-LDH/PVDF-HFP gelled electrolyte based DSSCs. The good performance in PCE achieved by the LDH based DSSCs was mainly attributed to the significant boost in open circuit voltages (Voc), from 0.74 V for both the liquid electrolyte and PVDF-HFP gelled electrolyte based DSSCs to 0.79V for the ZnAl-CO_3-LDH/PVDF-HFP gelled electrolyte based DSSC and 0.84V for the ZnAl-Cl-LDH/PVDF-HFP gelled electrolyte based DSSC. The boost in Voc was found to come from the positive shift in the redox potential of the electrolyte redox couple as revealed from a cyclic voltammetry analysis. As for the long term stability, a PCE retention rate of 89% after 624 h was achieved with the ZnAl-CO_3-LDH/PVDF-HFP gelled electrolyte based DSSC, much better than that of the liquid electrolyte based one. References:

1. A. Yella, H.-W. Lee, H.-N. Tsao, C. Yi, A.-K. Chandiran, M.-K. Nazeeruddin, Eric W.-G. Diau, C.-Y. Yeh, S.-M. Zakeeruddin, M. Grätzel, Science 334, (2011)
2. X.-Wang, S.-A. Kulkarni, B.-I. Ito, S.-K. Batabyal, K. Nonomura, C.-C. Wong, M. Gratzel, S.-G. Mhaisalkar, S. Uchida, ACS Appl. Mater. Interfaces, 5(2013)
3. C.-W. Tu, K.-Y. Liu, A.-T. Chien, M.-H. Yen, T.- H. Weng, K.-C. Ho, K.-F. Lin , Journal of Polymer Science: Part A: Polymer Chemistry, 46(2008)
4. C.-H. Lee , K.-Y. Liu, S.-H. Chang, K.-J. Lin, J.-J. Lin, K.-C. H, K.-Fu Lin, Journal of Colloid and Interface Science ,363 (2011)
5. Y. Geng, Y. Shi, L. Wang, B. Ma, R. Gao, Y. Zhu, H. Dong and Y. Qiu, Phys. Chem. Chem. Phys.,13 (2011)
6. Y.-H. Lai, C.-Y. Lin, J.-G. Chen, C.-C. Wang, K.-C. Huang, K.-Y. Liu, K.-F. Lin, J.-J. Lin, K.-C. Ho, Solar Energy Materials & Solar Cells ,94 (2010)

(375f) Boron-Modified TiO2 Microsphere With Enhanced Photovoltaic Performance

Rao, J., Nanjing University of Technology
Zhou, X., Nanjing University of Technology
Originally presented on: 11/5/2013 16:55:00 - 17:15:00

Nanotubes-assembled rutile TiO_2 microspheres were successful prepared by a facile two-step hydrothermal approach in the presence of sodium metaborate tetrahydrate. SEM and FESEM micrographs revaled that the addition of boron enrichs the assembled nanorodes with a reduced diameter during hydrothermal process, probably result from the increased crystal defects induced more reaction sites for crystal growth after introduction of lots of boron compound. A slight shift to the left and weakened intensity of the diffraction peaks appeared in XRD pattern without obvious peaks for diboron trioxide evidenced the obtained rutile TiO_2 accompanied with boron at interstitial sites. When applied to the photoanode of dye-sensitized solar cell, the photovoltaic device achieved a favorable photoelectricity conversion efficiency of 3%, which was 74% higher than that of the non-modified TiO_2. Further investigation on the cells' electrochemical properties were adopted by the impedance spectroscopy, which showed a longer electron lifetime and lower electron transfer rate at photoanode/dye/electrolyte interface compared to the pure TiO_2 sample.

(375g) Effect of Ionic Liquid Electrolytes in DSSCs With Titanium Dioxide (TiO2) Inverse Opal Structures

Ramesar, N., CUNY City College of New York
Kretzschmar, I., The City College of New York, The City University of New York
Originally presented on: 11/5/2013 17:15:00 - 17:35:00

Effect of Ionic Liquid Electrolytes in DSSCs with Titanium Dioxide (TiO_2) Inverse Opal Structures

Naomi Ramesar and Ilona Kretzschmar

Department of Chemical Engineering, CUNY City College, 160 Convent Avenue, New York, NY 10031

AIChE Conference 2013: ***Abstract***

Dye-sensitzed solar cells (DSSC) are low-cost alternatives to conventional solar cells that can work well in low light conditions. Despite considerable study on improving the efficiency of DSSCs, the current liquid electrolyte cell has plateaud at 11.1%. A major problem with these cells is the low viscosity and high volatility of the electrolyte, i.e., acetonitrile, which cause leakage and volatilization. We propose that using ionic liquids (ILs), which are more viscous, less volatile, and conductive, may be more suitable electrolytes. However, the higher viscosity of ILs may cause incomplete infiltration of the DSSC's nanoporous TiO_2 electrode. Here, we present a study of DSSCs with TiO_2 inverse opal electrodes of controlled pore sizes (0.1-1 mm) and ionic liquid derivatives of 1-alkyl-3-methylimidazolium tetrafluoroborate (alkyl: ethyl, butyl, and decyl) with viscosities ranging from 25.2 to 721 cP. The stability and functionality of the DSSC is tested using current-voltage characteristics.

(375h) Porous Nanowire Assemblies On Graphene for Dye-Sensitized Solar Cells

Charpentier, P. A., Western University
Alsharari, Q., Western University
Farhangi, N., Western University
Moula, G., University of Western Ontario
Ayissi, S., Western University
Originally presented on: 11/5/2013 17:35:00 - 17:55:00

Tremendous interest exists towards synthesizing nanoassemblies for dye-sensitized solar cells (DSSCs) using earth abundant and friendly materials with green scalable approaches.[1] Traditional approaches towards nanowires prevent multi-metallic nanowires, give low surface areas, and give a lack of surface chemistry for decorating nanoparticles. Supercritical CO_2 ($scCO_2$) is an enabling solvent to overcome several of these challenges for the synthesis of nanomaterials and nanodevices due to its low viscosity, high diffusivity, and "zero" surface tension [2]. This work focused on synthesizing titanium dioxide porous and metal-doped titanium dioxide nanowires from graphene surfaces using $scCO_2$. [3,4]. Graphene was examined as a catalyst mat for DSSC assemblies while quantum dots were decorated on the TiO_2 nanowires.

The results showed that undoped and doped (Fe, Zr, various quantum dots) TiO_2 nanowires were grown on the surface of graphene sheets using a sol-gel method in $scCO_2$, which helped exfoliate graphene sheets. TiO_2 nanowires less than 40 nm (when $scCO_2$ was used as a solvent) were uniformly decorated on the graphene sheets, in which surface -COOH functionalities acted as a template for nanowire growth. A variety of bi and trimetallic nanowires were synthesized, giving enhanced optical properties and photoefficiencies in DSSCs. Fe and Zr doped TiO_2 nanowires grown on graphene showed smaller crystal sizes, gave higher visible absorption and surface areas, and higher efficiency solar cells (>4% efficiency).[5] HRTEM images confirmed formation of porous nanowires with high aspect ratios and high metal dispersion using XPS. EELS mapping showed high dispersion of bimetallic and multi-metallic wires that could be connected directly to graphene assemblies. Higher surface areas of bimetallic and multi-metallic nanowires on graphene were found to enhance the electron-hole separation while reducing the band gap of the nanowires making them a potential for DSSCs. [3]. Theoretical band structures of assemblies were studied using the Vienna ab-initio Simulation Package (VASP) based on the Density Functional Theory (DFT). Functionalized graphene sheets decreased significantly the band gap but increase the binding energy values due to carboxylate adsorption sites, hence enhancing DSSC performance.[5]

1) Oudel, P., & Qiao, Q. (2012). One dimensional nanostructure/nanoparticle composites as photoanodes for dye-sensitized solar cells. Nanoscale, 4(9), 2826-2838.

2) R. Sui, Paul A. Charpentier*, Synthesis of Metal and Silicon Oxides Nanostructures by Direct Sol-Gel Reactions in Supercritical Fluids, *Chem. Rev.*, (2012), 112(6), 3057-3082..

3) Nasrin Farhangi, Yaocihuatl Medina-Gonzalez , Rajib Roy Chowdhury, and Paul A. Charpentier, Using Supercritical CO_2 for growing TiO_2/Graphene Nanoassembles: Synthesis and Photocatalytic Effect, *Nanotechnology* (2012), 23(29), 294005/1-294005/12.

4) Nasrin Farhangi, Rajib Roy Chowdhury, Yaocihuatl Medina-Gonzalez, Madhumita B. Ray and Paul A. Charpentier Visible light active Fe doped TiO2 nanowires grown on Graphene using Supercritical CO_2, Applied Catalysis B: Environmental 110 (2011) 25– 32.

5) Ayissi, Serge; Charpentier, Paul; Farhangi, Nasrin; Wood, Jeffery; Palotás, Krisztián; Hofer, WernerInteraction of Titanium Oxide Nanostructures with Functionalized Graphene Sheets: A DFT Study, J. Phys Chem. C (2013) Submitted.

(415a) Catalytic Conversion of Bio-Ethanol Into 1,3-Butadiene

Angelici, C., Utrecht University

Weckhuysen, B. M., Utrecht University

Bruijnincx, P. C. A., Utrecht University

Originally presented on: 11/6/2013 8:30:00 - 8:51:00

Catalytic conversion of bio-ethanol into 1,3-butadiene

Carlo Angelici, Bert M. Weckhuysen,* Pieter C. A. Bruijnincx*

Inorganic Chemistry and Catalysis, Debye Institute of Nanomaterials Science, Utrecht University, Utrecht, The Netherlands

*b.m.weckhuysen@uu.nl, p.c.a.bruijnincx@uu.nl

Introduction

The amount of bio-ethanol produced is rapidly growing as a result of the higher demand for biofuels and improved fermentation technologies. This increased availability also opens up opportunities for the use of bio-ethanol as an economically attractive feedstock for renewable bulk chemicals. [1] The Lebedev process, i.e. the conversion of ethanol to 1,3-butadiene, is a promising example of this approach. Indeed, the feasibility of this route has been recently positively assessed in a techno-economic study. [2] Importantly, The on-purpose production of butadiene through the Lebedev process presents an attractive solution to projected shortages in butadiene that result from the recent shifts to lighter, shale gas-based feedstocks for steam crackers. Here, we present our results on catalyst activity and stability for the ethanol-to-butadiene conversion with particular focus on the elucidation of structure-activity relationships, i.e. the influence of preparation method on morphology and amount/strength of the active sites.

Experimental details

Single and mixed oxides have been tested in the gas-phase synthesis of butadiene from ethanol (2% in N_2, total gas flow 100 mL/min, 425°C). Various SiO_2-MgO catalysts were obtained by (modified) co-precipitation (CP and CP-MOD) and wet-kneading procedures (WK, WK-MOD and WK-AER) Supported CuO/SiO_2-MgO catalysts were prepared via incipient wetness impregnation (IWI). Extensive characterization of fresh and spent catalysts was performed with to assess morphology (TEM), acid-base properties (CO_2- & NH_3-TPD, Pyridine-IR and titration) and amount of coke produced (UV-Vis and IR).

Results and Discussion

Of the various (combinations of) bulk oxides tested for butadiene synthesis, SiO_2-MgO mixed oxides proved the most interesting. The preparation method and the silica morphology were found to strongly influence both the activity and selectivity of the catalyst. The wet-kneaded samples performed best, giving the highest butadiene yield with little ethylene formation. The butadiene yield could be further boosted for all SiO_2-MgO samples by impregnation with CuO. The best result, i.e. a butadiene yield as high as 50%, was achieved over the CuO/WK-AER catalyst (Figure 1), a catalyst consisting of relatively small silica particles. Limited deactivation was observed with time on stream for all the catalysts.

TEM studies showed the morphology to be quite different for the various SiO_2-MgO samples (Figure 2); in particular, the wet-kneaded samples consist of silica spheres embedded in a thin MgO layer. Furthermore, these materials are less acidic than the others, in agreement with the lower amount of ethylene and coke produced for this catalyst. UV-Vis analysis provided interesting insights not only on coke formation, but also on the chemical nature of the CuO introduced as a promoter.

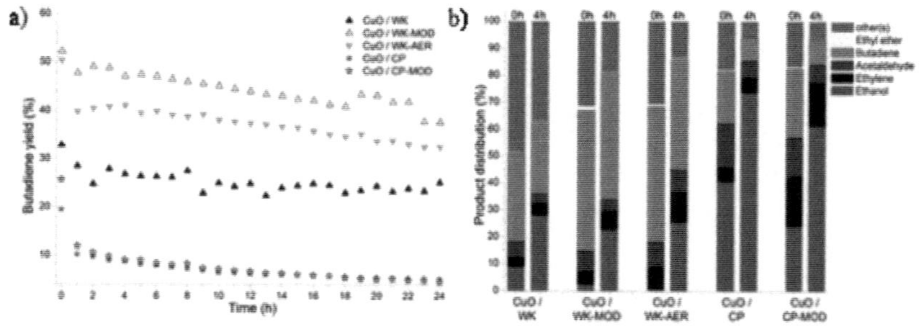

Figure 1. a) Butadiene yield with differently prepared CuO-containing catalysts with time on stream; b) product distribution at the start and after 4 h reaction.

Conclusions

The efficient and competitive synthesis of 1,3-butadiene from bioethanol would be of great economic importance. Our results show that catalyst preparation greatly influences butadiene yield and clear understanding of the underlying structure-function relation will aid in further improving the performance of these promising catalysts.

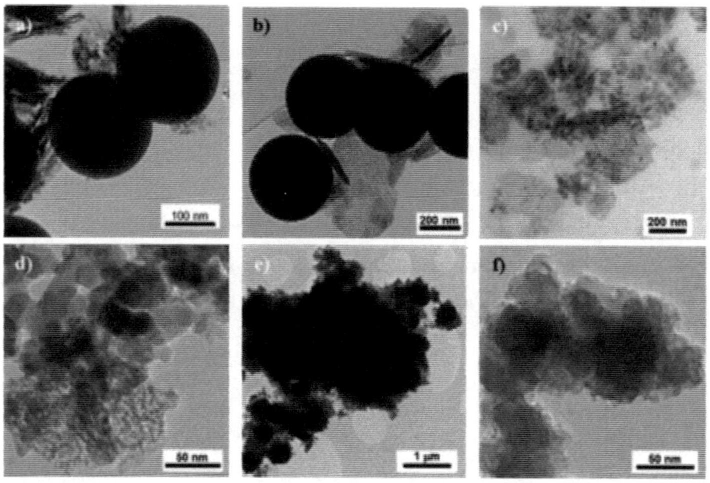

Figure 2. Morphology differences: a) physical mixture, b) WK, c) WK-MOD, d) WK-AER, e) CP, f) CP-MOD.

References

1. C. Angelici, B. M. Weckhuysen, P. C. A. Bruijnincx, *ChemSusChem* **2013**, DOI: 10.1002/cssc.201300214.

2. A. D. Patel, K. Meesters, H. den Uil, E. de Jong, K. Blok, and M. K. Patel, *Energy Environ. Sci.* **2012**, 5, 8430.

(415b) The Coupling of Ethanol to Butanol Over Calcium Hydroxyapatites

Hanspal, S.,
Davis, R. J., University of Virginia
Originally presented on: 11/6/2013 8:51:00 - 9:12:00

The use of corn-based bioethanol as a domestic and renewable transportation fuel alternative has led to a recent surge of ethanol production in the United States. This increased availability combined with decreasing market prices has made ethanol an economically viable platform feedstock for higher-value fuels and chemicals, such as butanol. Butanol is an important industrial chemical commodity and has recently generated interest as a potential gasoline fuel additive because of its high energy density and ability to improve fuel economy for biofuel blended gasoline.

The catalytic conversion of ethanol to butanol is thought to occur via the Guerbet reaction – a well-known industrial route for higher alcohol synthesis that ultimately couples two short-chain alcohol molecules to produce a longer chain saturated alcohol. The most commonly-accepted path for this reaction involves an initial dehydrogenation of ethanol to form acetaldehyde, which then undergoes an aldol condensation reaction followed by hydrogenation of the resulting unsaturated condensation products. Recent reports have demonstrated that the highest activity and butanol selectivity in ethanol coupling reactions can be obtained over hydroxyapatite (HAP) catalysts [1]; however the active site on these materials is unknown, which complicates catalyst design and optimization. This work explores the surface acid and base sites on calcium hydroxyapatite to determine their influence on catalytic performance during the Guerbet coupling of ethanol.

Ethanol conversion studies were performed at atmospheric pressure in a gas-phase, fixed-bed reactor system equipped with an on-line gas chromatograph. Acid and base sites of the catalytic materials were characterized using adsorption microcalorimetry of probe molecules such as NH_3 and CO_2, as well as diffuse reflectance FT-IR spectroscopy (DRIFTS) during stepwise temperature programmed desorption (TPD) and during reaction at elevated temperatures. The DRIFTS of ethanol on HAP revealed the presence of adsorbed ethanol. Results from TPD indicated that the ethanol is weakly held to the HAP surface compared to that on MgO, which is consistent with the affinity of the surface for other adsorbates as probed by adsorption microcalorimetry [2]. Results from characterization and reactivity testing will be used to determine important structure-function relationships for HAP materials.

References

[1] Tsuchida, Takashi, Shuji Sakuma, Tatsuya Takeguchi, and Wataru Ueda. "Direct synthesis of n-butanol from ethanol over nonstoichiometric hydroxyapatite." *Industrial & engineering chemistry research* 45, no. 25 (2006): 8634-8642.

[2] Birky, Theodore W., Joseph T. Kozlowski, and Robert J. Davis. "Isotopic transient analysis of the ethanol coupling reaction over magnesia." *Journal of Catalysis* 298 (2013): 130-137.

(415c) Catalytic Depolymerization of Lignin Extracted From Kraft Black Liquor
Carter, B., Clemson University
Originally presented on: 11/6/2013 9:12:00 - 9:33:00

Catalytic Depolymerization of Lignin Extracted from Kraft Black Liquor

Bethany D. Carter and David A. Bruce*

Department of Chemical and Biomolecular Engineering, Center for Advanced Engineering Fibers and Films, Clemson University, Clemson, SC, 29634-0909, USA.

* *Corresponding author. E-mail:* dbruce@clemson.edu

Lignin is the only renewable source of aromatic species, consisting of a network of guaiacyl, syringyl, and p-hydroxyphenyl units connected through various types of ether linkages forming an alkyl aryl ether backbone functionalized with methoxyl, benzyl alcohol, and phenolic hydroxyl species. Lignin conversion into lower molecular weight, monomeric type species with valuable functionality, such as phenols and aromatic acids, would shift dependence away from fossil fuels to a renewable source of aromatics. Lignin is a major component of woody biomass constituting 20 to 30 percent of woody biomass.

Lignin was extracted from Kraft black liquor via the continuous Sequential Liquid Lignin Recovery and Purification (SLRP) process whereby the pH of the incoming black liquor (pH ~ 13.5) was lowered to a pH ~ 9.5 by acidification with CO_2resulting in lignin precipitation. The precipitated, liquid lignin phase is then acidified with sulfuric acid until a final pH ~ 2.5 is achieved, whereby the lignin is fully acidified.

Catalytic conversion of SLRP lignin, as well as small model compounds, was investigated using a porous, heterogeneous solid acid catalyst containing noble a metal sites. The bifunctional nature of the catalyst is such that cleavage of the ether linkages occurs at the acid site, while product stabilization occurs via hydrogenation at the noble metal site. Hydrogenation of reaction products helps prevent repolymerization, which leads to improved conversion. The heterogeneous nature of the catalyst allows for separation and reuse of the catalyst.

Lignin depolymerization experiments were carried out in small batch reactor in the presence of the catalyst and under hydrogen pressure. Characterization of depolymerized lignin fractions was done using gas chromatography-mass spectrometry to analyze lighter molecular weight fractions. Higher molecular weight fractions, which are not easily volatilized, were analyzed using NMR spectroscopy.

(415d) Production of Biobased Phenol and Xylenols From Lignin

Mante, O., Brookhaven National Laboratory
Rodriguez, J. A., Brookhaven National Laboratory
Originally presented on: 11/6/2013 9:33:00 - 9:54:00

The production of high valued chemicals as part of a biorefinery is considered as one of the ways to enhance the economic viability of producing biofuels from lignocellulosic biomass. Lignin, which constitutes up to 35 wt% of biomass, is a phenolic polymer and is produced as a major byproduct in the process of making cellulosic ethanol and also from pulp and paper mills. Currently, catalytic cracking and hydroprocessing pathways such as hydrodeoxygenation (HDO), hydrogenation, and hydrocracking (HDC) are used for lignin conversion into hydrocarbons. Phenol is also an alternative important industrial chemical that can be produced from lignin. The goal of this work is to develop selective and inexpensive catalytic process for lignin depolymerization into phenol and xylenols that can be used for applications such as phenol/formaldehyde resins, epoxidized novolacs, adhesives, binders and other products. In the present study, TiO_2 catalyst was used to promote demethoxylation, dehydroxylation and demethylation of pyrolysis decomposition products from alkali lignin. The effect of catalyst-to-feed ratio and temperature on the conversion of monomeric phenolics into phenol, cresols and xylenols was investigated. The catalytic pyrolysis of kraft lignin was performed using a bench scale pyrolysis system (Py-GC/MS). The performance of TiO_2 was compared with ZrO_2, CeO_2, and MgO. The major products from the process were identified and quantified by gas chromatography mass spectroscopy. The catalysts were characterized using TEM, XRD, and NEXAFS. The results showed that TiO_2 (anatase) was effective at selectively removing methoxy, hydroxyl and carbonyl functionalities from the monomeric phenolic compounds.

(415e) On Route to Recyclable Carbon-Based Heterogeneous Catalysts for Xylan Conversion

Chung, P. W., University of California, Berkeley
Charmot, A., University of California, Berkeley
Katz, A., University of California Berkeley
Originally presented on: 11/6/2013 9:54:00 - 10:15:00

On Route to Recyclable Carbon-based Heterogeneous Catalysts for Xylan Conversion

Po-Wen Chung, Alexander Charmot, and Alexander Katz

Department of Chemical and Biomolecular Engineering,

University of California, Berkeley, CA 94720

Energy Bioscience Institute

2151 Berkeley Way, Berkeley, CA 94720

Lignocellulosic biomass feedstocks contain significant amount of fermentable carbohydrates, including cellulose and hemicellulose. Though the conversion of cellulose into glucose or chemicals has been extensively studied, there are relatively few studies of conversion of hemicellulose into pentose and chemicals. In light of our previous study, we have learned that mesoporous carbon material strongly adsorbs long-chain glucans from acid hydrolyzate likely due to carbohydrate-aromatic interaction between glucan and carbon surface. Here, we have demonstrated that functional mesoporous carbon material also strongly adsorbs long-chain xylans, which are extracted from dry *Miscanthus*. *Miscanthus* composed of 43 wt% cellulose, 22 wt% of hemicellulose and 21 wt% of lignin. We also demonstrated the conversion of those adsorbed xylans to xylose.

(415f) Bifunctional Catalysts With Improved Hydrothermal Stability for the Hydrolytic Hydrogenation of Cellulose

Galloza-Lorenzo, D. Z., University of Puerto Rico at Mayaguez
Contreras-Mora, J. L., University of Puerto Rico at Mayaguez
Mendez-Roman, R., University of Puerto Rico - MayagÃ¼ez
Cardona-Martinez, N., University of Puerto Rico - MayagÃ¼ez
Originally presented on: 11/6/2013 10:15:00 - 10:36:00

The production of glucose and/or sorbitol from the hydrolytic hydrogenation of cellulose, offers the potential to be used as a renewable feedstock in a biorefinery for the production of fuels and value added chemicals. Here we report the results of the study of the effect of cellulose crystallinity, cellulose/catalyst ratio, reaction temperature and reaction time on the catalytic performance for the conversion of cellulose into sorbitol. We studied a series of single, binary and ternary metal oxide supports with a range of surface acidity and phosphated niobic acid. Ru was supported using evaporative deposition on the supports. The catalytic materials were characterized using nitrogen adsorption, X-Ray Diffraction (XRD) and Fourier Transform Infrared Spectroscopy (FTIR). Ruthenium supported on $NbOPO_4$ displayed the best catalytic performance with hexitols yields of about 45% at 100% cellulose conversion. All materials suffer a decrease in their surface areas after treatment in hot water. However, treating $Ru/NbOPO_4$ in water at 210°C and 230 °C and 35 bar H_2 for 24 h or 48 h caused a slight decrease in cellulose conversion, but a significant increase in hexitol and hydrogenolysis products yield. Ru/Nb_2O_5 (HY 340) and the equivalent amount of phosphoric acid present in $Ru/NbOPO_4$ give essentially the same conversion as $Ru/NbOPO_4$ but a lower yield to sugar alcohols and a higher yield to hydrogenolysis products. Decreasing cellulose crystallinity and cellulose/catalyst ratio, increases the cellulose conversion and hexitols yield for $Ru/NbOPO_4$.

(415g) Selective Conversion of Cellobiose to Hexitols Over Solid Acid Supported Ru Catalysts Under Mild Conditions

Wang, H., Tianjin University
Ge, Q., Southern Illinois University Carbondale
Han, J., Tianjin University
Zhu, X., University of Oklahoma
Niu, Y., Tianjin University
Xie, X., Tianjin University
Originally presented on: 11/6/2013 10:36:00 - 10:57:00

Catalytic conversion of cellulose to hexitols under mild conditions has always been a great challenge because of the strong ß-1,4-glycosidic bond as well as intra- and inter- molecular hydrogen bonds in cellulose. A bifunctional catalyst, which offers balanced activity for hydrogenation and hydrolysis, is ideal to control the selectivity of the reaction. In the present study, we used two strategies to prepare catalysts containing both acid sites for hydrolysis and metallic sites for hydrogenation simultaneously. We then used cellobiose, a glucose dimer linked by the ß-1,4-glycosidic bond, to test the acidity of the supports and examine the overall hydrogenolysis performance of these solid acids supported Ru catalysts.

In the first approach, we synthesized an insoluble ionic liquid (IL)–HPA hybrid and used it as a support for Ru. We characterized the catalyst, $Ru/[Bmim]_3PW_{12}O_{40}$, using FT-IR spectra of pyridine adsorption and measured pH changes after treating the catalyst in H_2. The results confirmed the in-situ generation of Brønsted acidic sites as a result of hydrogen spillover from the Ru sites to the O sites of the support. The catalyst, which combines the Ru sites for hydrogenation and both Lewis and Brønsted acidic sites for hydrolysis, exhibited a superior selectivity toward sorbitol over the catalyst resulted from mixing $[Bmim]_3PW_{12}O_{40}$ and Ru/C. In addition to generate Brønsted sites on the oxygen sites, proton transfer of the hydrogen spilt over onto the oxygen sites acidifies the liquid product. These acidic active sites work synergistically with the supported Ru and contribute to the overall hydrogenolysis activity. The unreduced Ru(III) may also contribute to the observed hydrolysis activity through its Lewis acidity.

In the second approach, we use zirconia modified SBA-15 as a support for Ru. Herein, zirconia was used as a modifier to improve the acidity of the surface of the inner pores of SBA-15. In fact, coating of zirconia onto SBA-15 not only led to generation of acidic sites but also improved dispersion of Ru in the mesopore of Zr-SBA-15. The catalyst enables both hydrolysis and hydrogenation reactions to convert cellobiose into sorbitol. The reaction of cellobiose on Ru/Zr-SBA-15 follows the major reaction path of hydrogenolysis/hydrogenation of the C_1-O-C_5 ether bond in one glucose monomer of the cellobiose molecule and hydrogenates it to 3-β-D-glucopyranosyl-D-glucitol on Ru, hydrolysis of 3-β-D-glucopyranosyl-D-glucitol to sorbitol and glucose on acid sites, and hydrogenation of glucose to hexitols on Ru. Kinetic fittings suggested that hydrolysis of 3-β-D-glucopyranosyl-D-glucitol to sorbitol and glucose on the acid sites is the rating determining step of the overall reaction towards hexitols. We also showed that optimal Ru particles and well preserved mesopore with acid sites could be achieved with a Zr/Si ratio of 0.25, which allows the fast transportation of reactants to the active sites (both Ru and acidic sites). With Ru/Zr-SBA-15(0.25), full conversion of cellobiose and hexitols yield of 72.1% could be achieved in 1 hour at 140 Â¢ÂºC.

(416a) Use of Supported Group IB-Pd Bimetallic Catalysts Prepared By Electroless Deposition for the Selective Hydrogenation of Acetylene and 1,3-Butadiene

Zhang, Y., University of South Carolina

Williams, C. T., University of South Carolina

Monnier, J. R., University of South Carolina

Originally presented on: 11/6/2013 8:30:00 - 8:50:00

Use of Supported Group IB-Pd Bimetallic Catalysts
Prepared by Electroless Deposition for the Selective Hydrogenation of Acetylene
and 1,3-Butadiene

Yunya Zhang, Christopher T. Williams and John R. Monnier

Department of Chemical Engineering, University of South Carolina, Columbia SC 29208 USA

Ethylene is used as a starting point for many chemical intermediates in the petrochemical industry. The yearly demand for ethylene is over 150 million tons with a growth rate of 3.5% [1]. Ethylene obtained from steam cracking of higher hydrocarbons contains up to 2% acetylene, which acts as a poison for ethylene polymerization catalysts at even ppm concentrations (> 5 ppm) [2]. Thus, selective hydrogenation of acetylene to ethylene over low weight loading Pd catalysts is industrially used to purify ethylene streams. Due to inferior selectivity at high acetylene conversion and the formation of green oil during this reaction, various additives, such as Ag, Ni, Cu, Au, Pb, Tl, Cr and K have been added to Pd to improve the performance of current generation catalysts [3-6]. However, the bimetallic effects of the above additives are not yet well understood, possibly because the conventional methods of catalyst preparation (i.e., co-impregnation, successive impregnation) result in both monometallic and bimetallic particles with varying compositions. This in turn makes it difficult to determine the position of the two metallic components, since the bimetallic interaction occurs only when the two metallic components form bimetallic surface compositions instead of separate particles.

We use electroless deposition (ED) to prepare Ag- and Au-Pd/SiO_2 bimetallic catalysts with controlled and proximal contact between the two metals. ED uses a controlled redox chemical reaction to deposit Ag or Au onto Pd surface from a solution containing a reducible Ag or Au salt and a reducing agent. Because the reducing agent is catalytically activated by Pd surface atoms to produce an active hydrogen species, Ag or Au is only deposited on Pd surface atoms. Since ED is kinetically controlled, the final composition of a particular bimetallic catalyst can be controlled to give rather precise combinations of the two metallic components. Our goals are not only to discuss the improvements in catalysts selectivity of acetylene to ethylene, but also to illustrate the roles of Ag and Au additives during acetylene hydrogenation. If the promoting effect of Ag or Au is to modify the electronic structure of the catalytically-active Pd component, we should expect different behavior for the Ag-Pd and Au-Pd bimetallic catalysts. On the other hand, if Ag or Au only dilutes the Pd surface into different ensemble sizes that change the adsorption modes of acetylene on Pd to allow the hydrogenation of acetylene to ethylene and not ethylene to ethane, Ag-Pd and Au-Pd bimetallic catalysts should behave similarly.

Figure 1A shows that the conversion of acetylene for Ag-Pd/SiO_2 bimetallic catalysts decreased sharply at ¦È$_{Ag}$ > 0.7 while the selectivity for acetylene conversion to ethylene increased to 90% between ¦È$_{Ag}$ = 0.7 ¨C 0.95. It is proposed that at low coverages of Ag on Pd, there is an abundance of contiguous Pd surface sites where the strongly adsorbed acetylene and ethylene are adsorbed as ethylidyne. This surface species has been observed in the direct formation of ethane in the following sequential hydrogenation process: ethylidyne

°ú ethylidene °ú ethyl °ú ethane [7, 8]. As a result, this type of adsorption is not selective for hydrogenation to ethylene. At high coverages where Ag atoms cover most of the Pd surface, the remaining ensembles of Pd sites are too small to allow the formation of ethylidyne. In this case, acetylene is more weakly adsorbed as a ¦Ð-bonded species requiring only a single Pd surface site, which favors formation of ethylene.The turnover frequencies (TOF) results shown in Figure 1B are consistent with the above assumptions. At low Ag coverages, since ethylidyne requires more Pd surface sites than adsorbed ¦Ð-bonded species at high coverage, there are less C_2H_2 molecules reacted per Pd surface site. Thus, TOF of C_2H_2 conversion was lower at low Ag coverage. On the other hand, hydrogenation of these strongly bound ethylidyne species leads to the preferable formation of C_2H_6 while hydrogenation of ¦Ð-complex species prefers the formation of ethane, so TOF of C_2H_6 formation was higher at low Ag coverage.The kinetics was then investigated to verify the above hypotheses. In Figure 2, analyses for un-promoted Pd and ¦È$_{Ag\ on\ Pd}$= 0.92 gave C_2H_2 pressure dependences of ¨C0.67 and ¨C0.20, respectively, consistent with much more strongly adsorbed C_2H_2 for large Pd ensembles. Families of Au-Pd/SiO_2 bimetallic catalysts have similar performance trend as Ag-Pd/SiO_2 catalysts, indicating the bimetallic effect for these catalysts is not electronic but geometric in nature.

We are now using the same Ag- and Au-Pd/SiO_2 bimetallic catalysts to investigate the reaction mechanism of selective hydrogenation of 1,3-butadiene. Similar to acetylene, 1,3-butadiene, along with other diolefins also contaminates the ethylene streams and poisons the catalyst used for further polymerization of ethylene. In addition to evaluation and kinetics study, the mechanism of this reaction will be investigated by studying the adsorption of reactants and products on the surface of catalysts using Fourier transform infrared (FTIR) spectroscopy. The elucidation of reaction mechanism will advance our understanding of the catalytic phenomena and allow us to relate the variations in catalytic behavior to the properties of metals. In the future research, it will be interesting to study hydrogenation of 2-methyl-1,3-butadiene (isoprene), 1,3-pentadiene (isomer of isoprene) and 2,3-dimethyl-1,3-butadiene, which are electronically similar to 1,3-butadiene but have more steric hindrance. Comparison between the results of acetylene and diolefin hydrogenation will give us a better idea to understand how Ag- or Au-Pd/SiO_2 bimetallic catalysts affect the reaction pathways to control the product distribution during hydrogenation of unsaturated hydrocarbons.

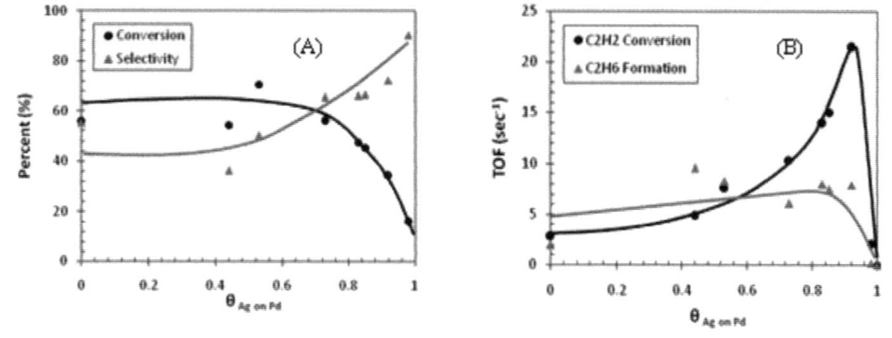

Figure 1. (A) Acetylene hydrogenation conversion and selectivity, (B) TOF for acetylene conversion and ethane formation as a function of coverage of Ag on Pd.

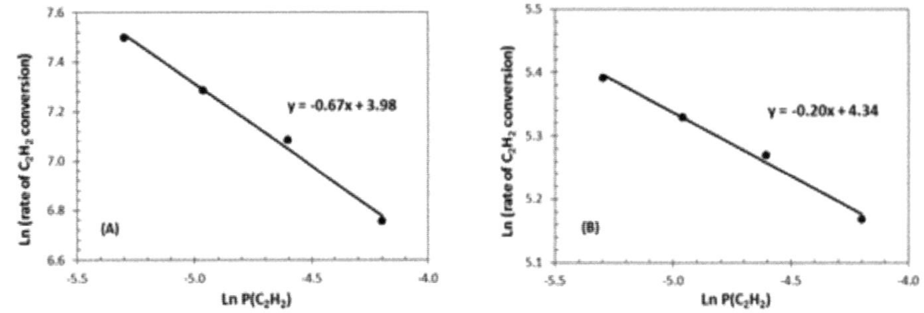

Figure 2. The C_2H_2 pressure dependence for (A) Pd/SiO_2 and (B) $Ag\text{-}Pd/SiO_2$ ($\theta_{Ag\ on\ Pd}$= 0.92) at 50°C with H_2 pressure of 0.05 atm.

References

[1] W. R. True, *Oil Gas J.* **110** (2012).

[2] H. Zea, K. Lester, A. K. Datye, E. Rightor, R. Gulotty, W. Waterman, *Appl. Catal., A* **282**, 237-245 (2005).

[3] D. C. Huang, K. H. Chang, W. F. Pong, P. K. Tseng, K. J. Hung, W. F. Huang, *Catal. Lett.* **53**, 155¨C159 (1998).

[4] P. Miegge, J. L. Rousset, B. Tardy, J. Massardier, J.C. Bertolini, *J. Catal.* **149**, 404-413 (1994).

[5] S. Leviness, V. Nair, A. H. Weiss, *J. Mol. Catal.* **25**, 131-140 (1984).

[6] V. H. Sandoval, C. E. Gigola, *Appl. Catal., A* **148**, 81-96 (1996).

[7] J. Wood, M. J. Alldrick, J. M. Winterbottom, E. H. Stitt, S. Bailey, *Catal. Today* **128**, 52-62 (2007).

[8] J. Margitfalvi, L. Guczi and A. H. Weiss, *React. Kinet. Cutal.Letr.* **15**, 475-479 (1980).

(416b) Decarbonylation and Decarboxylation of Heptanoic Acid Over Supported Pt and Pd Catalysts

Lopez-Ruiz, J. A.,
Davis, R. J., University of Virginia
Originally presented on: 11/6/2013 8:50:00 - 9:10:00

Deoxygenation refers to a broad class of chemical reactions involving the removal of oxygen from organic compounds that can be accomplished by dehydration, decarbonylation and decarboxylation. Linear carboxylic acids found in vegetable oils and animal fats can be readily converted to chemicals and fuels by decarbonylation and decarboxylation reactions. Recent work on decarbonylation and decarboxylation of carboxylic acids over transition metal catalysts is often performed in the presence of dihydrogen to inhibit catalyst deactivation, but the reactions are highly selective towards the production of paraffins because of the rapid hydrogenation of alkenes formed in the process. Since terminal alkenes formed by decarbonylation are appealing chemical intermediates in the polymer industry, a selective catalyst for decarbonylation of carboxylic acids that is stable in the absence of external dihydrogen is highly desirable.

In this work, we synthesized supported Pt and Pd nanoparticles for the decarbonylation of heptanoic acid with the goal of forming α-olefins. In particular, Pt and Pd nanoparticles were synthesized on Norit activated carbon by an incipient wetness impregnation method. The catalysts were tested for heptanoic acid conversion in a continuous flow fixed bed reactor operating at 573 K and 37 bar for liquid-phase operation and 1 bar for gas-phase operation. The effects of metal weight loading and conversion on catalyst stability and product selectivity will be discussed. Low conversion experiments in liquid-phase and gas-phase operation suggest that the main reaction path over Pt and Pd/Norit C was decarbonylation; however, decarboxylation products appeared at higher conversions presumably because of secondary side reactions such as water-gas shift (WGS) and hydrogenation. Metal sintering was detected by XRD and TEM.

(416c) Fast Hydropyrolysis and Hydrodeoxygenation of Biomass Model Compounds

Mehta, D. D., Purdue University
Smeltz, A. D., Purdue University
Nugent, N. J., Purdue University
Anderson, W. E., Purdue University
Delgass, W. N., Purdue University
Agrawal, R., Purdue University
Ribeiro, F. H., Purdue University
Originally presented on: 11/6/2013 9:10:00 - 9:30:00

Fast hydropyrolysis followed by in-line vapor-phase hydrodeoxygenation has been shown to have the potential to maximize carbon retention in bio-oil with modest hydrogen consumption [1]. In this work, both hydropyrolysis and vapor-phase hydrodeoxygenation have been studied in separate reactor systems with model compounds. Taking a cue from the existing hydrotreating process in petroleum refineries, both the reactor systems were designed to handle high pressure hydrogen.

The concept of the fast hydropyrolysis reactor was based on a modified hydrogen-oxygen torch igniter. This continuous feed reactor was designed to carry out hydropyrolysis on ca. twenty gram quantities of cellulose at ~30 bar hydrogen pressure with a residence time of ca.70 milliseconds. The key design goal of this novel reactor was to minimize secondary reactions by drastically reducing the residence time and diluting the feed in excess of hydrogen. At 500°C, the major product was levoglucosan (monomer) comprising about 50% of the product bio-oil, followed by glycolaldehyde and cellobiosan. Nearly 90% of the bio-oil was accounted for by products identified and quantified using a liquid chromatography – mass spectrometry technique. As the pyrolysis temperature was increased from 500 to 700°C, the yield of levoglucosan decreased by nearly 40%, compensated by an increase in proportion of C2 compounds like glycolaldehyde and formic acid, giving insights into the ring opening mechanism of levoglucosan.

A fixed-bed catalytic reactor was also built to study the kinetics of furfural hydrodeoxygenation over Pt-based catalysts supported on multi-walled carbon nanotubes. At 1 bar hydrogen pressure, the major pathway over the Pt catalyst was the decarbonylation to furan, which is consistent with literature reports on Pd catalyst [2]. However, at 20 bar hydrogen pressure, the major product was furfuryl alcohol from reduction of furfural, demonstrating the importance of hydrogen in retaining carbon in the liquid product.

References

[1] R. Agrawal, N. Singh, Synergistic Routes to Liquid Fuel for a Petroleum-Deprived Future, AIChE Journal, 55 (2009) 1898-1905.

[2] S. Sitthisa, T. Pham, T. Prasomsri, T. Sooknoi, R. Mallinson, D. Resasco, Conversion of furfural and 2-methylpentanal on Pd/SiO2 and Pd-Cu/SiO2 catalysts, Journal of Catalysis, 280 (2011) 17-27.

(416d) Hydrogenation of Biomass-Derived Furans Over Noble Metal Catalysts

Kang, J., University of Cincinnaty
Guliants, V., University of Cincinnati
Originally presented on: 11/6/2013 9:30:00 - 9:50:00

The production of furan derivatives from hemicellulose and cellulose is one of the major routes for biomass-based sustainable energy supply and valuable chemicals. Using cellulose will largely produce 2,5-dimethylfuran (DMF), while 2-methylfuran (2MF) results by hydrolysis, dehydration and hydrogenation reactions starting from hemicellulose. Although DMF and 2MF have been considered as next-generation biofuels, very little is known about their hydrogenation or hydrogenolysis reactions leading to higher-value added chemicals.

The hydrogenation and hydrogenolysis reactions of two methyl-substituted furans, DMF and 2MF, were studied over noble metal catalysts supported on carbon. We determined the conversion and product selectivity over the Pt and Pd catalysts for each furan under different reaction conditions. Over the entire temperature range of 50–300°C, the hydrogenated species, ring-opening products (alcohols and ketones), and cracked products were produced by hydrogenation of furans.

We further investigated the decomposition of hydrogenated methyl-substituted furans in the absence of hydrogen, because as the reaction temperature increases the decomposition of hydrogenated furans can affect the product selectivity. It is also possible that the furan hydrogenation reactions proceed via partial or full hydrogenation of the furan ring. Therefore, to probe the sequence of hydrogenation, hydrogenolysis and decomposition events during hydrogenation of these substituted furans, hydrogenation and decomposition of 2,5-dimethyltetrahydrofuran and 2-methyltetrahydrofuran were investigated over the same noble metal catalysts. The DMF and 2MF hydrogenation results showed that over a Pt/C catalyst the main reaction was furan ring opening and the primary hydrogenation products were ketones. We also observed various cracked hydrocarbons resulting from the hydrogenolysis reaction. These results indicated that the Pt/C catalyst was active in both the furan ring opening as well as ring saturation reactions. In contrast to the Pt/C catalyst, the Pd/C catalyst exhibited a completely different behavior in furan hydrogenation reactions. It showed a high selectivity to furan ring saturation products with only traces of other products. Also, the Pd/C catalyst showed deactivation due to carbon deposition, especially at high reaction temperatures, and the original catalytic activity could not be regained after attempted regeneration in hydrogen. We also observed that the Pt/C catalyst displayed high selectivity to 2-hexanol during DMF hydrogenation below 100°C.

In contrast to DMF, 2MF hydrogenation showed high cracking tendency even at low reaction temperatures. We propose that the steric hindrance caused by an additional methyl group in the furan ring of DMF affects the mode of furan ring adsorption on the catalytic surface causing a reduction in the cracking tendency. Therefore, this study sheds new light onto the nature of reaction intermediates and pathways during DMF and 2MF hydrogenation over noble metal catalysts.

(416e) Production of Dimethylfuran From Hydroxymethylfurfural Through Catalytic Transfer Hydrogenation With Ruthenium Supported On Carbon

Jae, J., University of Delaware
Zheng, W., University of Delaware
Lobo, R. F., University of Delaware
Vlachos, D. G., University of Delaware
Originally presented on: 11/6/2013 9:50:00 - 10:10:00

The hydrogenation of biomass-derived 5-(hydroxymethyl) furfural (HMF) to 2,5-dimethylfuran (DMF) is a key reaction in upgrading biomass derived furanic compounds into platform molecules toward the production of chemicals and liquid transportation fuels. For instances, DMF itself can be used as a gasoline additive or further upgraded into *p*-xylene through Diels-Alder reaction with ethylene. The hydrogenation of HMF to DMF has been studied over various metal catalysts (Cu, Pd, Ni, and Ru) using molecular hydrogen [1-4]. Copper-based catalysts, carbon supported copper-ruthenium and copper chromite, have been identified as the best metal for selective production of DMF (71% yield) [1]. However, hydrogenation utilizing petroleum-derived H_2 is neither entirely green nor economic because it requires high-pressure H_2 increasing process costs. In this study we introduce an alternative route for the conversion of HMF to DMF through catalytic transfer hydrogenation (CTH) where a hydrogen donor is used as a hydrogen source, instead of molecular H_2.

We have investigated the CTH of HMF using secondary alcohols and show that DMF is selectively produced with up to 80% yield over a carbon-supported Ru catalyst. Initial reaction kinetic studies were performed using 2-propanol (Isopropyl alcohol(IPA)) as a hydrogen donor in a batch reactor. It was found that the CTH reaction occurs at reaction temperatures lower than 373 K and the selectivity of the DMF increases with increasing reaction temperature. At low temperatures, the primary product is 2,5-bis(hydroxymethyl)furan (BHMF) and upon increasing the reaction temperature to 463 K, BHMF is completely converted with a selectivity to DMF of up to 81%. The main by-products detected are those formed through etherification of BHMF or 5-methyl furfuryl alcohol (MFA) with IPA. However, these ethers are slowly converted into DMF, especially at higher temperatures (e.g. 463 K). During the CTH reaction, H_2 needed for hydrogenation of HMF is supplied from dehydrogenation of IPA. We have found that dehydrogenation of IPA to acetone and H_2 is a faster reaction than hydrogenation of HMF. The acetone yield reaches chemical equilibrium with the carbon yield of 11.5% at 463 K after 1 hour while the corresponding H_2 partial pressure is 0.95 MPa in the closed reactor. Importantly, the hydrogen transfer hydrogenation of HMF to DMF can be carried out with other hydrogen donors, including 1-propanol, 1-butanol, and 2-butanol. XRD and TEM analyses of the fresh and spent catalysts showed the presence of Ru metal particles. The active site of the catalyst has been identified with a series of experiments. Overall, our work suggests that the catalytic transfer hydrogenation can be a vial route to upgrade other biomass-derived molecules in an efficient and sustainable way without utilizing petroleum-derived H_2.

[1] Y. Roman-Leshkov, C.J. Barrett, Z.Y. Liu, J.A. Dumesic, Nature, 447 (2007) 982-U985.

[2] S. Sitthisa, T. Sooknoi, Y.G. Ma, P.B. Balbuena, D.E. Resasco, J. Catal., 277 (2011) 1-13.

[3] S. Sitthisa, D.E. Resasco, Catal. Lett., 141 (2011) 784-791.

[4] J.P. Lange, E. van der Heide, J. van Buijtenen, R. Price, ChemSusChem, 5 (2012) 150-166.

(416f) Inhibition of Xylene Isomerization in the Cycloaddition of 2,5-Dimethylfuran and Ethylene for the Production of Renewable p-Xylene

Williams, C. L., University of Massachusetts Amherst
Chang, C. C., University of Massachusetts Amherst
Wiatrowski, M. R., University of Delaware
Fan, W., University of Massachusetts Amherst
Dauenhauer, P. J., University of Massachusetts Amherst
Originally presented on: 11/6/2013 10:10:00 - 10:30:00

Increasing demand for renewable and domestic energy and materials has led to an accelerated research effort to develop biomass-derived fuels and chemicals. One sustainable feedstock for these renewable chemicals is sugars produced by the saccharification of biopolymers (e.g., cellulose, hemicellulose). The catalytic conversion of these sugars to commodity chemicals, like *p*-xylene (used in the production of PET plastics), is currently a research area of great interest [1,2]. In previous work, the conversion of 2,5-dimethylfuran (DMF) and ethylene to *p*-xylene by a two-step reaction (Diels-Alder cycloaddition followed by dehydration), demonstrated a 75% yield of p-xylene [1]. This step completes the entire process of (a) glucose dehydration to hydroxymethylfurfural (HMF), (b) reduction of HMF to dimethylfuran, and (c) cycloaddition/dehydration of DMF to *p*-xylene.

In this work, diffuse reflectance infrared spectroscopy (DRIFTS) has shown competitive binding of DMF over *p*-xylene to the Brønsted acid sites located in the super and sodalite cages of H-Y faujasite. The persistence of these Brønsted acid sites after reaction has been shown through the use of ^{27}Al-NMR. Furthermore, the use of thermogravimetric analysis (TGA) indicates that DMF molecules can be adsorbed to Brønsted acid sites at a reaction temperature of 300 °C, while *p*-xylene will be completely desorbed. The combined use of DRIFTS, ^{27}Al-NMR, and TGA, in addition to detailed isomerization inhibition experiments, has allowed us to elucidate the source of *p*-xylene isomerization inhibition during the production of renewable aromatics from furans.

References

[1] Williams, C. L.; Chang, C.-chih; Do, P.; Nikbin, N.; Caratzoulas, S.; Vlachos, D. G.; Lobo, R. F.; Fan, W.; Dauenhauer, P. J. *ACS Catalysis* **2012**, *2*, 935-939

[2] N. Nikbin, P.T. Do, S. Caratzoulas, R.F. Lobo, P.J. Dauenhauer, D.G. Vlachos, "A DFT study of the acid-catalyzed conversion of 2,5-dimethylfuran and ethylene to p-xylene," *Journal of Catalysis*, **2013**, 297, 35-43

(416g) Hydrogenolysis of C-O Bond Over Ni-Based Bimetallic Nps: Application in Lignin Valorization

Yan, N., National University of Singapore
Zhang, J.,
Originally presented on: 11/6/2013 10:30:00 - 10:50:00

As lignin contains ca. 40% energy in lignocellulose biomass, its conversion to small molecules (mostly aromatics), which can serve as either fuels or chemicals, has received significant attention recently. Catalytic hydrogenolysis of the aryl ether bonds is a crucial step for lignin conversion. Previous studies are mostly focused on precious metal catalysts with a handful of others that use nickel-based catalysts. Nevertheless, identification of more efficient, yet affordable catalysts for C_{Ar}-O bond cleavage is still the bottleneck for lignin valorization via hydrogenolysis.

In this study, thirteen monometallic nanoparticle catalysts and twelve bimetallic nanoparticle catalysts were employed to catalyze the hydrogenolysis of 2-phenoxy-1-phenethanol, which is a typical lignin model compound with β-O-4 type linkage. All the nanoparticle catalysts were prepared by wet chemical reduction method with NaBH4 as the reductant and polyvinylpyrrolidone (PVP) as the stabilizer. The reactions were carried out at 100-160 °C and 1-20 bar H_2. The products were analyzed by GC, GC-MS and major products are also isolated by liquid chromatography and versified by NMR. Among catalysts being screened, NiPd, NiRu and NiAu exhibited superior catalytic performance over the other catalysts. In particular, NiAu nanoparticles were significantly more active than pure Au and Ni catalysts, and were effective in the hydrogenolysis of a variety of lignin model compounds and real lignin. TEM, XPS, XRD and EXAFS were employed to characterize the morphology, structure and electronics of the NiAu catalysts, revealing the interactions between Au and Ni atoms. The promotion effect of Au decorated Ni was proposed to be the electron donating of the slightly electron deficient Ni atoms on the NiAu nanoparticles that facilitating C-O bond break, providing implications on future catalyst design.

(436a) Hydrogen Production From Water

McDaniel, M. D.,
Ekerdt, J. G., University of Texas at Austin
Originally presented on: 11/6/2013 8:30:00 - 9:00:00

Silicon-based photocathodes show promising potential for photoelectrochemical (PEC) water splitting for renewable production of hydrogen. Trade-offs between stability and efficiency have limited the success of Si-based photoelectrodes. To overcome this issue, we utilize a thin layer (≈ 2 nm) of strontium titanate, $SrTiO_3$ (STO), which is grown epitaxially on p-doped Si (001) by molecular beam epitaxy. The STO layer greatly enhances the stability of the Si-based photocathode. In addition, the high quality STO layer and clean, low defect interface between STO(001) and Si(001) provides minimal trap states for charge carriers. Moreover, there is no conduction band offset between STO and Si. This allows the photo-generated electrons to easily move from the Si substrate to the STO surface, where hydrogen ions are reduced to form gaseous hydrogen. Based on our electrochemical measurements, we find that the STO-Si photocathode is both stable and exhibits high conversion efficiency. Studies with continuous Ti/Pt films for H_2 evolution yield a current density above 15 mA/cm^2; patterned metal layers are expected to give higher current densities. The results indicate that Si-based photocathodes for PEC water splitting can be improved using a thin STO layer, providing a viable method for clean hydrogen production.

(436b) Core-Shell Photoelectrochemical Electrodes for Water Splitting

Peng, Q., Duke University
Kalanyan, B., North Carolina State University
Hoertz, P.,
Miller, A.,
Kim, D. H., North Carolina State University
Hanson, K., UNC-Chapel Hill
Alibabaei, L., UNC-Chapel Hill
Liu, J., Duke University
Meyer, T., UNC-Chapel Hill
Parsons, G. N., North Carolina State University
Glass, J. T., Duke University
Originally presented on: 11/6/2013 9:00:00 - 9:18:00

Photoelectrochemical (PEC) processes offer one of most promising ways to convert solar energy into fuels (e.g. H_2, and CH_3OH), especially H_2. In order to synthesize high efficient and stable PEC water splitting electrodes, one must optimize the light absorption, charge separation, and charge transfer processes of the electrode simultaneously. Core-shell nanostructures with transparent conductive cores and semiconductor shells show great promise for optimizing PEC performance of photoactive materials, such as TiO_2. However, limited research has been performed in this field.

We will present the result of using transparent conductive network consisting of low cost, solution-processed antimony doped tin oxide nanoparticles (nanoATO) and tin doped indium oxide nanoparticles (nanoITO) for enabling core-shell TiO_2 PEC electrodes. In detail, we will present the synthesis of TiO_2 PEC electrodes by coating solution-processed nanoATO on FTO glass with TiO_2 through atomic layer deposition (ALD). The conductive, porous nanoATO film-supported TiO_2 electrodes, yielded a highest photocurrent density of 0.58 mA/cm^2 under AM 1.5G simulated sunlight of 100 mW/cm^2. This is approximately 3× the maximum photocurrent density of planar TiO_2 PEC electrodes on FTO glass. Transient photocurrent measurements showed that nanoATO films reduce charge recombination by accelerating transport of photoelectrons through the less defined conductive porous nanoATO network. Further, we will show that the effect of interfacial reaction during the TiO_2 ($TiCl_4/H_2O$) ALD on transparent nanoATO and nanoITO on their PEC performance. The scalable solution-processed porous transparent nanoATO and nanoITO films are promising as a framework to replace other conductive scaffolds (Si nanowire forest, CNTs, and $TiSi_2$ nanonets) for PEC electrodes. The results are not only valuable for PEC, but also informative for the related DSSC research.

(436c) Design of Photocatalysts Active in Infrared and Visible Range

Lee, D. C., Korea Advanced Institute of Science and Technology (KAIST)

Originally presented on: 11/6/2013 9:18:00 - 9:36:00

Infrared- and visible-active semiconductors are promising candidates for light-harvesting materials. When the semiconductor nanocrystals are within quantum confinement regime, they exhibit size tunable bandgap, enabling light-harvesting even more controllable. Furthermore, the dynamics of excited states can also be controlled through a design of heterostructure semiconductor nanocrystals. We synthesized PbSe/CdSe/CdS core/shell/shell nanocrystals with varying structural design, e.g., spheres, pyramid and tetrapod. These nanocrystals show varying electron-hole exchange interactions and radiative recombination lifetimes depending on their dimensions. We also prepared semiconductor/metal nanojunction and used the nanomaterials as photocatalysts for reduction of various dyes by growing metal tips at the corners of semiconductor nanocrystals. Interestingly, the morphology change leads to the difference in the tip growth pattern. In addition, the growth of metal particles affects the Fermi level of the entire hybrid composites, leading to the change in their photocatalytic activity. In the presentation, the implications of the heterostructure quantum dots and their hybrid composites with metal nanoparticles will be discussed in the context of energy applications.

(436d) Efficient Photoelectrochemical Water Splitting With Si-Based Metal-Insulator-Semiconductor Photoelectrodes

Esposito, D. V., University of Delaware
Lee, Y., National Instittue of Standards and Technology
Talin, A. A., Sandia National Laboratory
Moffat, T. P., National Institute of Standards and Technology
Originally presented on: 11/6/2013 9:36:00 - 9:54:00

Photoelectrochemical (PEC) water splitting is an attractive pathway for renewable production of hydrogen, but the efficiency and stability of semiconducting photoelectrodes must be improved. One promising approach to achieve both high efficiency and good electrochemical stability is the metal-insulator-semiconductor (MIS) photoelectrode architecture.[1,2] The MIS design consists of catalytic metal structures, or collectors, deposited on a 1-3 nm thick oxide-covered semiconductor. Of great importance to this design is the thin insulating oxide layer, which must protect the semiconductor from the potentially corrosive electrolyte while simultaneously mediating minority carrier tunneling between the semiconductor and collectors. In recently published work,[2] we have demonstrated significant improvement in the performance of Si-based MIS photocathodes through interfacial engineering. These planar photocathodes exhibit excellent stability and appreciable photoelectrode conversion efficiencies, up to 2.9%, but further improvement in efficiency must be achieved in order to make this process economically viable.

In this work, nano- and micro-structured Si-based MIS photoelectrodes fabricated by reactive ion etching and vapor-liquid-solid chemical vapor deposition are explored as a means to further increase the efficiency of Si-based photoelectrodes. The use of 3D-structured electrodes offer several potential advantages for achieving higher efficiency, including increased surface area for catalysis, improved collection of photo-generated carriers, and decreased optical reflection losses. However, the use of 3D photoelectrodes also introduces detrimental effects that create trade-offs between optical, catalytic, and electronic properties. In this talk, design principles for optimizing the performance of 3D-structured photoelectrodes are described and a comparison of the performance of planar and 3D-structured MIS photoelectrodes is presented. The incorporation of MIS photoelectrodes into stand-alone PEC water splitting devices will also be discussed. Calculations based on the method of detailed balance show that PEC devices utilizing Si-based MIS photoelectrodes are capable of achieving solar-to-hydrogen conversion efficiencies greater than the 2015 efficiency target of 15% set by the Department of Energy.[3]

References

[1.] H.J. Lewerenz, et al., *Electrochem. Acta*, **2011**, 56, 10726.

[2.] D.V. Esposito, I. Levin, T.P. Moffat, A.A. Talin, *Nature Materials*, **2013**, (in press).

[3.] Fuel Cells Technologies Office Multi-Year Research, Development and Demonstration Plan. US Department of Energy Efficiency and Renewable Energy. Available from: http://www1.eere.energy.gov/hydrogenandfuelcells/mypp/ (2012)

(436e) Graphene Oxide Photocatalysts for Water Splitting and Its Upconverted Photoluminescence

Yeh, T. F.,

Teng, H.,

Originally presented on: 11/6/2013 9:54:00 - 10:12:00

Graphene-Oxide Photocatalysts for Water Splitting and Its Upconverted Photoluminescence

Te-Fu Yeh[1] and Hsisheng Teng[1,2,*],

[1]*Department of Chemical Engineering and Research Center for Energy Technology and Strategy, National Cheng Kung University, Tainan 70101, Taiwan*

[2]*Center for Micro/Nano Science and Technology, National Cheng Kung University, Tainan 70101, Taiwan*

**E-mail:* *hteng@mail.ncku.edu.tw*

Abstract

Graphene oxide (GO) is a p-type semiconductor with a suitable band positions for photocatalytic watersplitting. GO sheets have a hydrophilic 2D carbon structure that allows for extensive chemical modification. This study presents strategies of tuning the conductivity type of GO from p-type to n-type for simultaneous evolutions of H_2 and O_2 under visible light irradiation. Further converting the n-type graphene oxide sheets to quantum dots results in stoichiometric evolution of H_2 and O_2 at 2-to-1 ratio. The quantum dots also serve as an energy-converting medium with upconverted photoluminescence to promote the activity of metal-oxide photocatalyts.

Introduction

Polymeric semiconducting molecules, which have a large contact area with water, attract significant attention as alternatives to metal-containing photocatalysts. Graphene oxide (GO) derived from graphite oxidation is a molecule-like semiconductor [1]. GO can be extensively dispersed in water to molecular scale and has a large exposed area. Graphene is a zero gap semiconductor whereas the GO gap increases with the oxidation level [2,3]. The bonding with electron-withdrawing oxygen functionalities results in the p-type conductivity of GO. N-type conductivity appears when graphene covalently bonds to electron-donating nitrogen functionalities [4-6].

In addition to chemical doping, we exploit the quantum-confinement effect to modulate electronic properties. Unlike two-dimensional graphene, the zero-dimensional graphene quantum dots (GQDs) have a finite band gap depending on their size [7]. The confined p-electron in sp^2 domain gives the quantized discrete levels [8]. With these features, GQDs exhibit upconvered photoluminescence (PL) [9].

This study presents the potential of GO as a photocatalyst for water splitting. P-type GO can catalyze H_2 evolution with the presence of electron donors. Ammonia-treated GO (NGO) exhibits n-type conductivity and promising photocatalytic activity for generating H_2 and O_2 simultaneously, but with the ratio below the stoichiometric value (that is, 2). Miniaturized N-doped graphene oxide quantum dots (NGO QDs) function as photocatalysts for stoichiometric H_2 and O_2 generation from water cleavage under visible light irradiation.

We also demonstrate the upconversion ability of NGO QDs.

Experimental

GO was prepared using a natural graphite powder through a modified Hummers' method. NGO was obtained through nitridation of the as-prepared GO performed at room temperature using a flow of NH_3 gas. NGO QDs were prepared from oxidation of nitrogen-doped graphene. Photocatalytic reactions were conducted at approximately 25 ¢XC in a gas-enclosed inner irradiation system or a side irradiation system.

Result and Discussion

We demonstrated the photocatalytic activity of GO in the gas-enclosed circulation system with mercury-lamp irradiation. Continuous H_2 evolution over metal-free GO was observed for reactions conducted in pure water and an aqueous methanol solution [1]. The total evolution of H_2 from aqueous methanol solution far exceeds the amount of H_2 obtained from pure water. GO cannot catalyze O_2 generation even with the presence of the electron scavenger Ag^+ ion.

We modified the as-synthesized GO photocatalyst by ammonia treatment to form NGO, which exhibited n-type conductivity. The change of conductivity type was contributed by the electron-donating property of the introduced amino groups. NGO effectively catalyzes O_2 evolution under mercury-lamp irradiation [6]. The n-type conductivity of NGO attracts hole migration toward the interface region for O_2 evolution. NGO gave rise to simultaneous evolutions of H_2 and O_2 from the side irradiated system under visible light irradiation [6]. The molar ratio of the evolved H_2 and O_2 was smaller than the stoichiometric ratio 2-to-1.

Quantum confinement effect elevates the conduction band edge, promoting the electron injection for H_2 evolution from water. We found stoichiometric evolutions of H_2 and O_2 from pure water over NGO QDs with visible-light irradiation. Figure 1 shows the PL spectra of NGO QDs excited by long-wavelength light (660-900 nm), with the upconverted emission located in the range of 450-530 nm. The upconverted PL emission of NGO QDs excited $NaTaO_3$ to catalyze H_2 evolution from an aqueous methanol solution under visible light irradiation.

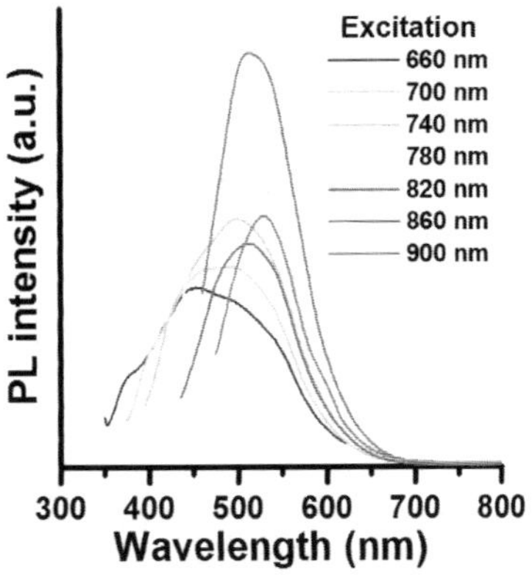

Figure 1. Upconverted PL from NGO QDs. The excitation wavelength ranges from 660 nm to 900 nm. The

upconverted emission located in the range of 450-530 nm.

Conclusions

The p-type GO catalyzed H_2 evolution from an aqueous methanol solution under mercury-lamp illumination. The introduction of nitrogen functionalities converted the p-type GO to an n-type NGO, which was capable of catalyzing simultaneous H_2 and O_2 evolutions from water under visible light. The NGO QD catalyst achieved overall water splitting ($H_2/O_2 = 2$) under visible light irradiation. The upconversion ability of NGO QDs made wide-gap photocatalysts active under visible-light illumination.

Acknowledgements

This research is supported by the National Science Council of Taiwan (101-2221-E-006-243-MY3 and 101-2221-E-006-225-MY3) and the "Aim for the Top-Tier University and Elite Research Center Development Plan" of National Cheng Kung University.

References

[1] T.F. Yeh, J.M. Syu, C. Cheng, T.H. Chang, H. Teng, *Adv. Funct. Mater.* **2010**, 20, 2255.

[2] T.F. Yeh, F.F. Chan, C.T. Hsieh, H. Teng, *J. Phys. Chem. C* **2011**, 115, 22587.

[3] T.F. Yeh, J. Cihlář, C.Y. Chang, C. Cheng, H. Teng, *Materials Today* **2013**, 16, 78.

[4] X. Wang, X. Li, L. Zhang, Y. Yoon, P. K. Weber, H. Wang, J. Guo, H. Dai, *Science* **2009**, 324, 768.

[5] C. C. Hu, H. Teng, *J. Phys. Chem. C* **2010**, 114, 20100.

[6] T.F. Yeh, S.J. Chen, C.S. Yeh, H. Teng, *J. Phys. Chem. C* **2013**, 117, 6516.

[7] L. A. Ponomarenko, F. Schedin, M. I. Katsnelson, R. Yang, E. W. Hill, K. S. Novoselov, A. K. Geim, *Science* **2008**, 320, 356.

[8] G. Eda, Y. Y. Lin, C. Mattevi, H. Yamaguchi, H. A. Chen, I. S. Chen, C. W. Chen, M. Chhowalla, *Adv. Mater.* **2010**, 22, 505.

[9] H. Li, X. He, Z. Kang, H. Huang, Y. Liu, J. Liu, S. Lian, C. H. A. Tsang, X. Yang, S. T. Lee, *Angew. Chem. Int. Ed.* **2010**, 49, 4430.

(436f) The Study of Cu2ZnSnS4 Nanocrystal/TiO2 Nanorod Arrays Heterojuction Photoelectrochemical Cell for Hydrogen Generation

Ho, T. Y.,
Chen, L. Y.,
Originally presented on: 11/6/2013 10:12:00 - 10:30:00

In this study, we reported TiO_2 nanorods arrays (TiO_2 NRAs) synthesized by hydrothermal method with controllable densities, diameters and lengths for photoelectrochemical hydrogen generation. The densities and diameters of the TiO_2 NRAs could be easily controlled by the concentration of titanium precursor (titanium isopropoxide, TTIP) and the amount of the HCl. With different reaction time, the lengths of TiO_2 NRAs could be increased from 2 μm to 9 μm. The photocurrent density and conversion efficient were increased with the lengths of TiO_2 NRAs. The maxima current density and conversion efficiency are 2.5mA/cm^2 at 1.23V vs RNHE and 0.95% with 4-5 μm in length under 100mA/cm^2 Xenon lamp illumination with AM1.5 filter, respectively. To improve the conversion efficiencies, various sensitizers, such as dye or quantum dot, are usually used to cover the absorption in the visible light range. Herein, Cu_2ZnSnS_4 (CZTS), which owns direct band gap semiconductor with 1.5eV and suitable band positions, was employed to extend the absorption range. In this study, CZTS nanoparticles were direct decorated on TiO_2 NRAs by solvothermal method with metal chloride and thiourea as precursor and ethanol as solvent to improve the conversion efficiency for hydrogen generation.

(436g) Development of New Generation of Sub-Nanometer Catalysts for Sustainable Energy Applications

Wu, Q., Stony Brook University/BNL
Shen, P., Stony Brook University / BNL
Zhao, S., Stony Brook University
Ramakrishnan, G.,
Su, D., Brookhaven National Laboratory
Li, Y.,
Orlov, A., Stony Brook University / BNL
Originally presented on: 11/6/2013 10:30:00 - 10:48:00

Developing novel heterostructured nanomaterials for energy applications as water splitting under visible light has enormous potential for energy production. Currently the existing materials have very low quantum efficiency, which indicate that new approaches are required in catalysts design. In this work we have developed and characterized sub nanometer Au and Pt nanoparticles, which greatly enhanced rate of hydrogen production of CdS. These heterostructured materials were characterized by TEM, STEM, XRD, UV-Vis spectroscopy and DRIFTS. We have also conducted comprehensive studies of these catalysts in photocatalytic reactors, where we observed a substantial hydrogen production under visible light.

(436h) ZrO2 Incorporated ZnO/TiO2 Mesoporous Photocatalyst for H2 Generation

Hussein, A., South Dakota Shool of Mines and Technology
Shende, R., South Dakota School of Mines and Technology
Originally presented on: 11/6/2013 10:48:00 - 11:06:00

The present research deals with the synthesis of ZrO_2 incorporated coupled ZnO/TiO_2 photocatalyst and its effectiveness on H_2 generation from aqueous methanol solution. Ordered mesoporous ZrO_2 supported coupled TiO_2/ZnO materials were synthesized using surfactant as structure directing agent, titanium isopropoxide, zinc acetate dihydrate and zirconium n-propoxide precursors. The ZrO_2 incorporated coupled ZnO/TiO_2composites were characterized by XRD, UV-Vis spectra, SEM/TEM, XRF, Raman and N_2 adsorption-desorption isotherm. Photocatalytic hydrogen evolution was carried out in a 7 mL quartz reactor at ambient temperature. The focused intensity and area on the quartz cuvette were 0.60 W and 2 cm^2, respectively. In a typical experiment, 2.0 mg of ZrO_2/ZnO-TiO_2 photocatalyst was dispersed in 3.5 mL of H_2O containing 0.5 mL methanol or 10 mg biomass as a sacrificial agent. A 300 W xenon lamp (Oriel) equipped with a UV-filter (240 nm - 400 nm) was employed as the light source. Prior to irradiation, the suspension of the catalyst was degassed with ultrapure nitrogen for 30 min to completely remove the dissolved oxygen and to ensure the reactor in an anaerobic condition. From 3 mL reactor overhead space, 0.1 mL of gaseous samples were withdrawn periodically and analyzed using the gas chromatograph (GC), from which photocatalytic effectiveness towards H_2 generation was evaluated and compared with coupled TiO_2/ZnO material.

(450a) Bench Scale Shale Shaker Performance

Chase, G. G., The University of Akron
Benis, S., The University of Akron

Originally presented on: 11/6/2013 8:30:00 - 8:55:00

Shale shakers separate coarse solid particles from drilling muds in well bore-hole drilling operations. Drilling muds are engineered with rheological properties needed for efficient transport of coarse solid particles generated by the drill cutting into the soil and rock. Experimental measurements on full scale equipment are difficult, expensive, and inconvenient.

In this work experimental measurements are taken from a bench scale shale shaker. Experimental results show the effects of vibration frequency and amplitude, deck angle, mud flow rate, and mud concentration. Experimental results are compared with a shale shaker model.

(450b) Composition of Heavy Oil From Speed of Sound Data

Riazi, M. R., Kuwait University
Dashti, H., Kuwait University
Originally presented on: 11/6/2013 8:55:00 - 9:20:00

As reserves for light and conventional oils are in decline, demand on heavy and unconventional oils is on the rise. Production and processing of such oils require knowledge of their composition for proper characterization and design of related processing equipment. Heavy oils are usually rich in asphaltene which upon precipitation may cause significant flow problems in production and processing units. In this presentation we examine several measurable properties such as speed of sound and dipole moment which may be used for characterization of heavy oils.

Speed of sound in oils can be accurately measured under different conditions of temperature and pressure. In this presentation we show some recently measured data for the speed of sound in pure hydrocarbons, crude oil and deasphalted oils. Analysis of data show that sonic velocity may be used to determine amount of asphaltene and other heavy components in oils. In addition a predictive method to determine speed of sound from bulk properties is presented for cases that data on the speed of sound is not available.

(450c) A Case Study Of Debottlenecking Strides In A Petrochemical Plant

Okechukwu, C., INDORAMA ELEME PETROCHEMICALS LTD
Shrivastava, B.,
Dada, E. A., ChemProcess Technologies (CPT), LLC
Originally presented on: 11/6/2013 9:20:00 - 9:45:00

Process plant debottlenecking is aimed at increasing the plant's efficiency, optimizing raw material consumption, energy and other utilities, capacity utilization, on-stream availability, safety rating, competitiveness and profitability among others. To be competitive, the promoters of a process plant has to pro-actively adapt to ever changing realities and templates caused by equipment obsolescence, raw materials vulnerability, technological changes and even stringent local legislations and environmental laws.

Nigerian's government quest to attract foreign and local direct investment into key sectors of the economy in general and the oil and gas sector in particular to create jobs, contribute to GDP growth, facilitate the exploitation of vast natural resources are among the driving force for debottlenecking of existing facilities in the Petroleum and Gas sector for optimization and process improvement.

Indorama Eleme Petrochemical Ltd (IEPL), sited in the Niger Delta region of Nigeria has pursued and implemented a robust and integrated debottlenecking master plan to modernize its plants and facilities. In this presentation, we shall examine the giant debottlenecking strides made by IEPL Eleme which is fast transforming Nigeria into a large petrochemical hub in Africa. These debottlenecking strides include the modification for catalysts separation and regeneration. We shall also discuss some of the resources added by IEPL to consolidate the robust maintenance programs, upgrading of manufacturing facilities, streamlining of feedstock availability constraints and implementing a safety culture all of which have led to capacity improvements at a cost effective and profitable manner.

(450d) Experimental Measurements and Modelling of the Dissociation Conditions of Clathrate Hydrates for (Refrigerant + NaCl + Water) Systems

Ngema, P. T.,
Petticrew, C. N., University of KwaZulu-Natal
Naidoo, P., University of KwaZulu-Natal, Howard College Campus
Mohammadi, A. H., MINES ParisTech
Ramjugernath, D., University of KwaZulu-Natal
Originally presented on: 11/6/2013 9:45:00 - 10:10:00

Abstract

Globally, gas hydrates have been proposed as a potential means for many positive applications, *e.g.* wastewater treatment and desalination, CO_2 capture and separation, separation of close-boiling point compounds, hydrogen/methane storage, refrigeration and air conditioning industry, food industry[1,2,3,4,5]. Javanmardi and Moshefeghian[4]; Chun *et al.*[6]; Seo and Lee,[7] and Eslamimanesh *et al.*[8] have reported that the use of refrigerants for the formation of gas hydrates in water desalination processes is promising when compared to traditional desalination processes. The use of refrigerants to form gas hydrates is attractive because hydrates as host molecules may occur at ambient conditions. According to Eslamimanesh *et al.*[9], the dissociation of gas hydrate results in the production of pure water and the release of refrigerant which can be recycled. However, there is a lack of research regarding the use of hydrate technology for the treatment of industrial wastewater and desalination processes, particularly using fluorinated refrigerants as the hydrate former.

The focus of this study is to generate accurate hydrate dissociation data which will be used to design a wastewater treatment and desalination processes using gas hydrate technology. In this study, three refrigerants 1,1,1,2-tetrafluoroethane (R134a), (0.5 mass fraction difluoromethane + 1,1,1,2,2-pentafluoroethane) (R410a), and (0.5 mass fraction 1,1,1-trifluoroethane + 1,1,1,2,2-pentafluoroethane) (R507)were investigated in the absence and presence of a NaCl aqueous solution at various salt concentrations. Hydrate dissociation data was measured using an isochoric pressure-search method.

The binary systems, in this study, which consisted of {R134a, or R410a, or R507} + water were measured in the temperature range between (276.2 to 291.8) K and pressures ranging from (0.114 to 1.106) MPa. The ternary system R134a + water + NaCl, at three salt concentrations of (5, 10, and 15) %, was measured in the temperature range between (268.1 to 280.6) K and pressures ranging from (0.086 to 0.340) MPa. For the ternary systems comprising of {R410a or R507} + water + NaCl, at two salt concentrations of (5 and 10) %, measurements were undertaken in the temperature range between (275.1 to 290.3) K and pressures ranging from (0.269 to 1.170) MPa.

Modeling of the measured data was undertaken using a combination of the solid solution theory of van der Waals and Platteeuw[10] for the hydrate phase, the Aasberg – Petersen *et al.*[11] model for the electrolyte aqueous system, and the Peng-Robinson[12]equation of state with classical mixing rules for the liquid and vapour phases. The correlated results show good agreement with the experimental dissociation data.

The results of this study show that the presence of the NaCl in the aqueous solutions has a thermodynamic inhibition effect on refrigerant gas hydrates. The {R134a or R507} Hydrate–Liquid water–Vapour ($H–L_w–V$) equilibrium phase boundary is shifted to lower dissociation temperatures compared to the dissociation temperatures and pressures for R22 + water + NaCl systems measured by Chun et al.[6]. Also, it is higher than propane + water + NaCl systems measured by Chun et al.[13]. The hydrate dissociation temperatures for R410a + water system are close to ambient conditions. In addition, quadruple points at which the four phases ($H–L_w–L_{R134a}–V$) coexist were determined for the water + R134a + NaCl system. Further research is being investigated to identify suitable promoters with the systems discussed to determine feasible operating conditions the desalination processes.

***Keywords**:*Gas hydrate, Refrigerant, Salt, Dissociation data, Model, Desalination.

***Corresponding authors**: Deresh Ramjugernath, e-mail: *ramjuger@ukzn.ac.za*,

Amir H. Mohammadi, e-mail: *a.h.m@irgcp.fr*

References

(1) Mohammadi, A. H.; Richon, D. Phase equilibria of methane hydrates in the presence of methanol and/or ethylene glycol aqueous solution, *Ind. Eng. Chem. Res,* **2010,**49, 925-928.

(2) Tumba, K.; Reddy, P.; Naidoo, P.; Ramjugernath, D.; Eslamimanesh, A.; Mohammadi, A. H.; Richon, D. Phase equilibria of methane and carbon dioxide clathrate hydrates in the presence of aqueous solutions of tributylmethlyphosphonuim methylsulfate ionic liquid, *Journal of Chemical and Engineering Data,* **2011**, 56, 3620-3629.

(3) Sloan, E. D.; Koh, C. A. *Clathrate hydrates of natural gases*, 3rdEdition, CRC Press, Taylor & Francis Group, London, New York, 2008.

(4) Javanmaardi, J.; Moshefeghian, M. Energy consumption and economic evaluation of water desalination by hydrate phenomenon, *Appl. Therm. Eng*, **2003**, 23, 845-857.

(5) Huang, C. P.; Fennema, O.; Poweri, W. D. Gas hydrates in aqueous-organic systems: II. Concentration by gas hydrate formation, *Cryobiology*, **1966**, 2, 240-245.

(6) Chun, M-K.; Lee, H.; Ryu, B. J. Phase equilibria of R22 ($CHClF_2$) hydrate systems in the presence of NaCl, KCl and $MgCl_2$, *J. Chem. Eng. Data*, **2000**, 45, 1150-1153.

(7) Seo, Y.; Lee, H. A new hydrate-based recovery process for removing chlorinated hydrocarbons from aqueous solutions, *Environ. Sci. Technol***. 2001**, 35, 3386-3390.

(8) Eslamimanesh, A.; Mohammadi, A. H.; Richon, D. Thermodynamic model for predicting phase equilibria of simple clathrate hydrates of refrigerants, *Chemical Engineering Science*, **2011**, 66, 5439-5445.

(9) Eslamimanesh, A.; Mohammadi, A. H.; Richon, D.; Naidoo, P.; Ramjugernath, D. Application of gas hydrate formation in separation processes: A review of experimental studies, *Journal of Chemical Thermodynamics*, **2012**, 46, 62-71.

(10) Van der Waals, J. H.; Platteeuw, J. C. Clathrate Solutions, *Adv. Chem. Phys,* **1959**, *2*, 1-57.

(11) Aasberg-Petersen, K.; Stenby, E.; Fredenslund, A. Prediction of high pressure gas solubilities in aqueous mixtures of electrolytes, *Ind. Eng. Chem. Res*, **1991**,30, 2180-2185.

(12) Peng, D. Y.; Robinson, D. B. A new two constant equation of state. *Ind. Eng. Chem. Fundam,* **1976**, 15, 59-64.

(13) Mohammadi, A. H.; Afzal, W.; Richon, D. Gas hydrate of methane, ethane, propane and carbon dioxide in the presence of single NaCl, KCl and $CaCl_2$ aqueous solutions: Experimental measurements and predictions of dissociation conditions, *J. Chem. Thermodynamics,* **2008,** 40, 1693-1697.

(473a) Selective Dehydration of Polyols to Commodity Chemicals

Nolan, M., Georgia Institute of Technology
Wanninayake, U., Iowa State University
Kraus, G., Iowa State University
Shanks, B. H., Iowa State University
Originally presented on: 11/6/2013 12:30:00 - 12:51:00

Synthesis of commodity chemicals from biorenewable sources is an important component of a sustainable society. Due to the very different chemical nature of bio-based feedstocks, a fundamental understanding of the chemical conversion of bio-derived molecules, and the mechanisms thereof, will be necessary for production of biorenewable chemicals. One reaction of interest for biorenewable chemicals is the selective removal of oxygen from polyols. In this presentation, the dehydration of 1,2,ω triols and 1,3,ω triols will be discussed. The dehydration reactions were carried out by pulsing solutions of a given triol through a gas chromatogram (GC) liner packed with amorphous silica-alumina, and observing selectivity changes as a function of material introduced versus catalyst weight.
The role of acid site density and deactivation of the catalyst will be discussed in the presentation, along with the role in hydroxyl group location on the triol and selectivity toward dehydration of primary versus secondary alcohols. Further, the role of hydroxyl group location and carbon chain length on selectivity to either the elimination of a hydroxyl or the formation of a furan, pyran, or oxepan ring will also be discussed in this presentation. Overall, a set of reactivity rules, including observed exceptions to the traditionally expected results for a dehydration reaction, will be proposed.

(473b) Selective Oxidation of Guaiacol to Produce Organic Acid Over TS-1 With H2O2

Su, J., University of Nevada Reno
Yang, L., University of Nevada Reno
Liu, Y., University of Nevada Reno
Lin, H., University of Nevada, Reno
Originally presented on: 11/6/2013 12:51:00 - 13:12:00

Biomass has drawn an increasing attention in recent years as a renewable feedstock in response to the depletion of fossil fuels and the increasing concern over environmental protection. Lignin, with a highly branched polymer of substituted phenolic monomeric entities, is a important part of biomass (15-30 wt%). [1] Due to lignin's amorphous structure and high stability of its monomers (benzene ring), the utilization of lignin become a challenge of biomass industry.[2] Hydrodeoxygenation is a common method to produce fuel hydrocarbons, such as benzene, cyclohexane or methyl-cyclohexane form the model compounds of lignin.

[3]

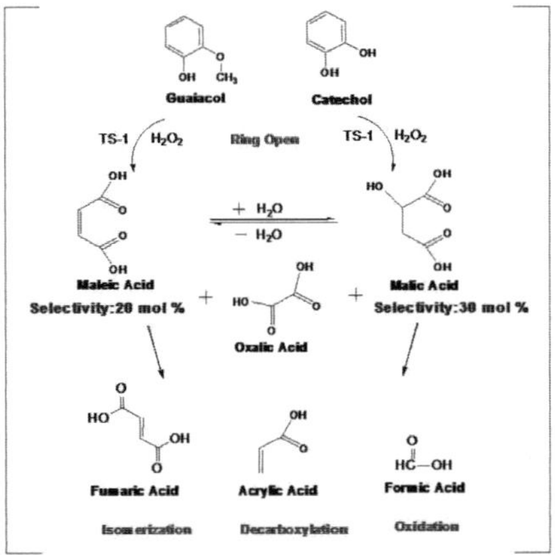

In this work we introduce an alternative way to utilization of lignin for production of useful chemaicals. We use H2O2 as the oxidants to selectively oxidize lignin model compounds such as guaiacol, catechol or syringol to organic acid. As we know that, organic acids are high value added polymer monomers or food additives.

As shown in scheme 1, when use H2O2 solutions as the oxidant and titanium silicalite zeolite (TS-1) as the catalyst, guaiacol and catechol can high selectively oxidized to maleic acid (selectivity: 20 mol%) and malic acid (selectivity:30 mol%) respectively. The byproducts are oxalic acid, fumaric acid, acrylic acid and formic acid.

In this work, the different oxidation pathway of guaiacol and catechol, as well as the catalysis mechanism of this selectively oxidation reaction will be discussed. Our work shown that, the utilization of lignin to produce high value added chemicals is feasible.

Reference

[1] Crawford, R. L. Lignin Biodegradation and Transformation; Wiley-Interscience: New York, 1981.

[2] Huber, G. W.; Iborra, S.; Corma, A. Chem. Rev. 2006,106, 4044--4098.

[3]a) Ind. Eng.Chem. Res. 1993, 32, 1535--1541; b) Angew. Chem., Int. Ed. 2007, 46, 7164--7183.

(473c) Phthalic Anhydride Production From Renewable Carbon Sources

Mahmoud, E.,
Lobo, R. F., University of Delaware
Mahmoud, E.,
Originally presented on: 11/6/2013 13:12:00 - 13:33:00

Increasing exploitation of fossil fuel feedstocks for the production of energy and materials has led to an increase in the price of crude oil as well as increasing levels of CO_2 in the atmosphere [1]. Therefore, there has been an accelerated research effort invested in the conversion of renewable biomass feedstocks into fuels and chemicals. One important chemical at risk for future shortages is phthalic anhydride because of its dependence on petroleum production. Phthalic anhydride is used for the production of plasticizers, polyesters, and resins. In 2000, the worldwide production of phthalic anhydride was 3.23 million tons [2]. The objective of this study is to investigate a renewable route for the production of phthalic anhydride from lignocellulosic biomass.

A renewable route to phthalic anhydride from biomass-derived furan and maleic anhydride is investigated using quantitative NMR spectroscopy to identify products and byproducts. Furan and maleic anhydride can be converted to phthalic anhydride in two steps: Diels Alder cycloaddition followed by dehydration. Here the two reactions were investigated independently. Good yields of the Diels-Alder reaction between furan and maleic-anhydride were obtained at room temperature and solvent-free conditions (SFC) to yield ~81% to the cycloadduct after 1 hr of reaction. The dehydration of the Diels-Alder cycloadduct was carried out using a number of promoters and reaction conditions. Ultimately, optimal yields for this dehydration were identified under reaction conditions containing mixed sulfonic-carboxylic anhydrides at room temperature as the key reagent. The NMR investigations show a yield of 57% to phthalic anhydride (81% conversion, 70% selectivity) after 1 hour of reaction can be obtained.

References:
[1] Climent, M., Corma, A., & Iborra, S. (2011). Green Chemistry , 520-540.
[2] Phthalic Anhydride. Retrieved from http://www.inchem.org/documents/sids/sids/85449.pdf

(473d) Reaction Kinetics of the Diels-Alder Cycloaddition of Dimethylfuran and Ethylene for Renewable p-Xylene

Williams, C. L., University of Massachusetts Amherst
Green, S. K., University of Massachusetts Amherst
Chang, C. C., University of Massachusetts Amherst
Fan, W., University of Massachusetts Amherst
Dauenhauer, P. J., University of Massachusetts Amherst
Originally presented on: 11/6/2013 13:33:00 - 13:54:00

Diels-Alder cycloaddition of dimethylfuran (DMF) and ethylene to produce *p*-xylene is a promising method of sustainably producing valuble plastic precursors [1,2]. This step completes the entire process of (a) glucose dehydration to hydroxymethylfurfural (HMF), (b) reduction of HMF to dimethylfuran, and (c) cycloaddition/dehydration of DMF to *p*-xylene. The DMF/ethylene reaction proceeds via two steps: a symmetry-allowed [4+2] Diels-Alder cycloaddition of DMF and ethylene to produce an oxa-norborene cycloadduct intermediate, and subsequent dehydration of the intermediate to form *p*-xylene. In this work, the kinetics of the combined cycloaddition/dehydration chemistry was experimentally evaluated, and the rate-limiting phenomena were identified. The apparent activation energy for the formation of Diels-Alder products from DMF and ethylene was experimentally measured as 17±3 kcal/mole using DMF in heptane with an H-Y catalyst in the temperature range of 200-250°C. Research also focused on characterizing the effects of catalyst loading and determining reaction orders with respect to DMF, ethylene and *p*-xylene.

References

[1] Williams, C. L.; Chang, C.-chih; Do, P.; Nikbin, N.; Caratzoulas, S.; Vlachos, D. G.; Lobo, R. F.; Fan, W.; Dauenhauer, P. J. *ACS Catalysis* **2012**, *2*, 935-939

[2] N. Nikbin, P.T. Do, S. Caratzoulas, R.F. Lobo, P.J. Dauenhauer, D.G. Vlachos, "A DFT study of the acid-catalyzed conversion of 2,5-dimethylfuran and ethylene to p-xylene," *Journal of Catalysis*, **2013**, 297, 35-43

(473e) Ruthenium-Catalyzed Hydrogenation of Levulinic Acid: Influence of the Support On Selectivity and Stability

Luo, W.,

Deka, U.,

Beale, A. M.,

van Eck, E. R. H., Radboud University Nijmegen

Bruijnincx, P. C. A., Utrecht University

Weckhuysen, B. M., Utrecht University

Originally presented on: 11/6/2013 13:54:00 - 14:15:00

Ruthenium-Catalyzed Hydrogenation of Levulinic acid: Influence of the Support on Selectivity and Stability

Wenhao Luo,[1] Upakul Deka,[1] Andrew M. Beale,[1] Ernst R. H. van Eck,[2] Pieter C.A. Bruijnincx,[1*] Bert M. Weckhuysen[1*]

[1]*Inorganic Chemistry and Catalysis, Debye Institute for Nanomaterials Science, Utrecht University, Utrecht, The Netherlands*

[2]*Institute for Molecules and Materials, Radboud University Nijmegen, Nijmegen, The Netherlands*

* p.c.a.bruijnincx@uu.nl; b.m.weckhuysen@uu.nl

Introduction

Levulinic acid (LA) has been identified as a promising, renewable platform molecule that can be converted by catalytic hydrogenation/deoxygenation to other valuable chemicals, such as γ-valerolactone (GVL), methyltetrahydrofuran (MTHF), pentanoic acid (PA) or its esters (PE) [1,2]. Here, we present our results on the influence of supports of varying acidity on the selectivity and stability of Ru catalysts in the hydrogenation of LA under increasingly harsh conditions in three solvents (dioxane, the LA mimic 2-ethylhexanoic acid (EHA), and neat LA) [3].

Experimental details

1wt% ruthenium catalysts on TiO_2, Nb_2O_5, HZSM-5 and H-Beta were prepared by wetness impregnation. The fresh and spent catalysts were extensively characterized by NH_3-TPD, TEM, N_2 physisorption, XRD, AAS, TGA, IR after pyridine adsorption, and solid state NMR. Catalytic tests were run at 40 bar H_2 and 200 °C in a Parr batch autoclave system.

Results and Discussion

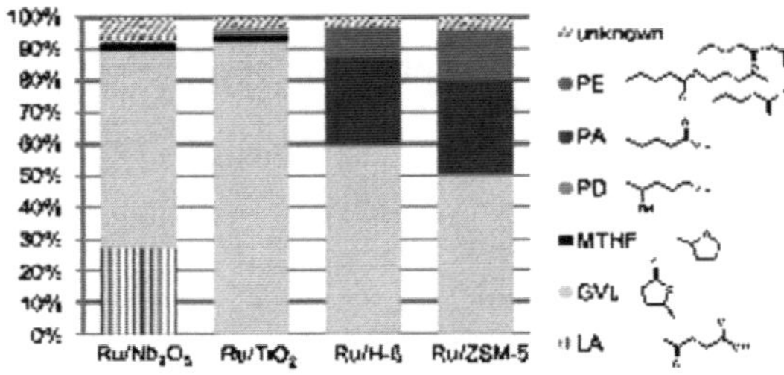

Figure 1. Catalytic hydrogenation of 10 wt% LA in dioxane.

Large differences in activity and selectivity are observed with the different catalysts (Figure 1). Notably, the zeolite-supported Ru-catalysts are the first examples of catalysts for the direct, one-pot conversion of LA to pentanoic acid (PA) and its derivatives (max. yield 45.8 % in dioxane) at this fairly low temperature. The acid sites on the support accelerate both the dehydration step from LA to GVL as well as the GVL ring opening, which is the most difficult step on the pathway to PA. Indeed, reactions with the intermediates GVL and pentenoic acid (PEA) in EHA showed that the acid-supported catalysts are capable of catalyzing the further conversion of GVL.

Spent catalysts were extensively studied to assess stability and detect any changes in either the metal phase or the support. The Ru/TiO_2 catalyst was highly selective to GVL and proved remarkably stable, even under the most severe condition (i.e. in neat LA). Gradual deactivation of the zeolite-supported catalysts was observed in EHA and neat LA. Coke build-up resulted in a decrease in surface area and pore volume for the niobia- and zeolite-supported catalyst, but not for the titania-supported one. TGA measurements showed that carbon build-up to decrease in the order H-ZSM5 > H-Beta > Nb_2O_5 > TiO_2 and that angelicalactone is involved in coke formation. Remarkably, coke build-up on the Ru/HZSM-5 catalyst preferentially occurred only in the zigzag channels, as evidenced by systematic shifts in the XRD patterns. While limited leaching and sintering of ruthenium does occur after long runs under severe conditions, dealumination is considered to be the main reason for deactivation, as demonstrated by pyridine IR studies and 2D ^{27}Al MQ MAS NMR (Figure 2).

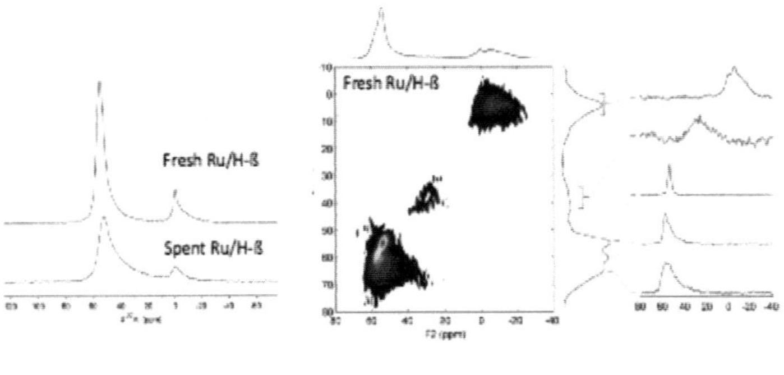

Figure 2. 1D and 2D ^{27}Al MQ MAS NMR spectra of fresh and spent Ru/H-Beta.

Conclusions

The obtained results provide further insight into the required properties for selective levulinic acid hydrogenation and show that Ru catalysts on strongly acidic supports are capable of the direct conversion of levulinic acid to pentanoic acid. The insights obtained in the catalyst deactivation mechanisms can be used for further improvement of these first promising examples.

References

1. J.Q. Bond, D.M. Alonso, D. Wang, R.M. West, and J.A. Dumesic, *Science* **2010**, 327, 1110

2. J.-P. Lange, R. Price, P.M. Ayoub, J. Louis, L. Petrus, L. Clarke, and H. Gosselink, *Angew. Chem. Int. Ed.* **2010**, 49, 4479.

3. W. Luo, U. Deka, A.M. Beale, E.R.H. van Eck, P.C.A. Bruijnincx, and B.M. Weckhuysen, *J. Catal.* **2013**, 301, 175.

(473f) Catalyst Design for the Integration of Heterogeneous Catalysis With Biocatalysis

Schwartz, T. J., University of Maine

Johnson, R. L.,

Cardenas, J., University of California, Irvine

Da Silva, N. A., University of California, Irvine

Schmidt-Rohr, K., Iowa State University

Dumesic, J. A., University of Wisconsin-Madison

Originally presented on: 11/6/2013 14:15:00 - 14:36:00

A promising strategy for the production of chemicals from biomass uses heterogeneous catalysts to upgrade platform chemicals that are produced biologically. Such a scheme exploits the high efficiency of heterogeneous catalysis and couples it with the ability of biocatalysis to generate molecules not easily accessible by other means. An important challenge for this approach is that the product solutions of biological processes may contain biologically-active compounds, such as amino acids, biomacromolecules, or vitamins, and these biologically-active species may act as poisons or inhibitors of heterogeneous catalysts used in subsequent processing steps. In this work, we have studied the hydrogenation of triacetic acid lactone (TAL, Figure 1) as a probe reaction to examine catalyst inhibition by biogenic impurities and to explore strategies to overcome such inhibition.

1 **2**

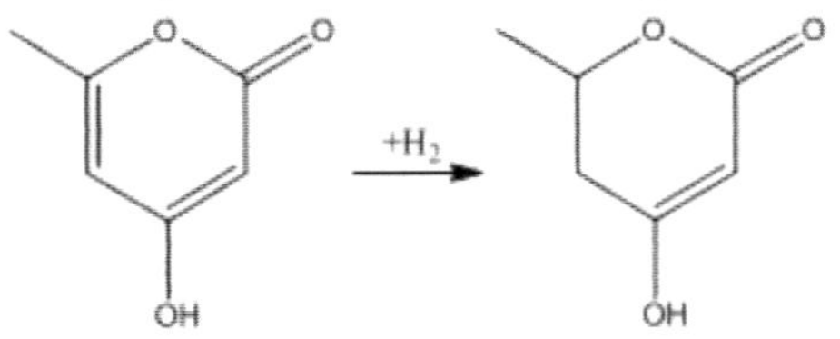

Figure 1. Hydrogenation of triacetic acid lactone (TAL, **1**) to 5,6-dihydro-4-hydroxy-6-methyl-2*H*-pyran-2-one (**2**). Sorbic acid can be produced from TAL by further hydrogenation of **2** and subsequent dehydration and ring opening.

TAL can be produced from glucose through polyketide biosynthesis, and it can be subsequently upgraded to sorbic acid. The Pd-based hydrogenation catalyst used in this process is particularly susceptible to inhibition by amino acids. In particular, we have observed that sulfur-containing amino acids are some of the most inhibitory compounds that are present in spent cell culture medium, with methionine causing an 80% decrease in catalyst activity at low concentration (0.01 mM). Additionally, the same concentration of tryptophan (an aromatic amino acid) or alanine (which has only a methyl side chain) results in 37% or 30% decreases in activity, respectively. As such, we have investigated some of the aspects of catalyst design that can make a catalyst resistant to this type of inhibition.

We have taken advantage of the different chemical properties of amino acids compared to the targeted cell-culture product (i.e., TAL) to design catalysts that are resistant to inhibition by amino acids. We have tailored the local chemical environment surrounding the metal nanoparticles by crosslinking poly(vinyl alcohol) inside the pores of a Pd/γ-Al_2O_3 hydrogenation catalyst, thus making the adsorption of polar compounds unfavorable. Using hydrogenation of TAL as a probe reaction, we studied this catalyst for resistance to the same three amino acids that inhibited the non-overcoated catalyst. This novel polymer-overcoated catalyst showed good resistance to inhibition by all three, with no measurable decrease in reaction rate after aging the catalyst for 14 hours in 0.01 mM methionine, tryptophan, or alanine. We therefore suggest that polymeric overcoating is a viable strategy for imparting amino acid tolerance to hydrogenation catalysts.

(473g) Reactions of Aliphatic and Aromatic Oxygenates On Ruthenium

Chiu, C. C., Agency for Science, Technology and Research
Genest, A., Agency for Science, Technology and Research
Borgna, A., Institute of Chemical and Engineering Sciences
Roesch, N., Technische UniversitÃ¤t MÃ¼nchen
Originally presented on: 11/6/2013 14:36:00 - 14:57:00

Biomass derived oxygenates can be used as feedstock for the production of alkanes or H_2 in aqueous phase processing over various metal catalysts [1,2]. While late transition metals, e.g., Pt or Pd, typically favor reforming yielding hydrogen, ruthenium catalysts were reported to show high selectivity for alkane formation [2], which demonstrates their potential as hydrodeoxygenation catalyst. We examined computationally transformations of selected model oxygenates over Ru(0001), applying a DFT-GGA periodic slab model approach.

We explored the decomposition of ethanol [3], the smallest alcohol containing a C-C bond, focusing on the competition between C-O and C-C bond cleavage steps. Preference of C-C bond cleavage has been associated with reforming, preference of C-O bond cleavage has been suggested to favor the formation of hydrocarbons via hydrodeoxygenation. We found that dehydrogenating ethanol yielding unsaturated centers is required for kinetically accessible C-C and C-O cleavage. In part, this is related to a stronger substrate-surface interaction, stabilizing transition structures. For most intermediates studied, C-O cleavage is both kinetically and thermodynamically favored over C-C scission. The latter becomes kinetically favored only for highly dehydrogenated species CH_kCO ($k = 1, 2$). Based on the calculated energetics we suggest a decomposition pathway of ethanol on Ru via formation of ketene CH_2CO and subsequent C-C cleavage, to yield methylene and CO. The crucial barrier, 77 kJ mol^{-1}, is associated with the initial dehydrogenation at the saturated ethyl group [3].

We went on to explore reactions of the aromatic oxygenate guaiacol (*2*-methoxyphenol) on Ru. Due to its hydroxyl, alkoxyl and aromatic functionalities, guaiacol may serve as a model system for understanding the chemistry of lignin, one of the major components of woody biomass. Again we started with the evaluation of preferred adsorption complexes, addressing especially the interaction of aromatic functionalities with Ru. We also studied the hydrodeoxygenation of guaiacol by evaluating pertinent reaction energies and activation barriers.

References:

[1] D.M. Alonso, J.Q. Bond, J.A. Dumesic, *Green Chem.* **2010,** 12, 1493-1513.

[2] R.R. Davda, J.W. Shabaker, G.W. Huber, R.D. Cortright, J.A. Dumesic, *Appl. Catal. B-Environ.* **2005**, 6,171-186.

[3] C.-c. Chiu, A. Genest, N. Rösch, *Top. Catal.* **2013**, DOI: 10.1007/s11244-013-0051-0.

(478a) Enhanced Yield of Bio-Gas Production With Modified Themo-Reactor Configuration and Separation of CO2

Tahir, M. S., University of Gujrat
Saleem, M., Institute of Chemical Engineering and Technology, University of the Punjab, Lahore, Pakistan
Originally presented on: 11/6/2013 12:30:00 - 12:50:00

Pakistan is currently facing severe energy crisis with significant shortfall. A number of short-term technical solutions are available but are not adequate to solve the issue in the longer run. Production of bio-gas with modified thermo-reactor from the animal waste at regional level shows potential of reducing burden on total natural gas resource in Pakistan.

This research study investigates the effects of solar thermo-rector configuration on the overall yield and composition of bio-gas from dung. The designed capacity of the bio-gas plant is 10 m^3. The solar thermo-reactor employs internal heating system with coils using warm water heated from the solar energy. The bio-gas reactor was operated over the range of 25-55 ^{0}C to observe effects on bio-gas yield and composition. It was observed that the increase in temperature within the thermo-reactor resulted in significant increase in yield of bio-gas i.e. thermo-reactor operated at 50-55 ^{0}C (thermophilic). Operation time and plant size of state of the art process can be reduced.

The calorific value of bio-gas obtained at 50-55 ^{0}C was slightly higher than calorific value of bio-gas obtained at mesophilic condition (25-35°C) due to increased production of CH_4. Thermophilic operation of bio-gas plant also enhances the methane content in the bio-gas i.e. 57-60 % vol of methane. Bio-gas produced was purged from bottom and cold water from top a packed column and decreases in CO_2 was also recorded. The results were scaled up from pilot scale capacity to regional level in order to estimate potential of bio-gas production

(478b) Microwave-Assisted Catalytic Glycolysis of Poly(ethylene terephthalate) By Using Ionic Liquids

Liu, Y., Donghua University
Xing, Y., Donghua University
Originally presented on: 11/6/2013 12:50:00 - 13:10:00

Poly(ethylene terephthalate) (PET) was widely used for beverage bottles, electrical/electronic instruments and household wares. As a consequence of dramatically increasing consumption, recycling of post-consumer PET products has become an important environmental opportunity for sustainable usage in society. The presented study is aiming to investigate the use of ionic liquids as catalysts for the glycolysis of PET. A series of novel ionic liquids using 1-butyl-3-methylimidazolium chloride (BMIMCl) and zinc halides (ZH) as precursor with different molar ratios was first synthesized. The glycolysis of PET by microwave-assisted method using the synthesized ionic liquids as catalysts and glycol as solvent was also studied. The influences of the type of zinc halide, the molar ratio of BMIMCl:ZH, the dosage of ionic liquid and solvent, reaction time and reaction time on the yield of recovered terephthalic acid were investigated. The synthesized ionic liquids and the glycolysis product were characterized by X-ray diffractometer, infrared spectrometric analyzer, ^{1}H NMR, differential scanning calorimeter and DSC. The results showed that the synthesized ionic liquids exhibited higher catalytic activity for the glycolysis of PET by microwave-assisted method□The halogen anion has a key influence on the glycolysis of PET. A study on the recycling of the ionic liquid shows that it could be used repeatedly. A mechanism of the degradation of PET was proposed and the kinetics of the reaction was needed further investigated.

(478c) Degradation of Rhodamine-B By Heteropolyoxometalate Ionic Liquid Under Visible Light

Liu, Y., Donghua University
Chen, Y., Donghua University
Xing, Y., Donghua University
Originally presented on: 11/6/2013 13:10:00 - 13:30:00

The photo-catalytic degradation of Rhodamine-B(RhB) dye was successfully achieved using different heteropolyoxometalate phosphonium ionic liquids(PWPILs) under visble light irradiation. A series of heteropolyoxometalate phosphonium ionic liquids were first prepared from phosphotungstic acid and different quaternary phosphonium salts trihexyl(tetradecyl)phosphonium bromide, tributyl(octyl)phosphonium bromide and tetrabutyl phosphonium bromide with different molar ratioa. The melting points of PWPILs, all lower than 100°C,were characterized by differential scanning calorimeter(DSC). The results show that the characterized peaks of phosphotungstic anion in IR spectra of PWPILs are consistent with that of phosphotungstic acid. This indicates that the quaternary phosphonium cations do not change the Keggin structure of heteropoly acid anion. The influences of experimental parameters, such as the carbon chain length in phosphoniums, the number of free hydrogen in phosphotungstic anions ,addition of different catalysts, irradiation time and the concentration of RhB solution on the photo-catalytic degradation of RhB were investigated. The photo-degradation efficiency was characterized by decolourization rate of RhB. The results indicated that $[P(C_{14}H_{29}(C_6H_{13})_3]_3[PW_{12}O_{40}]$ showed the best photo-catalytic degradation efficiency.
The RhB solution (2×10-4mol/L) was almost decomposed catalyzed by $[P(C_{14}H_{29})(C_6H_{13})_3]_3[PW_{12}O_{40}]$ under UV-Vis light irradiation for 3 hours. The degradation rate was more than 95%.

(478d) Metal-Containing Ionic Liquid As An Effective Catalyst for the Degradation of Poly(ethylene terephthalate) in Ethylene Glycol

Lu, X., Institute of Process Engineering, Chinese Academy of Sciences
Wang, Q., Institute of Process Engineering, Chinese Academy of Sciences
Zhang, S., Institute of Process Engineering, Chinese Academy of Sciences
Originally presented on: 11/6/2013 13:30:00 - 13:50:00

The vast emissions of poly(ethylene terephthalate) (PET) wastes have caused serious environmental pollution and resource waste. Its effectively recycling has become a global problem which must be urgently solved.

Chemical recycling is an attractive process because PET can be degraded into monomers which can be used as raw materials to produce virgin plastic products. Glycolysis, as an important chemical recycling method, has received extensive concern due to its advantages of less volatile solvents, high conversion of PET, giving monomers and easily continuous production. But there are some drawbacks, such as high temperature, high pressure, low reaction rate and difficult separation between catalyst and product. Many researchers devoted to developing new catalysts to overcome the drawbacks above. Ionic liquids have been reported that could effectively catalyze the degradation of PET to give BHET monomers with high purity and make the reaction condition mild. But the conversion of PET and the selectivity of monomer bis(2-hydroxyethyl terephthalate)(BHET) are both low, and the reaction rate is slow. Therefore, the development of new catalysts is still critical.

It has been found that synergetic effect of catalysts can speed up a chemical reaction or make reaction condition mild.[1,2] Yamamoto et al. found that combinations of organic and inorganic molecules bearing acidic groups could result in higher catalytic activity than individual catalysts for acid-catalyzed reactions.[2] Wang et al. and Zhou et al. found that the synergetic effect of cation and anion of ionic liquid catalyst could make the degradation condition of PET milder.[3,4] Therefore, we infer that a new catalyst containing two active components maybe has highly catalytic activity.

In this study, a metal-containing ionic liquid catalyst including two active components was synthesized and used for the PET degradation in ethylene glycol. The results showed that the synthesized catalyst has better low-temperature catalytic activity. The conversion of PET and the selectivity of BHET are 100 % and 78.7 % respectively at 175 Â°C for 1.25 h. The high catalytic activity is mainly attributed to the synergetic effect of two components. Moreover, this catalyst has the advantages of mild reaction condition, fast reaction rate, products being easily separated from catalyst and being of high purity.

[1] P. R. Schreiner. Science, 2010, 327, 965.

[2] H. Yamamoto and K. Futatsugi. Angew. Chem. Int. Ed., 2005, 44, 1924.

[3] H. Wang, R. Y. Yan, Z. X. Li, X. P. Zhang and S. J. Zhang. Catal. Commun., 2010, 11, 763.

[4] X. Y. Zhou, X. M. Lu, Q. Wang, M. L. Zhu and Z. X. Li. Pure Appl. Chem., 2012, 84, 789.

(478e) Anaerobic Digestion Technology:The Energy,Environment and Health of Developing Nations

Adetule, O., Cornell University

Originally presented on: 11/6/2013 13:50:00 - 14:10:00

Anaerobic digestion is one of the most cost effective and established technologies available for managing organic wastes for energy production,environmental management and pathogen reduction. In developing countries such as Nigeria, the application of this technology at household level as a method for gas generation and safe organic waste management has not been adequately studied. In light of the above, a pilot scale digester was designed for the production of Biogas which at the same time manages waste and reduced pathogens through anaerobic digestion process. The pilot scale digester was 10 litres and manages 9 litres of waste.The digester was fed with different types of waste:Cattle manure, Chicken litter, Rumen of cattle and Water hyacinth. After some weeks, the wastes produced Biogas, reduced pathogens and reduced the amount of wastes in the environment. A prototype digester was later designed and constructed which can accommodate more waste and produced more biogas. The prototype digester is a $5m^3$ underground fixed dome biodigester made of concrete and was designed, constructed and used for digesting cattle dung and water hyacinth. The digester was operated by mixing two parts of cattle dung (900kg) and one part of chopped water hyacinth (100kg) in 2,000 liters of water for two weeks following standard procedures. The gas produced was collected and piped to a kitchen for household use. It was also observed that there was reduction in pathogens and sanitization of the environment through waste management. This study shows that anaerobic digestion can be maximized for waste reduction, energy production,pathogen reduction and conversion of green house gases into usable forms.

(478f) Phosphorus Recovery From Municipal Wastewater Treatment Sludge

Lamichhane Upadhyaya, K., Mississippi State University
French, W. T., Mississippi State University
Hernandez, R., University of Louisiana at Lafayette
Originally presented on: 11/6/2013 14:10:00 - 14:30:00

Phosphorus recovery from municipal wastewater treatment sludge

Kamal Lamichhane Upadhyaya[1], W. Todd French[1] and Rafael Hernandez[2]

[1]Dave C. Swalm School of Chemical Engineering, Mississippi State University, MS, 39762, USA

[2]Department of Chemical Engineering, University of Louisiana at Lafayette, Lafayette, LA, 70503, USA

Abstract

Phosphorus is an essential nutrient for plants, animals and humans. Phosphorus is primarily used in agriculture, as it is one of the important components in fertilizer. Besides, phosphorus is also used in laundry detergent, matchstick, steel production and pesticides. The United States phosphate reserves are declining, which could lead to phosphorus shortage in near future. The United States now imports 10% of its phosphorus supply after exporting for several decades. Hence, there is need to search for renewable and sustainable source of phosphorus such as municipal wastewater sludge. Municipal wastewater sludge contains up to 10 wt% phosphorus and recovering these could significantly decrease the phosphorus shortage in the United States. In this study, the potential of phosphorus recovery from primary and secondary sludge through acidic and basic leaching processes will be investigated. The wastewater treatment sludge will be incinerated to obtain sewage sludge ash (SSA) and subjected to acidic and basic leaching at different acid and base concentration and pH levels. After leaching process, the supernatant will be separated from residue through filtration. Phosphorus will then be recovered from the supernatant mixture through the addition of calcium chloride in basic medium. The recovered phosphorus ($Ca_5(PO_4)_3OH$) has various agricultural and industrial usage as it resembles the phosphate mineral ore (apatite).

(478g) Techno- Economic Analysis of Wastewater Biosolids Gasification

Lumley, N. P. G., Colorado School of Mines
Porter, J. M., Colorado School of Mines
Originally presented on: 11/6/2013 14:30:00 - 14:50:00

Wastewater treatment in the U.S. produces over eight million dry tons of sludge per year. Wastewater sludge in its raw form is a noxious dilute suspension of organic matter in up to 95% water. Management and disposal is an expensive challenge for wastewater treatment plants (WWTPs), where the conventional approach to sludge management is to dewater the material to varying degrees and dispose by landfilling or application as a soil amendment. Such an approach requires energy and financial expense both in drying and ultimate disposal. Dried wastewater sludge has a lower heating value (LHV) close to that of low- rank coal. Thermochemical conversion of wastewater sludge to energy and/ or fuel product may allow for a more value- added approach to sludge management in addition to concentrating and stabilizing the waste stream.

Many WWTPs currently valorize a portion of the energy in sludge via anaerobic digestion. The biological process allows bacteria to operate on dilute sludge in an anaerobic environment producing a combustible bio- gas. Anaerobic digestion is an established technology but allows for only about 20% recovery of the energy in sludge. Anaerobic digestion does not significantly reduce the amount of sludge requiring downstream disposal. Additionally, bio -gas contaminants, especially siloxanes, are a problem for downstream combustion apparatus.

Thermochemical conversion (TCC) technologies subject sludge to chemical processes at elevated temperatures producing heat and/or combustible fuel products. TCCs applicable to wastewater treatment include direct combustion, pyrolysis, and gasification. TCC offers a great advantage over digestion by allowing nearly full utilization of sludge energy and a similarly great reduction in disposable waste volume. Each of these TCCs has been successfully applied to biomass. Wastewater sludge however is fundamentally different from most biomass sources. Raw wastewater sludge may be up to 95% water and 30% ash (dry basis). Balancing moisture removal with reactor operating conditions is the most important consideration in whether TCCs can be net energy- and net economically-positive. Ash concentration and composition present certain challenges to reactor design. Additionally, wastewater sludge organic composition can be variable leading to uncertainty in fuel output.

This work reviews the current state of practice of various TCCs with application to sewage sludge. We discuss technical and economic challenges restricting greater implementation of TCC within the wastewater process. A thermodynamic model of a sludge gasification- electric generator system is built using performance characteristics of a generic, commercially available small biomass gasification platform. We perform energy and techno- economic analyses on sludge gasification to determine lower- limit WWTP capacity and as- delivered sludge moisture content for both air- blown and steam- blown gasification processes.

(499a) SiC Photoanodes for Solar Water Splitting

Bohn, C., National Institute of Standards and Technology
Sharifi, F.,
Kang, M. G.,
Szalai, V., National Institute of Standards and Technology
Originally presented on: 11/6/2013 12:30:00 - 12:49:00

AIChE 2013 Session: Nanomaterials for Hydrogen Production and Fuel Cells

Title: SiC Photoanodes for Solar Water Splitting

Authors: C.D. Bohn, M.G. Kang, F. Sharifi, V.A. Szalai

SiC ceramics offer a rich field for exploration for solar water splitting.Â Polytypes include 4H-SiC with a bandgap of 3.23 eV, 6H-SiC with a bandgap of 3.05 eV and 3C-SiC with a bandgap of 2.36 eV.Â Single crystal substrates of some of these materials were obtained from commercial vendors.Â The effect of nanoscale morphology and the addition of Pt nanocrystals at the interface between the SiC and electrolyte were systematically investigated in a pH=7 potassium phosphate buffer under AM 1.5 illumination.Â Gas chromatography provided insight into the chemical products formed by the n-type SiC photoanodes.Â Additionally, linear sweep voltammetry, impedance spectroscopy and scanning electron microscopy were used to monitor the structural evolution of the semiconductor surface.Â Results are discussed in the broader context of photoelectrochemical hydrogen production and recommendations for improving the efficiency of water splitting with SiC are proposed.

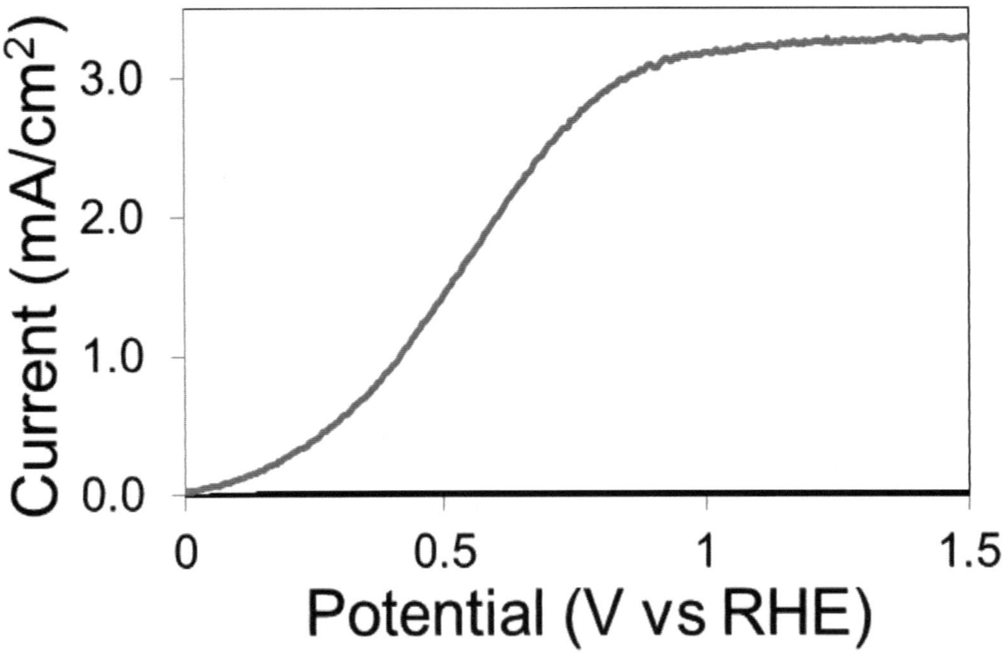

Keywords: solar water splitting, SiC, photoanode

(499b) Advanced Electrocatalysts From Organic Solution Synthesis

Wang, C., Johns Hopkins University

Originally presented on: 11/6/2013 12:49:00 - 13:08:00

Alloy nanoparticles (NPs) have attracted increasing interest due to their superior performance in various catalytic applications. For example, Pt alloys with transition metals (PtM with M = Fe, Co, Ni, etc.) have been found to be highly active for the oxygen reduction reaction in fuel cells. A lot of efforts have been dedicated to the synthesis of alloy catalysts, which are usually in the form of alloy NPs dispersed in high surface area carbon matrix. The approaches mostly include co-precipitation of metal salts in aqueous solution, impregnation, and electrodeposition. In spite of the successful preparation of various types of alloy catalysts, it yet remains challenging to synthesize monodisperse and homogeneous alloy catalysts with robust control over size and composition. Here we introduce our organic solvothermal approach towards high-quality alloy nanomaterials for electrocatalytic applications. Control over particle size, alloy composition and alloy homogeneity has been achieved by matching the different reduction and growth rates of the alloying elements, which has enabled systematic studies of the effects of these parameters on electrocatalysis. Our studies thus provide clues for the design and optimization of alloy nanomaterials for catalytic applications.

(499c) Nanostructured Ti –Fe2O3/Cu2O Bilayered Thin Films for Photoelectrochemical Water Splitting

Sharma, D., Dayalbagh Educational Institute,Dayalbagh

Upadhyay, S., Dayalbagh Educational Institute

Choudhary, S., Dayalbagh Educational Institute

Satsangi, V. R., Dayalbagh Educational Institute

Shrivastav, R., Dayalbagh Educational Institute

Dass, S., Dayalbagh Educational Institute

Originally presented on: 11/6/2013 13:08:00 - 13:27:00

Nanostructured Ti-Fe_2O_3/Cu_2O Bilayered Thin Films for Photoelectrochemical Splitting of Water

Dipika Sharma, Sumant Upadhyay, Surbhi Choudhary, Vibha R. Satsangi[?],

Rohit Shrivastav and Sahab Dass,

Department of Chemistry, [?]Department of Physics & Computer Sciences

Faculty of Science, Dayalbagh Educational Institute (Deemed University),

Agra-282110 (India)

Nanostructured thin films of pure Fe_2O_3, Ti ?Fe_2O_3, Cu_2O, and bilayered Fe_2O_3/Cu_2O, Ti-Fe_2O_3/Cu_2O were deposited on ITO glass substrate using spray pyrolysis method. Among various metal oxide semiconductors, cuprous oxide Cu_2O [Eg=2.4eV] and Iron oxide Fe_2O_3 [Eg=2.1eV] were
chosen for the present study on account of being abundant, non-toxic, environmentally safe, having low band gap and possess staggered type band edges which is suitable for efficient charge separation in water splitting process.

All samples were characterized using XRD, SEM and UV?Vis spectrometry. Photo electrochemical properties were also investigated in a three-electrode cell system. UV-Vis absorption spectrum for pure Fe_2O_3, Ti ?Fe_2O_3, Cu_2O, and Fe_2O_3/Cu_2O, Ti-Fe_2O_3/Cu_2O bilayered thin films shows absorption in visible light region. Nanostructured thin films for all the samples were used as photoelectrode in the Photoelectrochemical cell for water splitting reaction. Maximum photocurrent density of 2.22 mA/cm^2 at 0.95 V/SCE was exhibited by 454 nm thick Ti-Fe_2O_3/Cu_2O bilayered photoelectrode (Fig.1). Increased photocurrent density and enhanced photoconversion efficiency, exhibited by the bilayered thin films may be attributed to improved conductivity and efficient separation of the photogenerated charge carriers at the

Ti-Fe_2O_3/Cu_2O interface.

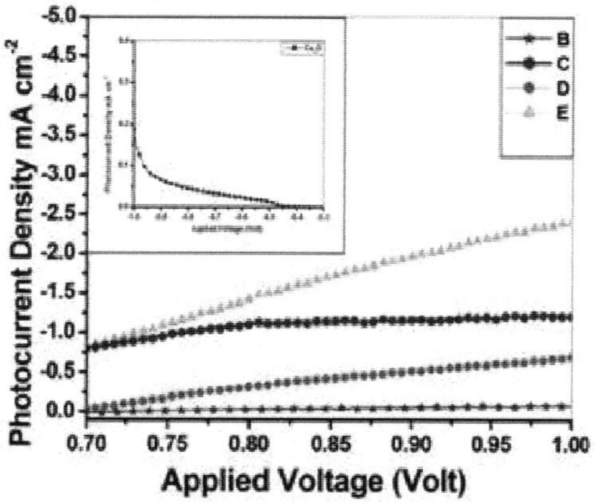

Fig. 1 Photocurrent density vs. Applied potential curve for pure Cu_2O(A), Fe_2O_3(B), Ti ?Fe_2O_3(C), and Fe_2O_3/Cu_2O(D), Ti-Fe_2O_3/Cu_2O(E) bilayered thin films respectively ,under visible light illumination in 0.1M NaOH electrolytic solution using 150W visible light source of irradiation 150 mW cm^{-2} at the position of sample.

Keywords: *Photoelectrochemical, Water splitting, Bilayered, Cu_2O, Fe_2O_3, Ti- Fe_2O_3, Spray-Pyrolysis*

(499d) Synthesis and Characterization of Mn2O3-ZnO Nanocomposites for Hydrogen Generation Via Photoelectrochemical and Photocatalytic Splitting of Water

Singh, N., Dayalbagh Educational Institute
Choudhary, S., Dayalbagh Educational Institute
Satsangi, V. R., Dayalbagh Educational Institute
Dass, S., Dayalbagh Educational Institute
Shrivastav, R., Dayalbagh Educational Institute
Originally presented on: 11/6/2013 13:27:00 - 13:46:00

Synthesis and Characterization of Mn_2O_3-ZnO Nanocomposites for Hydrogen Generation *via* Photoelectrochemical and Photocatalytic Splitting of Water

Nirupama Singh, Surbhi Choudhary, Vibha R. Satsangi[#], Sahab Dass and Rohit Shrivastav*

Department of Chemistry, [#]Department of Physics & Computer Science,

Dayalbagh Educational Institute, Dayalbagh, Agra-282 005, India

*E-mail: rohitshrivastav_dei@yahoo.co.in

The growing consumption of energy, which is now mainly produced from fossil fuels leads to a further depletion of the limited fossil reserves, increased dependency on fossil fuel imports from often politically unstable counties and furthermore gives rise to an increased concentration of green house gases mainly CO_2 in the atmosphere. The report deals with the synthesis of Mn_2O_3-ZnO nanocomposites, their characterization, and possible use in photoelectrochemical and photocatalytic splitting of water for hydrogen generation. Powdered samples of nanocomposites were prepared by co-precipitation at varying Mn/Zn molar ratios. XRD analysis revealed the evolution of wurtzite ZnO and Mn_2O_3 phases in samples, while optical absorption threshold was found shifted towards higher wavelength with increasing Mn amount. The Zn/Fe molar ratio in nanocomposites was optimized, yielding significantly enhanced photoelectrochemical current and photocatalytic hydrogen yield. Plausible explanations for observed change in photoresponse of samples have been offered.

(499e) High Performance Fluorine Doped Oxygen Evolution Reaction (OER) Electro-Catalysts for PEM Based Water Electrolysis

Kadakia, K., University of Pittsburgh
Datta, M. K., University of Pittsburgh
Velikokhatnyi, O.,
Kumta, P. N., University of Pittsburgh
Originally presented on: 11/6/2013 13:46:00 - 14:05:00

Identification of electro-catalysts containing non noble metal or significantly reduced amounts of expensive noble metals (e. g. RuO_2) is highly desirable. Development of such a catalyst with comparable electrochemical performance to the standard noble metal oxide counterpart for the proton exchange membrane (PEM) based water electrolysis would constitute a major advancement in generation of hydrogen by water electrolysis. We have accordingly developed a two-pronged theoretical first principles and experimental approach to generate a nano-structured solid solution of SnO_2:10wt.% F containing only 20at.% RuO_2 [*e.g.* $(Sn_{0.80}Ru_{0.20})O_2$:10F] that displays a remarkably similar electrochemical activity and moreover, comparable or even much improved electrochemical stability and durability compared to pure the noble metal counterpart, RuO_2. Results of these studies will be presented and discussed.

(499f) Multiwalled Carbon Nanotubes Drive the Activity of Metal@Oxide Core-Shell Catalysts in Modular Nanocomposites

Cargnello, M., University of Pennsylvania

Originally presented on: 11/6/2013 14:05:00 - 14:24:00

Multiwalled Carbon Nanotubes Drive the Activity of Metal@oxide Core-Shell Catalysts in Modular Nanocomposites

M. Cargnello,[1,2] M. Grzelczak,[1] B. Rodriguez-Gonzalez,[3] Z. Syrgiannis,[1] K. Bakhmutsky,[4]

V. La Parola,[5] L. M. Liz-Marzán,[3] R. J. Gorte,[4] M. Prato,[1] and P. Fornasiero[1]

[1] *Department of Chemical and Pharmaceutical Sciences, ICCOM-CNR, Consortium INSTM, University of Trieste, via L. Giorgieri 1, 34127 Trieste, Italy*

Â [2] *Department of Chemistry, University of Pennsylvania, 231 S. 34th Street, Philadelphia, PA 19104, USA*

[3] *Departamento de QuÃmica FÃsica, Universidade de Vigo, 36310 Vigo, Spain*

[4] *Department of Chemical and Biomolecular Engineering, University of Pennsylvania, 311A Towne Building, 220 S. 33rdÂ Street, Philadelphia, PA 19104, USA*

[5] *Istituto per lo Studio dei Materiali Nanostrutturati (ISMN-CNR), Via Ugo La Malfa 153, Palermo I-90146, Italy*

The ability to build hierarchical structures by arranging different building blocks with nanometer-scale precision is one of the most useful aspects of nanotechnology. This concept can produce novel materials with properties that are different from those expected from the simple sum of the individual blocks. Indeed, the interactions between the constituent parts in nanometer-scale ensembles can lead to novel electronic, optical, or catalytic properties not available in the initial building blocks. In the field of heterogeneous catalysis in particular, the interactions between the active phase, supports, and promoters is critical for obtaining high performance materials. These interactions can be both electronic and geometric. Furthermore, the demands in terms of geometry and binding energy of the active sites for different catalytic reactions can be very different.

In this contribution, rational nanostructure manipulation has been used to prepare nanocomposites in which multiwalled carbon nanotubes (MWCNTs) were embedded inside mesoporous layers of oxides (TiO2, ZrO2, or CeO2), which in turn contained dispersed metal nanoparticles (Pd or Pt). We show that the MWCNTs induce the crystallization of the oxide layer at room temperature and that the mesoporous oxide shell allows the particles to be accessible for catalytic reactions. In contrast to samples prepared in the absence of MWCNTs, both the activity and the stability of core-shell catalysts is largely enhanced, resulting in nanocomposites with remarkable performance for the water-gas-shift reaction, photocatalytic reforming of methanol, and Suzuki coupling. The modular approach shown here demonstrates that high-performance catalytic materials can be obtained through the precise organization of nanoscale building blocks.

(499g) Separation of Sulphur Dioxide and Oxygen in Thermochemical Hydrogen Production

Elder, R. H., University of Sheffield
He, G., University of Sheffield
Cumming, D. J.,
Allen, R. W. K., University of Sheffield
Originally presented on: 11/6/2013 14:24:00 - 14:43:00

Dense Yttria-stabilised-zirconia (YSZ) membranes have been shown to have great potential for separating SO_2 and O_2 in the sulphur family of thermochemical cycles for hydrogen production. Use of a catalytic YSZ membrane reactor in the sulphuric acid decomposition stage of these cycles could drive the reaction forward and significantly increase the yield of SO_2 and O_2. Whilst SO_2/O_2 separation has been shown to be feasible at temperatures between 750°C and 950°C, significant performance degradation is seen over time. This performance degradation is shown to be due to sulphur poisoning (from both sulphur adsorption and sulphide formation) of the triple phase boundaries. A similar effect is seen in solid oxide fuel cells, where ppm concentrations of sulphur severely diminish cell performance.

Here we investigate the effect of voltage pulsing on membrane performance recovery. Significant performance recovery is seen after the application of a short voltage pulse. The optimum pulse duration and magnitude were found to be 0.08 s and 10 V respectively. The lower the magnitude of voltage pulse, the longer the duration required to achieve maximum recovery. Impedance spectroscopy and SEM results are presented, showing the contribution of different resistances and the effect of the voltage pulsing on the membrane cell.

Consideration of oxygen reduction mechanisms at the triple phase boundary shows that as voltage across the membrane increases, the rate controlling step changes from oxygen adsorption and surface diffusion, to oxygen availability in the feed. The higher voltage pulses are beneficial to performance recovery as polarisation resistance and charge transfer resistance decrease due to the removal of sulphur from the surface. Pulses that have too great a magnitude, or are too long, dramatically promote oxygen reduction and lead to concentration polarisation due to lack of oxygen in the feed.

Longer term testing (20 hours) showed that repeated, regular voltage pulsing can significantly improve membrane performance compared to those that receive no voltage pulsing.

(499h) Chemisorption, Physisorption, and Hysteresis of Hydrogen On Carbon Nanotubes

Barghi, S., University of Southern California

Tsotsis, T. T., University of Southern California

Sahimi, M., University of Southern California

Originally presented on: 11/6/2013 14:43:00 - 15:02:00

We used a magnetic suspension balance, coupled with a residual gas analyzer, to investigate hydrogen adsorption/desorption on carbon nanotubes (CNTs) for pressures up to 100 bars at 25 ℃. Such a combination of gravimetric and mass analysis systems, used for the first time, is capable of overcoming the shortcomings of the common hydrogen adsorption measurement techniques. The most crucial drawbacks of the conventional measurement methods are as follows. (i) The inability to distinguish between hydrogen adsorption and chemisorptions; (ii) the relatively large errors caused by the leakage of hydrogen out of the system, and (iii) the difficulty of measuring hydrogen uptake capacity under operating conditions that differ from those of hydrogen storage systems used in practice.

In the experiments multi-walled carbon nanotubes (MWCNTs) were exposed to hydrogen, and were subsequently kept under vacuum for 2 consecutive cycles, in order to study hydrogen adsorption/desorption in the nanostructured materials. The difference between the first and the second hydrogenation cycles was in the initial preparation of the MWCNTs. For the first cycle, the MWCNTs were first degasified at an elevated temperatures (120 ℃). For the second cycle the MWCNTs were not degasified at the elevated temperatures and, therefore, hydrogen residue from the first adsorption/desorption cycle was still adsorbed on the MWCNTs at the beginning of the second cycle.

The first hydrogenation cycle was started after pretreatment of the CNT at the elevated temperature. Hydrogen adsorption kinetics for the cycle indicates that the adsorption equilibrium occurs within 4 - 5 hours. Careful examination of the recorded adsorption kinetics revealed the fascinating fact that a considerable part of hydrogen adsorption in the MWCNTs takes place *instantaneously*. The measured equilibrium data revealed some hysteresis for hydrogen desorption in this cycle, which does not vanish, even when the system's pressure approaches zero. This indicates that hydrogen adsorption in the first hydrogenation cycle is a combination of physisorption and chemisorption.

After the first cycle was finished, the CNTs were exposed to hydrogen for the second time without any pretreatments at the elevated temperature. Measurements indicated no hysteresis for during the second cycle. Therefore, only physisorption took place during the second cycle. Comparison between the first and the second hydrogen isotherms, together with the measured hydrogen adsorption kinetics, made it possible to quantitatively distinguish between the physisorption and chemisorption portions of hydrogen adsorption in the MWCNTs. According to our data, for pressures less than 20 bars physisorption and chemisorption are both important, with the former becoming the dominant mechanism as the pressure increases. To the best of our knowledge, this is the first time that hydrogen physisorption and chemisorption on carbon nanotubes have been measured separately.

(519b) Biorefinery: Liquefaction of Pyrolysis Char to Biofuel

Feiner, R., Graz University of Technology
Schwaiger, N., Graz University of Technology
Pucher, H., Graz University of Technology
Pucher, P., BDI BioEnergy International AG
Siebenhofer, M., Graz University of Technology
Originally presented on: 11/6/2013 12:52:00 - 13:14:00

Lignocellulosic biomass is the most abundant renewable material [1] in the world with coverage of 95% of all plant biomass [2] which makes it a feasible feedstock for chemical conversion to biofuels. For successful substitution of fossil fuel there are still obstacles to overcome. State of the art technologies of the petrochemical and chemical industry are not directly applicable for converting biomass to biofuels. This is due to the fact that biomass is a highly oxygenated raw material. Behrendt et al. specify the chemical composition of dry wood as $CH_{1.4}O_{0.7}$ [3] figuring out nearly one atom of oxygen per carbon atom, underlining the indispensable need of hydrogen for converting lignocellulosic biomass to biofuels. By these means simple and cheap pre-treatment is needed to eliminate oxygen in the feedstock. Liquid-phase pyrolysis is a feasible route for thermo-chemical elimination of oxygen from biomass. Products of liquid-phase pyrolysis are (1) *non-condensable gases*, (2) *water*, (3) *pyrolysis oil* of poor combustion properties and (4) biochar with an amount of 40wt% on dry biomass feed basis. The formation of liquid-phase pyrolysis products was investigated and reported [4], [5].

Biochar is a major product of pyrolysis processes, such as torrefaction and hydrothermal carbonization [6], [7]. Biochar from liquid-phase pyrolysis mainly consists of biogenous carbon with an approximate amount of 99 wt% [4]. This makes biochar a considerable feedstock for direct liquefaction in order to produce biofuels. In principle the direct liquefaction of biochar may well base on the experience from fossil coal liquefaction with hydrogen and/or hydrogen donor solvents at elevated temperature and pressure. The products of direct coal liquefaction are highly aromatic and suitable for usage as high-octane gasoline or feedstock for aromatic chemicals [8].

During coal liquefaction, while thermal decomposition ruptures the coal bonds, hydrogen from the gaseous phase or from a hydrogen donor solvent stabilizes the fragments and eliminates heteroatoms like oxygen. Stabilization ensures the breakdown of macromolecules into smaller hydrocarbon fractions. At low hydrogen availability fragments are not stabilized and will undergo re-polymerization reactions to form tar [9].

Liquefaction of biochar was investigated in a stirred 450ml batch reactor between 370°C and 425°C. The feed product was obtained from pilot scale liquid-phase pyrolysis from BDI (OMV refinery, Vienna). Biochar was extracted with n-hexane, dried and milled to <200microns. For each experiment 30g of dried biochar batches, mixed with tetralin, were prepared. The mixture was fed to the reactor and pressurized with 50bar of hydrogen. Temperature was elevated from ambient to final operation temperature with a heating rate of 15 °C/min. In order to distinguish between the contribution of different process phases to conversion and product formation the *heating period* was investigated separately prior to investigation of the *isothermal phase.* Reaction time for the *isothermal stage* was limited to 30min. Experiments were conducted with continuous hydrogen supply at two hydrogen pressure levels. Upon completion of the experiment the heating jacket was removed and the reactor was cooled with compressed air. The gas amount as well as the gas composition were determined when reactor pressure was relieved.

The lumped parameters gas (G), oil (O), asphaltenes (A), pre-asphaltenes (PA) and residue (R) were chosen for evaluation of the progress of liquefaction. The parameters were calculated from the results of sequential extraction of filtered liquefaction residues after each experiment. The liquid product phase was subjected to GC-MS, GC-FID, and elemental Analysis as well as size exclusion chromatography (SEC) in order to generate data for product quality, product formation, donor-solvent consumption as well as product size distribution.

Experiments with hydrogen donor at 425°C resulted in biochar conversion (C) of 84% and an oil yield of 72% after 30min of isothermal reaction, with the initial heating phase contributing to biochar conversion (C) to an extent of 37%. Biochar is converted into oil (O), gas (G), pre-asphaltenes (PA) and asphaltenes (A). Pre-asphaltenes and asphaltenes are still high molecular intermediates, formed during conversion of biochar [8], [10]. Liquefaction of biochar at greater extent starts at 415°C.

[1] X. Zhao, K. Cheng, and D. Liu, "Organosolv pretreatment of lignocellulosic biomass for enzymatic hydrolysis.," *Applied microbiology and biotechnology*, vol. 82, no. 5, pp. 815–27, Apr. 2009.

[2] R. Rinaldi and F. Schüth, "Design of solid catalysts for the conversion of biomass," *Energy & Environmental Science*, vol. 2, no. 6, p. 610, 2009.

[3] F. Behrendt, Y. Neubauer, M. Oevermann, B. Wilmes, and N. Zobel, "Direct Liquefaction of Biomass," *Chemical Engineering & Technology*, vol. 31, no. 5, pp. 667–677, May 2008.

[4] N. Schwaiger, R. Feiner, K. Zahel, A. Pieber, V. Witek, P. Pucher, E. Ahn, P. Wilhelm, B. Chernev, H. Schröttner, and M. Siebenhofer, "Liquid and Solid Products from Liquid-Phase Pyrolysis of Softwood," *BioEnergy Research*, vol. 4, no. 4, pp. 294–302, Jun. 2011.

[5] N. Schwaiger, V. Witek, R. Feiner, H. Pucher, K. Zahel, A. Pieber, P. Pucher, E. Ahn, B. Chernev, H. Schroettner, P. Wilhelm, and M. Siebenhofer, "Formation of liquid and solid products from liquid phase pyrolysis.," *Bioresource technology*, vol. 124, pp. 90–94, Nov. 2012.

[6] D. Ciolkosz and R. Wallace, "A review of torrefaction for bioenergy feedstock production," *Biofuels, Bioproducts and Biorefining*, vol. 5, pp. 317–329, 2011.

[7] A. Funke and F. Ziegler, "Hydrothermal carbonization of biomass: a summary and discussion of chemical mechanisms for process engineering," *Biofuels, bioproducts and biorefining*, vol. 4, no. 2, pp. 160–177, 2010.

[8] T. Kaneko, F. Derbyshire, E. Makino, D. Gray, and M. Tamura, "Coal Liquefaction," in *Ullmann's Encyclopedia of Industrial Chemistry*, Wiley-VCH Verlag GmbH, 2005, pp. 1–83.

[9] M. W. Haenel, "Catalysis in Direct Coal Liquefaction," in *Handbook of Heterogeneous Catalysis*, Wiley-VCH Verlag GmbH, 2008, pp. 3023–3036.

[10] X. Li, H. Hu, S. Zhu, S. Hu, B. Wu, and M. Meng, "Kinetics of coal liquefaction during heating-up and isothermal stages," *Fuel*, vol. 87, pp. 508–513, Apr. 2008.

(519c) Mechanistic Insight Into Sonochemical Biodiesel Synthesis Using Heterogeneous Base Catalyst

Choudhury, H. A., Indian Institute of Technology Guwahati
Moholkar, V., Indian Institute of Technology, Guwahati

Originally presented on: 11/6/2013 13:14:00 - 13:36:00

Abstract

The beneficial effect of ultrasound on transesterification reaction is well known. Heterogeneous (or solid) catalysts for biodiesel synthesis have merit that they do not contaminate the byproduct of glycerol. In this paper, we have attempted to identify the mechanistic features of ultrasound–enhanced biodiesel synthesis with the base–catalyst of CaO. A statistical design of experiments (Box–Behnken) was used to identify the influence of temperature, alcohol to oil molar ratio and catalyst loading on transesterification yield. The optimum values of these parameters for the highest yield were identified through Response Surface Method (with a quadratic model) and ANOVA. These values are: temperature = 62°C, molar ratio = 10:1 and catalyst loading = 6 wt%. The activation energy was determined as 82.3 kJ/mol, which is higher than that for homogeneous catalyzed system (for both acidic and basic catalyst). The experimental results have been analyzed vis–à–vis simulations of cavitation bubble dynamics. Due to 3–phase heterogeneity of the system, the yield was dominated by intrinsic kinetics, and the optimum temperature for the highest yield was close to boiling point of methanol. At this temperature, the influence of cavitation bubbles (in terms of both sonochemical and sonophysical effect) is negligible, and ultrasonic micro–streaming provided necessary convection in the system. The influence of all parameters on the reaction system was found to be strongly inter–dependent.

(519d) Energetic and Environmental Assessment of Biomass Processing in the Sugar-Cane Industry

Kahwaji Janho, M., Cleveland State University
Shuster, Q. A., Cleveland State University
Gatica, J. E., Cleveland State University
Hernandez, M. R., Universidad Nacional de Tucuman
Mele, F. D., Universidad Nacional de Tucuman
Colombo, M., Universidad Nacional de Tucuman
Originally presented on: 11/6/2013 13:36:00 - 13:58:00

Bio-fuels have come to be recognized as possible, environmentally friendly alternatives to fossil fuels. This work studies a Bio-ethanol production process complementing a sugar mill, through computer simulations, using ASPEN PLUS® as the simulation engine.

The process flowsheet examined includes three major processing stages typical of distillery installations for fuel-grade production in typical sugar-cane processing facilities: (1) the fermentation stage, (2) the separation stage, and (3) the dehydration stage.

Using plant data from a bio-ethanol pilot plant in Northwestern Argentina (NOA) a kinetic model is formulated for a fed-batch molasses fermentation process. The model for this unit, along with a user-defined separation module, is then integrated into a commercial process simulator. Different schemes meeting equivalent performance standards in terms of purity and recovery are examined in this paper.

Particular attention is given to energetic integration by performing a pinch-analysis to each of the fundamental stages to find a global optimum. Optimization is achieved by minimizing a given objective function that accounts for bio-ethanol recovery, separation efficiency, and minimization of effluents. Life Cycle Analysis is used to complete the assessment of operating parameters/conditions on environmental effects.

(519e) Investigation of Solvent System for the Production of Biodiesel From Sludge Palm Oil (SPO) By Enzymatic Transesterification

Alam, M. Z., International Islamic University Malaysia
Nasaruddin, R. R., International Islamic University Malaysia
Jami, M. S., International Islamic University Malaysia
Uddin, M. S., International Islamic University Malaysia
Originally presented on: 11/6/2013 13:58:00 - 14:20:00

Biodiesel is a non-toxic, renewable and environmental friendly fuel that can replace the fossil fuels which have issues on toxic emissions, unstable escalation of price and high depletion in reserve. The high cost of common feedstocks of biodiesel can be reduced by using low cost and abundant waste oils such as sludge palm oil (SPO) especially in Malaysia which produces 40 million tons SPO per year. In this study, SPO with a very high FFA content (51.64 ± 0.59% as palmitic acid) was used to produce biodiesel using locally-produced *Candida cylindracea* lipase from a low-cost based medium, palm oil mill effluent (POME). The inhibition of lipase activity is reduced by studying the effects of several factors to the conversion of free fatty acids (FFA) into biodiesel and the total yield of biodiesel (Y_T).

In this study, sludge palm oil (SPO), a low grade and by-product of palm oil mill industry was used to produce biodiesel by direct enzymatic transesterification using locally produced *Candida cylindracea* lipases from palm oil mill effluent (POME). The solvent system was investigated by comparing the effect of methanol and ethanol. The yield of biodiesel and conversion of free fatty acids (FFAs) into biodiesel for ethanol were higher than the methanol. Ethanol was selected for the based on the findings of ethanol-to-SPO molar ratio, water content, and additional organic solvent. The optimum factors of ethanol solvent system were 4:1 ethanol/SPO molar ratio in the presence of small water content (1:15-1:10 water/SPO volume ratio). The optimum factors of process conditions were 10 U, 40^0C, 250 rpm and 24 hours for enzyme loading, reaction temperature, mixing speed and reaction time respectively. The highest conversion and Y_T of the optimal conditions from OFAT method was 12.0% and 62.3 % (w/w SPO) respectively.

(519f) Update to Biodiesel Production From Activated Sludge and Economic Analysis

Coker, A. T., Mississippi State University
Hernandez, R., University of Louisiana at Lafayette
Revellame, E. D., Mississippi State University
Holmes, W. E., University of Louisiana at Lafayette
French, W. T., Mississippi State University
Originally presented on: 11/6/2013 14:20:00 - 14:42:00

Biodiesel is a renewable fuel that can supplement petroleum fuel supply. One of the reasons why biodiesel is not yet cost-competitive with petroleum fuels is due to the high cost of feedstock - up to 70 -95% of biodiesel production cost. A feedstock with high potential for less expensive biodiesel production is activated sludge obtained from wastewater treatment plants. However, drying of sludge prior to oil extraction and reaction is a major operating cost of this process since it is so high in water content. Previous studies by this group have shown that the lipids present in sludge can be converted to biodiesel. In this case, supercritical methanol was used as a reactant and catalyst for biodiesel production.

A study was conducted to determine the optimum and cost-efficient yield of biodiesel produced from reacting wet sludge using only supercritical methanol. The goal of this investigation is to address the challenge of reducing sludge moisture content by investigating the feasibility of using activated sludge for cost-efficient biodiesel production with little or no drying of the feedstock. A model system of oleaginous yeast, *Rhodotorula glutinis*, was used to evaluate the production of biodiesel in a system similar to sludge. The combination of a fluidized sand bath and 46ml tubular stainless steel reactors were used for the supercritical methanol reactions. Two methods of this non-catalytic process, 1-step and 2-step processes, that could produce high FAME yields were studied and compared in terms of FAME yields and kinetic rate constants. With these results, plant schematics were proposed and economic analyses were conducted to investigate the cost efficiency of both non-catalytic methods. One was recommended with great potential for producing biodiesel from activated sludge at a cost-efficient price.

(519g) An Intensified Integrated Biodiesel Manufacturing Process for Conversion of Waste Oils and Alcohols Into Biodiesel to Power Campus Bus System

Sunol, A. K., University of South Florida
Cerniga, Z., University of South Florida
Mcgranaghan, E., University of South Florida
Originally presented on: 11/6/2013 14:42:00 - 15:04:00

A novel biodiesel manufacturing process is described. The continuous process manufactures biodiesel from waste oils and waste alcohols from local community and aims to power the University bus system. The process utilizes supercritical alcohols in transesterification reactions with residence times of few minutes. The autothermal reactor is heat integrated with the waste alcohol distillation process. The presentation reports the pilot process operation, process scale up, transporatation system powering implications and resulting life cycle analysis.

(529a) Lignocellulosic Biomass Pyrolysis Product Comparison From a Micropyrolyzer, Batch and Continuous Reactor Systems

Garrido, R., Villanova University
Hammer, N. L., USDA
Zmiewski, A., Villanova University
Satrio, J. A., Villanova University
Originally presented on: 11/6/2013 15:15:00 - 15:31:00

Lignocellulosic Biomass Pyrolysis product comparison from a micropyrolyzer, batch and continuous reactor systems

Rene Garrido, Nicole L. Hammer, Alexander Zmiewski, and Justinus A. Satrio

Department of Chemical Engineering, Villanova University

ren.garrido@villanova.edu, nhammer@villanova.edu: justinus.satrio@villanova.edu

This project aims at advancing the current state of knowledge of biomass pyrolysis by developing an understanding of the pyrolytic reaction mechanisms of different feedstocks, and to find common grounds between different lignocellulosic biomass and system configurations.

To understand these differences three successive scales of pyrolysis equipment have been implemented; Stage I, a micro-pyrolyzer, Stage II, a small-scale tubular batch reactor, and Stage III, a fluidized-bed continuous reactor. Understanding the differences on reaction mechanisms for each of the stages is a key element to properly identify and determine the products obtained from pyrolysis. To separate primary from secondary reactions, a μg scale pyroprobe system combined with gas chromatography and mass spectrometry (Py-GCMS) is used to study the chemical composition of volatile products from previously hydrolyzed and torrefied biomass. Effects of the severity of hydrolysis pretreatment and torrefaction on the chemical composition and yield distribution of the fast pyrolysis products will be evaluated. The chemical composition study of pyrolyzed biomass would then aid in the understanding of the chemistry behind the different compounds to further enhance the process of upgrading bio-crude oil.

For a batch reactor, a tubular furnace is used to pyrolyze the samples, and the products will be compared with a continuous fluidizing bed reactor. Although the reaction mechanisms are different, and the biomass particles in the fluidization medium have greatly enhanced heat and mass transfer in fluidized beds reactors. This is the first step to understand and find common ground between the different feedstocks, and to move from lab scale to a pilot scale reactor.

This study investigates common Reed Grass *(Phragmites Australis)*, and Pinewood as examples of different sources of lignocellulosic biomass as potential feedstock for bio-oil production.

Phragmites represents grassy biomass; it is a prevalent waste on Villanova's campus and around the world. It is potentially a valuable energy and chemical feedstock due to its high yield (18-28 tons of biomass per acre per year). Pinewood represents hardwood biomass with high cellulose content and for our study it is the control system in which all other biomasses will be related to.

Based on the results obtained by analytical pyrolysis this study will focus on obtaining bio-crude oil from *Phragmites australis* and Pinewood using a batch reactor and compare the results from a fluidized bed continuous reactor. Through this study, groundwork is established as a preliminary attempt to understand how various feedstocks can be used for biomass conversion to more useful energy sources.

(529b) Spatio-Temporal Carbohydrate Characterization of Wood Particles During Fast Pyrolysis

Paulsen, A. D., University of Massachusetts Amherst
Hough, B.,
Schartz, D.,
Pfaendtner, J., University of Washington
Dauenhauer, P. J., University of Massachusetts Amherst
Originally presented on: 11/6/2013 15:31:00 - 15:47:00

Fast pyrolysis provides an inexpensive chemical process for converting lignocellulosic biomass including wood fibers and grasses to a liquid intermediate called bio-oil which can be transported and upgraded to chemicals and fuels. The complexity of fast pyrolysis arises from the feedstock which is comprised of several biopolymers (cellulose, hemicellulose, and lignin) integrated within a cellular microstructure. During rapid heating, cellulose and biopolymers rapidly degrade to a mixture of short-chain oligomers and eventually volatile components and aerosols which are collected as bio-oil [1,2,3]. When wood fibers are heated in pyrolysis, heat and mass transfer limitations within the cellular lignocellulosic structure produce dramatic temperature gradients, which ultimately dictate the local reaction conditions and the distribution of pyrolysis products. The complexity of characterizing the reacting particle chemistry and transport phenomena requires a large number of kinetic parameters and has been identified as one of the major fundamental challenges in research of fast pyrolysis of biomass [4].

In this work, we aim to improve reaction-transport pyrolysis models of reacting woody biomass by demonstrating a new experimental technique that characterizes the composition of woody biomass as it pyrolyzes to bio-oil. Using diffuse reflectance spectroscopy, the carbohydrate composition of Yellow Poplar (*Liriodendron tulipifera*) is characterized in position and time through a particle exhibiting heat transfer dominant in only one spatial dimension. Compositional profiles exhibit a propagating reaction front consistent with heat-transfer limited (high Bi) pyrolysis conditions. A range of reaction conditions are examined, and the resulting data are described using a one-dimensional reaction-transport model using existing lumped-kinetic reaction chemistry. Sensitivity analysis of the resulting model and comparison of the prediction with experimental spatio-temporal compositional data allows for robust characterization of wood fiber pyrolysis.

References

[1] M.S. Mettler, D.G. Vlachos, P.J. Dauenhauer, *Energy & Environmental Science* **2012**, 5, 7797-7809.

[2] V. Agarwal, P.J. Dauenhauer, G.W. Huber, S.M. Auerbach, *Journal of the American Chemical Society* **2012**, 134(36), 14958-14972.

[3] A.R. Teixeira, K.G. Mooney, J.S. Kruger, C.L. Williams, W.J. Suszynski, L.D. Schmidt, D.P. Schmidt, P.J. Dauenhauer, *Energy & Environmental Science* **2011**, 4, 4306.

[4] M.S. Mettler, D.G. Vlachos, P.J. Dauenhauer, *Energy & Environmental Science* **2012**, 5, 7797-7809.

(529c) Numerical Investigation of Multiphase Dynamic Effects in Catalytic Upgrading of Biomass Pyrolysis Vapor

Capecelatro, J. S., Cornell University

Desjardins, O., Cornell University

Pepiot, P., Cornell University

Jarvis, M., National Renewable Energy Laboratory

Foust, T. D., National Renewable Energy Laboratory

Originally presented on: 11/6/2013 15:47:00 - 16:03:00

A recurring challenge among the variety of existing biomass-to-biofuel conversion technologies is the need to ensure optimal and homogeneous contact between the various phases involved. One example can be found in the catalytic upgrading of bio-oil vapors, a bi-phasic process critical in pyrolysis-based conversion, typically performed in circulating fluidized bed (CFB) reactors. While the upgrading process is expected to strongly depend on the contact time between raw gas and catalysts, and on catalysts deactivation over time, the formulation of robust design rules from an empirical standpoint alone remains difficult, due to the wide range of granular flow regimes coexisting within the reactor, from near close-packing at the bottom of the reactor to more dilute conditions in the riser. The issue is amplified by the highly unsteady and turbulent nature of gas-solid flows. In particular, clusters of particles are known to form in risers, which can significantly decrease the overall catalytic activity in the reactor. In this work, a Lagrangian particle tracking tool recently developed in the large eddy simulation (LES) framework of NGA is used to quantify the role of cluster dynamics on biomass pyrolysis vapor catalytic upgrading in risers. This numerical framework has been extensively validated for a range of particle-laden flows, in particular, cluster fall velocities in risers of CFB reactors have been shown to match experimental correlations (Fig.1). Two configurations are investigated and compared: an homogeneous system for which the catalyst-pyrolysis vapor contact time is statistically uniform, and a pseudo-two-dimensional riser, such as shown in Fig.2, combining all granular flow regimes of interest.

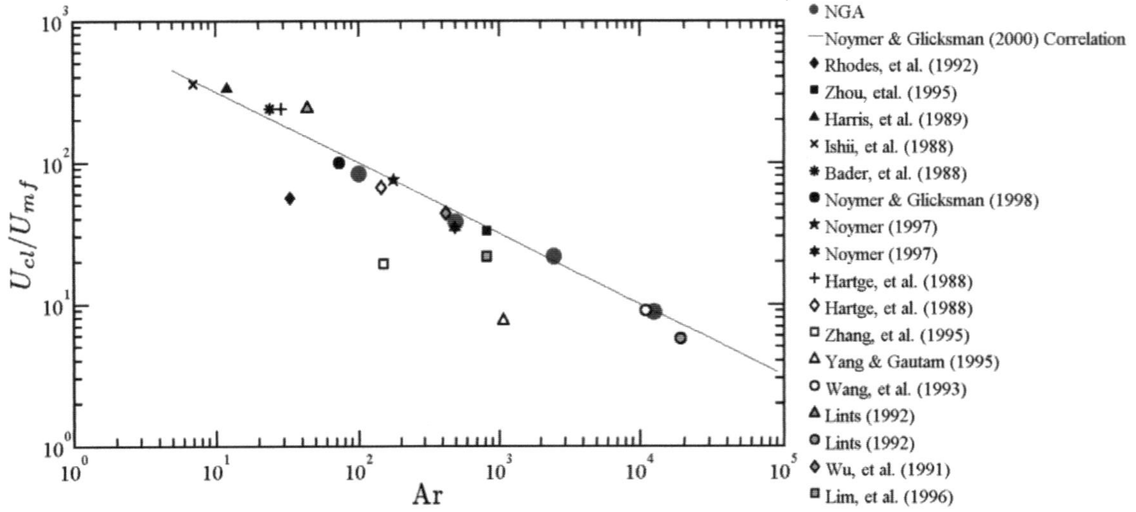

Figure 1: Mean cluster fall velocity normalized by the minimum fluidization velocity as a function of Archimedes number, a non-dimensional number that characterizes the flow inertia due to density differences between the gas and solid phases. The solid line indicates a correlation developed from a range of experimental studies by Noymer & Glicksman (2000).

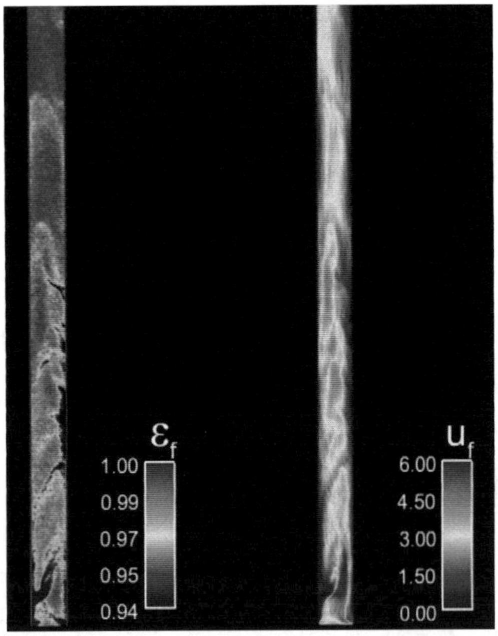

Figure 2: Pseudo two-dimensional simulation of a riser showing gas volume fraction (left) and gas velocity magnitude (right). Particles are injected at the bottom where the gas inlet is located.

(529d) Catalytic Effects of MgSO4 and ZnCl2 On Fast Pyrolysis of Pine for the Production of Bio-Oils

Yuan, L.,
Liu, Y. Q., Chevron Technology Ventures LLC
Lv, D. C., Xiamen University
Wang, D., Xiamen University
Originally presented on: 11/6/2013 16:03:00 - 16:19:00

The effects of $MgSO_4$ and $ZnCl_2$ as catalysts on fast pyrolysis of pine were investigated in a bench-scale bubbling fluidized bed reactor. The study showed that both $MgSO_4$ and $ZnCl_2$ could shift the pyrolysis of pine to a lower temperature zone, that is, most of pyrolytic reactions for cellulose and hemicellulose would happen at lower temperature while lignin at higher temperature, which results in different composition of bio-oils from that without adding any of these two salts. Compared to the bio-oils obtained without catalysts (base case), glucopyranose in bio-oils was increased to 20.69 % from less than 1% (base case) when 20 wt% of $MgSO_4$ was added; while ketones decreased from 26.71 % (base case) to 8.41 %, 2-Propanone, 1-hydroxy from 9.94 % (base case) to 0.62 %. When 30 wt% of $ZnCl_2$ is added, glucopyranose in bio-oils hiked to 20.21 % from less than 1% (base case); while phenols reduced to 6.25% from 27.09 % (base case). The pH of bio-oils rises whether $MgSO_4$ or $ZnCl_2$ was added. This study suggests that by adding specific salts into biomass, the composition of bio-oils obtained from fast pyrolysis can be of greatly different. Therefore, we can increase the content of certain chemicals in bio-oils on purpose, which could make the extraction of value-added chemicals from bio-oils more feasible and economical.

(529e) Pyrolysis of Alkali Pre-Treated Biomass and Biomass Constituents

Case, P., University of Maine
Stemmler, E., Bowdoin College
DeSisto, W. J., University of Maine
Wheeler, M. C., University of maine
Originally presented on: 11/6/2013 16:19:00 - 16:35:00

Although upgrading of pyrolysis oil has been extensively studied, it is still plagued with low product yields and significant catalyst deactivation. These issues are largely because of the instability of oxygenated species in pyrolysis oil, which leads to considerable char formation by polymerization at the elevated temperatures that are required for hydrodeoxygenation. In the current study, we eliminate compounds that cause polymerization by pre-treating sawdust with various alkali compounds before pyrolyzing it in a fluidized bed reactor. The resulting stabilized oils have significantly lower oxygen content than conventional pyrolysis oil. The higher heating values of the oils are in the range of 40 MJ/kg, and the energy yields exceed those commonly achieved in catalytic fast pyrolysis. Detailed analysis of conventional and pre-treated pyrolysis oils is presented and potential reaction mechanisms are discussed.

(529f) Study On the Aging Mechanism of Pyrolysis Bio-Oil

Meng, J., North Carolina State University
Tilotta, D., North Carolina State University
Adhikari, S., Auburn University
Kelley, S., North Carolina State University
Park, S., North Carolina state university
Originally presented on: 11/6/2013 16:35:00 - 16:51:00

Session 16004 Biomass Pyrolysis I

AIChE 2013

November 3 ~ 8, 2013

San Francisco, CA

Study on the Aging Mechanism of Pyrolysis Bio-oil

Jiajia Meng*,[1] David Tilotta,[1] Sushil Adhikari,[2] Stephen Kelley,[1] and Sunkyu Park[1]

[1] Department of Forest Biomaterials, North Carolina State University, Raleigh, NC 27695

[2] Biosystems Engineering Department, Auburn University, Auburn, AL 36849

*Presenting author: jmeng@ncsu.edu

The unstable nature of pyrolysis oil produced from biomass fast pyrolysis is one of the key issues that limit its application and upgrading, as aged bio-oil typically shows increased water content, viscosity, and phase separation. The aging phenomena are considered due to the polymerization of bio-oil components under acidic and thermal conditions catalyzed by the mineral compounds contained in char particles in the bio-oil. Current research on bio-oil aging is focused on characterizing the behavior together with seeking methods to slow down the aging speed. However, little research was reported on investigating the molecular-level polymerization mechanism of pyrolysis bio-oil during the aging. In this study, accelerated aging of the bio-oil produced from loblolly pine was performed with different aging time and temperature and with or without additives to access the bio-oil thermal stability by measuring water content, viscosity, acidity, molecular weight distribution, and chemical composition. With the chemical analysis of aged bio-oils, it was found that ether soluble fraction of the bio-oil was significantly reduced during the aging process and glycolaldehyde is the most reactive compound in that fraction. The pyrolytic lignin fraction, especially high molecular weight lignin, was found to gradually increase, which suggests the reaction between the compounds of ether soluble fraction and the lignin fraction. Because of the great reduction of glycolaldehyde concentration during the aging, synthetic bio-oil with different concentration of glycolaldehyde was mixed with pyrolytic lignin and aged at 80 °C to further study the role of glycolaldehyde on bio-oil aging chemistry. To elucidate the polymerization chemistry between ether soluble fraction and pyrolytic lignin fraction, model compounds, such as ^{13}C labeled glycolaldehyde and ^{13}C labeled phenolic compounds (phenol and guaiacol type), were reacted in a similar condition to that of accelerated aging. In addition, other suggested mechanisms of bio-oil aging will be reviewed and discussed in detail.

(529g) Effects of Post-Condensation Centrifugal Filtration On Pyrolysis Oil

Varadarajan, A., Mississippi State University
Walters, K. B., Mississippi State University
Originally presented on: 11/6/2013 16:51:00 - 17:07:00

The effects of post-condensation filtration of the crude pyrolysis oil with relatively high water content and solids content was studied. Tests were performed using a pilot plant-scale centrifuge filtration system. The centrifuge can be used as a three-phase separator [light liquid-heavy liquid-solids] or a two-phase clarifier [liquid-solid]. The testing was performed by varying the parameters such as separation time and displacement time which affects the separation efficiency of the centrifuge. In addition, the impact of increased temperature prior to and during the filtration process was examined. Effect of feed temperature on the separation efficiency of the centrifuge and the consequent physical and chemical property change on the filtered bio oil was performed. The filtered samples were collected and characterized. To determine the water content, the Karl Fischer titration method was utilized. Fourier transform infrared (FTIR) spectroscopy was used to determine chemical composition. Gel permeation chromatography was used to determine molecular weight distribution. Particle size distribution in the filtered samples was found using optical microscopy.Parameters were examined for the centrifugal filtration of post-condensation pyrolysis oil to optimize the particulate loading, processing time, and physicochemical stability for downstream processing, including upgrading

(529h) Transportable Fast-Pyrolysis Process for Distributed Conversion of Waste Biomass to Renewable Liquid Fuels

Yelvington, P. E., Mainstream Engineering Corporation
Schwartz, N., Mainstream Engineering Corporation
Zastrow, D., Mainstream Engineering Corporation
Amundsen, T., Mainstream Engineering Corporation
Originally presented on: 11/6/2013 17:07:00 - 17:23:00

In Florida alone, sugarcane bagasse, orange residue from juicing, and logging/mill residues are produced at a rate exceeding 16,000 dry tons per day. However, these agricultural residues are often spread over large areas and are difficult to access, particularly the logging residues. Rather than trucking large quantities of low-density biomass to a massive centralized biofuel production plant, the distributed conversion of agricultural residues into bio-oil is an attractive alternative for sustainable biofuel production. Opportunities also exist for the conversion of mixed waste to biofuel at forward-deployed military outposts, where the true delivered cost of fuel often exceeds $15 per gallon. Mainstream Engineering (MEC) recently commissioned a highly instrumented fast-pyrolysis pilot plant (1 ton/day) to demonstrate small-scale conversion of waste organic material to bio-oil. The pilot reactor was designed for flexibility and tight process control, as we explore the product yield and quality for a variety of feedstocks. Testing to date has used pine sawdust or a granulated mixed waste surrogate consisting of corrugated cardboard, office paper, and several plastics.

MEC is working to commercialize the portable process at a 10 ton/day scale for commercial applications and at the <1 ton/day scale for military applications. Our efforts have focused on improving the reactor yield by minimizing secondary cracking and improving the energetics with tight heat integration. Both improvements are geared towards achieving viable process economics at a small thoughput. We have also experimentally investigated the pre-processing (size reduction and drying) and end-use utilization of raw bio-oil in a burner/boiler. Results will be presented on our pilot reactor testing, techno-economic analysis of the portable reactor approach, and initial results from our efforts to mildly upgrade raw bio-oil to a renewable residual-grade fuel oil.

(529i) Biomasspyrolysisrefinery: Product Stability of Intermediates and Products

Schwaiger, N., Graz University of Technology
Feiner, R., Graz University of Technology
Pucher, H., Graz University of Technology
Steiner, L., Graz University of Technology
Ellmaier, L., BDI-Bioenergy International AG
Pucher, P., BDI BioEnergy International AG
Siebenhofer, M., Graz University of Technology
Originally presented on: 11/6/2013 17:23:00 - 17:39:00

The BiomassPyrolysisRefinery is a dual step concept for biomass liquefaction and hydrogenation, intending production of liquid energy carriers. First step of this concept is liquid phase pyrolysis (LPP). During pyrolysis lignocellulose is dehydrated to form biochar and pyrolysis oil (Schwaiger et al., 2011, 2012) The LPP process has been scaled up to a capacity of 100 kg/h biomass (BDI bioCRACK facility) at OMV refinery Schwechat. The second step of the BiomassPyrolysisRefinery is product upgrading. Pyrolysis oil can pass a dewatering step to raise the calorific value. It may also be deoxygenated and hydrogenated. Biochar is separated from the heat carrier after liquid phase pyrolysis and then hydrogenated.

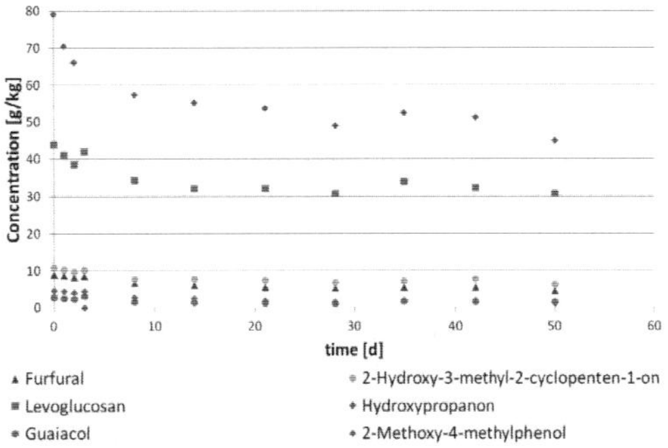

Figure 1: Stability of selected substances from liquid phase pyrolysis during storage at ambient temperature

During liquefaction biogenous material passes several heating and storing steps. Therefore sufficient chemical stability of intermediates and products is needed. Stability of pyrolysis oil for storing and upgrading purposes may be very limited (Fahmi et al., 2008; Kim et al., 2012), because of polymerization. The molar mass, viscosity and water content may rise significantly. The storing properties of liquid phase pyrolysis oil, upgraded liquid phase pyrolysis oil and liquified biochar at ambient temperature was monitored by GC-MS, GC-TCD, SEC and pH, and TAN analysis. Density and viscosity were recorded.

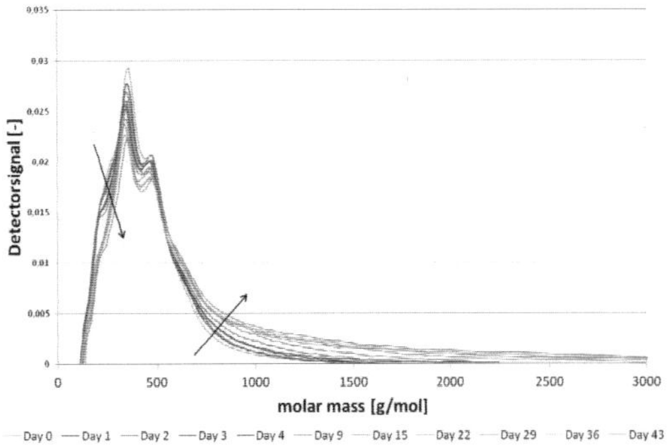

Figure 2: Liquid phase pyrolysis oil; molar mass versus storage time

Figure 1 shows the concentration of liquid key substances from liquid phase pyrolysis. During the first week of storage nearly one third of highly oxygenated substances vanish, afterwards degradation may cease. The same trend of property changes in the first days after pyrolysis is recorded with SEC-analysis. Figure 2 shows stabilization of molecular mass within day 15.

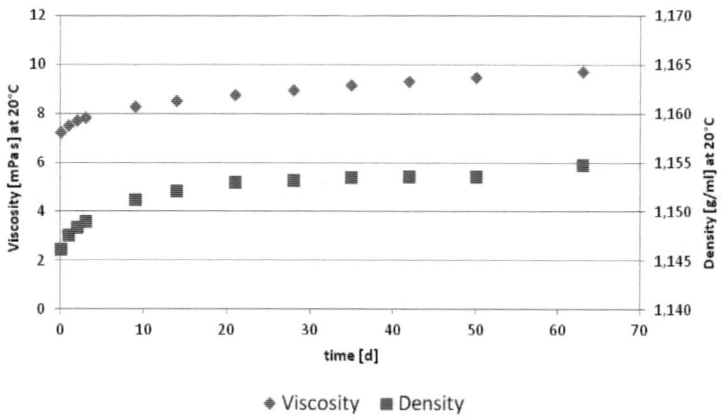

Figure 3: Viscosity and density at T = 20Ã‚Â°C of liquid phase pyrolysis oil

Viscosity and density trends, shown in figure 3, confirm the findings. Liquid phase pyrolysis oil undergoes ageing through repolymerization within the first 15 days of storage. After upgrading stability of pyrolysis oil and liquified biochar is significantly better. Compared with flash pyrolysis oil the stability of liquid phase pyrolysis oil is seemingly achieved through separation from char based particulate matter and inorganics.

Fahmi, R., Bridgwater, A., Donnison, I., 2008. The effect of lignin and inorganic species in biomass on pyrolysis oil yields, quality and stability. Fuel 87, 1230--1240.

Kim, T.-S., Kim, J.-Y., Kim, K.-H., Lee, S., Choi, D., Choi, I.-G., Choi, J.W., 2012. The effect of storage duration on bio-oil properties. Journal of Analytical and Applied Pyrolysis 95, 118--125.

Schwaiger, N., Feiner, R., Zahel, K., Pieber, A., Witek, V., Pucher, P., Ahn, E., Wilhelm, P., Chernev, B., SchrÃ¶ttner, H., Siebenhofer, M., 2011. Liquid and Solid Products from Liquid-Phase Pyrolysis of Softwood. BioEnergy Research 4, 294--302.

Schwaiger, N., Witek, V., Feiner, R., Pucher, H., Zahel, K., Pieber, a, Pucher, P., Ahn, E., Chernev, B., Schroettner, H., Wilhelm, P., Siebenhofer, M., 2012. Formation of liquid and solid products from liquid phase pyrolysis. Bioresource technology 124, 90--4.

(529j) Biomass Gasification: Effect of Sulfur Compounds On Catalytic Tar Removal Activity Using Nickel-Clay Catalysts

Vutukuri, J., University of Louisiana at Lafayette
Buchireddy, P., University of Louisiana at Lafayette
Bricka, R. M., Mississippi State University
Guillory, J., University of Louisiana at Lafayette
Bajpai, R., University of Louisiana at Lafayette
Zappi, M. E., University of Louisiana at Lafayette
Originally presented on: 11/6/2013 17:39:00 - 17:55:00

Biomass gasification is one of the thermochemical routes of converting biomass to fuel gas. This fuel gas can be used in various possible applications such as generation of heat or electricity, synthesis of liquid transportation fuels, production of hydrogen, synthesis of chemicals and generation of electricity in fuel cells. However, this fuel gas needs to be cleaned from the impurities like particulates, ammonia, hydrogen sulfide, hydrogen chloride, alkali metals, metals, and tars. The presence of tars in the fuel gas is one of the main technical barriers in the biomass gasification development. Tar buildup cause cracking in the pores of filters, forming coke and plugging the filters, and condensing in the cold spots and plugging the lines, resulting in serious operational interruptions.

Nickel (Ni) supported clays have been developed and evaluated for its activity towards tar removal at Mississippi State University. The results from these studies suggested that Ni-Clay catalysts have very good activity for tar removal. However, the influence of sulfur compounds, although, present in very low concentrations has not been evaluated. Hence, this study focuses on the effect of sulfur compounds on the catalyst performance for removal of tars from syngas. Literature suggests that the presence of sulfur in the fuel gas decreases the catalytic activity of the Ni catalysts due to strong chemisorption of sulfur on the Ni surface forming Nickel Sulfide, which is inactive. Although, the presence of sulfur has shown to have an inhibiting effect on the catalytic activity, the effect of sulfur on tar removal using Ni-Clays has not been investigated till date. This study investigates the influence of sulfur compounds on tar reforming activity of Ni Clay catalysts. A plug flow reactor has been designed and fabricated to perform these studies. Experiments are being conducted using naphthalene as a tar model compound and hydrogen sulfide as a sulfur compound in syngas. The effect of hydrogen sulfide partial pressure, temperature(650-850° C), steam to carbon ratio, and residence time on naphthalene removal activity is being evaluated, the results of which will be presented. Also, the influence of catalyst support, clay in this study is being investigated and will be compared with commercial silica/alumina supports.

(534a) Plate Food Waste: A Valuable Source of Energy

Jung, S., Iowa State University

Originally presented on: 11/6/2013 15:15:00 - 15:35:00

Post-consumer food waste, because of its quantity and composition, could be a good feedstock for biofuel production. The effect of glucan loading (2 and 6%), addition of an enzyme cocktail, pretreatment (grinding/drying or sterilizing), and food waste composition (breakfast vs. lunch, presence of used napkins) on ethanol concentration was reported. The potential for sorting and converting used napkins into ethanol was also assessed. At a 2% glucan content, ethanol concentration of up to 16 g/l was obtained after saccharification/fermentation of breakfast and lunch food waste. Increasing glucan content to 6% significantly increased ethanol concentration to 54 g/L, while there was no significant effect of food waste composition on carbohydrate conversion efficiency. When converting ethanol concentration from g/L to L/g of waste, concentrations of 0.06, 0.06 and 0.40 were obtained for food waste, food waste w/o napkins, and used napkins, respectively, identifying used napkins as an excellent source of bioethanol production.

(534b) Biofuel Production and Kinetics Study of Catalytic Microwave Pyrolysis of Douglas Fir Pellet Over Activated Carbon Supported Metal Catalyst

Bu, Q., Washington State University
Lei, H., Washington State University
Wang, L., Washington State University
Tang, J.,
Originally presented on: 11/6/2013 15:35:00 - 15:55:00

The presented study is aiming to improve the quality of bio-oils by catalytic upgrading of pyrolysis oil from microwave pyrolysis of biomass using activated carbon (AC) supported metal catalyst. A central composite experimental design (CCD) was used to optimize the reaction condition of this process. The effects of reaction temperature and reaction time on product yields were investigated. GC/MS analysis indicated that the main chemical compounds of bio-oils were furans, phenols, guaiacols, and ketones/ethers which were in the range of 80-87% of bio-oils for most of the samples. The ketones/ethers mainly composed by 2-cyclopen-1-one and straight chain ethers were increased up to about 38% of bio-oil, and the phenols increased significantly accompanying with the decrease of guaiacols in comparison with raw biomass pyrolysis. Thermogrametric analysis (TGA) was used to analyze the thermal behavior or Douglas fir pyrolysis, and the kinetic parameters were investigated using thermogrametric analysis (TGA) data and compared with catalytic microwave pyrolysis of Douglas fir in this process, the activation energy of Douglas fir pyrolysis in this process was remarkable lower than conventional TGA pyrolysis of Douglas fir. And the reaction mechanism of this process was analyzed.

(534c) Molecular-Level Modeling of Municipal Solid Waste Gasification

Horton, S. R., University of Delaware
Zhang, Y., Air Products and Chemicals, Inc.
Bennett, C. A., University of Delaware
Klein, M. T., University of Delaware
Petrocelli, F., Air Products and Chemicals, Inc

Originally presented on: 11/6/2013 15:55:00 - 16:15:00

Molecular-level Modeling of Municipal Solid Waste Gasification

Scott R. Horton*, Craig A. Bennett, Michael T. Klein

Department of Chemical & Biomolecular Engineering, University of Delaware

Yu Zhang, Francis P. Petrocelli

Air Products and Chemicals, Inc.

In 2010, there was 250 million tons of municipal solid waste (MSW) in the United States. Over half of this waste ended up in landfills which fail to recover the energy stored within MSW. The most basic waste-to-energy technology is incineration. While this process recovers energy, it also poses problems to the environment due to SO_x, NO_x, dioxin, and furan emissions. Plasma-arc gasification has nonstoichiometric amounts of oxygen and reduces the production of these harmful oxygenates while recovering energy from waste via syngas production. Furthermore, due to the extremely high temperatures of the plasma torch, the final by-product is a salable glass-like slag. The objective of this study is to aid process design by developing a molecular-level kinetics model of plasma-arc gasification.

The MSW is typically described at the lumped level, i.e., biomass, food waste, and plastics. In the present work, each of these lumped components was described at the molecular-level. Biomass was modeled as cellulose, hemicellulose, and lignin. In addition to these molecules, lipids, common amino acids, and sugars were used to model food waste. The modeled plastics were polyethylene, polyethylene terephthalate, and polyvinyl chloride. These feeds were all represented at the molecular level using Flory statistics for nearly linear polymers and lattice theory for lignin.

The feed stream was converted into a reaction network using an in-house software tool, the Interactive Network Generator (INGen). There were two major categories of reactions: devolatilization/pyrolysis and gasification. The major pyrolysis reaction families were thermal cracking, decarboxylation, and decarbonylation. For gasification, the dominant reaction type was reaction with oxygen. The generated reaction network was solved and optimized using another in-house software, the Kinetic Modeling Editor (KME).

For integration with process simulation, the optimized KME model was integrated into ASPEN PLUS using user-defined models. In the case of gasification, the KME model contains reactors in series and rate laws employing linear-free energy relationships for reaction families. The process of KME-Aspen integration was generalized allowing for the automatic implementation of any reaction model into an ASPEN worksheet.

The KME model could also be coupled to a computational fluid dynamics (CFD) solver and take equipment level information into account in the design. Once the products and kinetics of the pyrolysis and gasification reactions are obtained from KME, the reaction network could be converted into one or several global pyrolysis/gasification reactions by kinetic model reduction. The simplified mechanism could be easily integrated into any CFD solver, such as Ansys FLUENT. On the other hand, equipment information obtained from the CFD simulation, such as flow and temperature pattern, provides guidance for choosing reactors in ASPEN PLUS process simulation.

(534d) Performance Analysis of Rdf Gasification in a Two Stage Fluid Bed - Plasma Process

Materazzi, M., University College London

Lettieri, P., University College London

Chapman, C., Advanced Plasma Power

Taylor, R.,

Originally presented on: 11/6/2013 16:15:00 - 16:35:00

PERFORMANCE ANALYSIS OF RDF GASIFICATION IN A TWO STAGE FLUID BED - PLASMA PROCESS

M. MATERAZZI* **, P. LETTIERI*, R. TAYLOR **, C. CHAPMAN **

** Department of Chemical Engineering, University College London, Torrington Place, London WC1E 7JE, UK*

*** Advanced Plasma Power, South Marston Business Park, Swindon, SN3 4DE, UK*

Tar generation and ash disposal represent the strongest barrier for use of single stage fluid bed gasification for waste treatment, whereas sufficing for both is only possible with expensive cleaning systems and further processing. Waste fuels have, in general, physical and chemical characteristics much different from fossil fuels, so that the behaviour of these two feedstock types in high temperature fluid beds is very different one from the other. The high content of ashes and inorganic materials in RDF (Refuse Derived Fuel) dictates the thermal conditions in the fluid bed. A relatively low temperature has to be used to prevent agglomeration and sintering of bed material. For this reason, the gas that is produced by a standard fluid bed gasifier (FBG) has tars and other condensable organic species that are technically difficult and costly to remove. Furthermore, the bottom ash that is generated in the reactor may contain high levels of carbon, heavy metals and organic pollutants which lower the conversion efficiency of the process and limit any secondary usage.

The use of plasma systems has increasingly been applied with thermal waste treatment for its ability to completely decompose the input waste material into a tar-free synthetic gas and an inert, environmentally stable, vitreous material known as slag. The principal advantages that plasma offers to thermal conversion processes, besides the already mentioned tar/ash related issues absence, are a smaller installation size for a given waste throughput, and the use of electricity as energy source, characteristics which permit the technology to treat a wide range of low calorific value materials including liquids and solids. In applying the plasma technology to the gaseous products from a FBG, an advanced two-stage thermal process is able to achieve efficient conversion of the chars and the complex organics to the primary syngas constituents whilst limiting the electrical energy demand of the process (Figure 1).

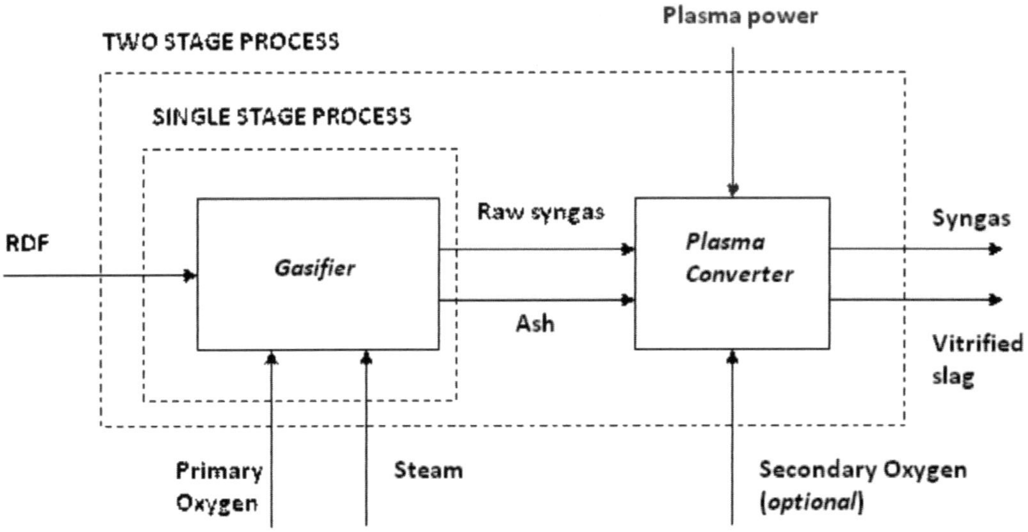

Figure 1: Schematic of Gasplasma two-stage thermal process.

This study focused on the thermodynamic assets of using a two-stage (FBG + plasma) thermal process over the conventional single-stage approach. These include, for example, the fact that the primary thermal waste decomposition is performed in conditions of optimal stoichiometric ratio for the gasification reactants. Furthermore, staging the oxidant injection in two separate intakes significantly improves the efficiency of the system, reducing the plasma power consumption. The aim of the present work is also to determine the effects of the thermal parameters so as to be able to select optimum gasification conditions in a two stage process and to clarify the mechanism of thermal decomposition of carbonaceous materials in the FBG and in the plasma converter.

A flexible model capable of providing reliable quantitative predictions of product yield and composition after the two-stage process has been developed. The method has a systematic structure that embraces atom conservation principles and equilibrium calculation routines, considering all the conversion stages that lead from the initial waste feed to final products. The model was also validated with experimental data from a demonstration plant in Swindon (UK) (Table 1).

	Case 1	**Case 2**	**Case 3**	**Case 4**	**Case 5**
Description:					
O2/fuel ratio (w/w)	0.51	0.59	0.59	0.79	0.50

Bed temperature (°C)	770	720	795	720	800
Proximate analysis, % (w/w)					
Fixed carbon	6.4	12.2	11.6	8.5	22.8
Volatile matter	59.6	50.2	64.8	47.6	68.0
Ash	19.1	23.2	12.1	8.9	0.5
Moisture	14.9	14.4	11.5	35.0	8.7
Ultimate analysis, % (w/w)					
C	41.0	47.0	43.0	31.5	45.2
H	5.7	6.3	5.6	4.1	6.46
O	17.5	6.9	26.6	19.7	45.38
N	1.2	1.74	0.61	0.4	0.26
S	0.2	0.15	0.25	0.17	0.01
Cl	0.4	0.31	0.34	0.23	0.25
GCV, MJ/kg (dry basis)	22.1	26.4	21.0	21.0	22.0

Table 1: Experimental parameters and characteristics of solid wastes (as received).

The study effectively demonstrated that the two-stage gasification system significantly improves the gas yield of the system and the carbon conversion efficiency, which are crucial in other single stage FBG systems, whilst maintaining high energy performances.

The most important outcome of this analysis is that the important parameters that define the performance of the fluid bed-plasma process are the oxidants (oxygen and/or steam) streams and the plasma power. These parameters are function of the type of waste and the detailed aim of the project. Furthermore, the two-stage performance can be quite accurately predicted by assumption of thermodynamic equilibrium.

THE GRAPHIC ON THIS PAGE WAS NOT AVAILABLE ON THE E-MEDIA VERSION OF THIS PAPER.

Figure 2: Comparison of model results with experiment data from the (left) FBG gasifier; (right) FBG + Plasma converter

There is close correlation between the observed and predicted values of the syngas exiting the plasma converter, although the actual level of CO is still ~1% below the theoretical value while the CO_2 level is 1.5% higher. The noted differences are likely due to flow and RDF composition uncertainties rather than from the predictive model. Given the size and configuration of the demonstration plant and the inherent variability of real RDF uncertainties are to be expected. The layout of syngas heating value (LHV) and cold gas efficiency (CGE) have been calculated and analyzed. It was found out that high LHV and CGE values are maintained in different power and oxygen conditions. The reason is that addition of plasma power into the converter decreases the amount of secondary oxygen required for complete gasification and produces larger amounts of CO and H_2 in the product gas. The optimizing direction for the two-stage process can only be determined after considering the detailed situation on different projects.

(534e) Process Modeling of Continuous Catalytic Gasification As a Waste to Energy Alternative

Tillie, C. F., Cleveland State University
Riedthaler, M. J., Cleveland State University
Reeves, S. A., Cleveland State University
Adhikari, S., Cleveland State University
Gatica, J. E., Cleveland State University
Originally presented on: 11/6/2013 16:35:00 - 16:55:00

Presently, space exploration waste is either carried onboard for the duration of the mission or discarded into a trash module which is sent back into Earth's atmosphere for destruction. On the other hand, with an estimated cost of $10,000 per pound, the cost of delivering mass to locations beyond low earth orbit (LEO) locations is prohibitive.

Since approximately 15 pounds of waste are produced daily by a crew of four during a mission, the intrinsic value of this waste can be greatly enhanced by waste to energy technologies. Moreover, conceptualizing waste gasification as an in-situ alternative for processing of trash is also an efficient route to control waste, while maintaining a healthy habitable environment, during long-duration missions (Santiago-Maldonado et al., 2010).

The gasification of long-chain polymers is hypothesized to follow a mechanism of four combined reactions —two parallel oxidation reactions which are kinetically controlled, accompanied by two series reaction where oxidation products (carbon dioxide and carbon monoxide) are converted into methane and hydrogen. Integrating this process into spaceflight missions would greatly reduce the onboard mass requirements, as these products can be used as propellant fuel. Further, under certain conditions the process could be also be adapted to produce oxygen and water, two essential life support species for space missions.

This paper explores scale-up alternatives to implement low-temperature continuous gasification processes as potential waste management processes. A kinetic model formulated from batch gasification experiments is integrated into a process simulator along with a detailed thermodynamic description.

Emphasis is placed on the preparatory treatment of the waste slurries, the reactor module, and equilibrium stages for recycling reactants and recovering the catalyst. Detailed characterization of low and mid fidelity simulants is presented.

(534f) Performance Evaluation of Accelerated Carbonation of Basic Oxygen Furnace Slag Via a Rotating Packed Bed: Modeling, Analysis and Maximization

Pan, S. Y., National Taiwan University
Chiang, P. C., National Taiwan University
Chen, Y. H., National Taipei University of Technology
Chang, E. E., Taipei Medical University
Tan, C. S., National Tsing Hua University
Originally presented on: 11/6/2013 16:55:00 - 17:15:00

Performance of accelerated carbonation by basic oxygen furnace slag (BOFS) in a rotating packed bed (RPB) was evaluated in this investigation. The design of experiments (DOE) was introduced to evaluate the effect of different operating parameters including reaction time, particle size of BOFS, temperature, liquid-to-solid (L/S) ratio, and rotating speed on carbonation conversion of BOFS. A quadratic response surface was established statistically to yield maximum carbonation conversion based on experimental data. In addition, both the fresh and carbonated BOFS were characterized by the thermogravimetry and differential scanning calorimetry (TG/DSC), X-ray diffraction (XRD), and scanning electron microscopy equipped with X-ray energy dispersive spectroscopy (SEM-XEDS) to confirm the results of accelerated carbonation of BOFS in an RPB.

Furthermore, by combining the process chemistry, reaction kinetics, and mass transfer, a prediction model for carbonation of BOFS within an RPB, with an error less than 10%, was well-developed. The enhancement factors for mass transfer coefficient in case of carbonation of BOFS within an RPB were determined accordingly. In comparison with the developed model and the RSM results, a standard operation procedure (SOP) was proposed for maximizing the carbonation conversion of BOFS. It was thus concluded that the accelerated carbonation of BOFS using an RPB is a promising process for CO_2 capture due to its higher carbonation conversion achieving in a short reaction time.

(534g) Production of Alumina From Fly Ash By a Mild and Cleaner Hydrometallurgical Process

Zhong, L., Institute of Process Engineering, Chinese Academy of Sciences
Duan, D., Institute of Process Engineering, Chinese Academy of Sciences
Zhou, E.,
Han, H.,
Originally presented on: 11/6/2013 17:15:00 - 17:35:00

Fly ash is a large volume solid waste generated during the coal combustion by coal-fired power plants, wherein the valuable alumina content could be up to 30-50%. A mild hydrometallurgical process to produce alumina from fly ash only by alkaline solution at comparatively low pressure and temperature was investigated. This novel process is also a cleaner process, where the excess alkaline solution reactant is recycled to the leaching of fly ash, followed by the crystallizing of sodium aluminate hydrate and then its seed decomposing to prepare aluminum hydroxide, calcined to produce alumina. By controlling the phase transformation of the multiphase complex components in the fly ash, overall more than 90% of alumina was recovered from fly ash. The final residues could be utilized as feedstock in construction materials. Furthermore, the reaction mechanism of the process is investigated. This mild and cleaner process could also be appropriate for the produce of alumina from other high Al-containing solid wastes.

(575a) Direct Land Use Change Greenhouse Gas Emissions Associated With Forest-Based Biofuels and Bioenergy Production in Michigan

Fan, J., Michigan Technological University
Froese, R. E., Michigan Technological University
Shonnard, D. R., Michigan Technological University
Handler, R. M., Michigan Technological University
Originally presented on: 11/6/2013 15:15:00 - 15:37:00

Michigan has one of the largest amounts of timber net growth in excess of removals of any state (135.6 million cubic feet per year). Some of this excess timber growth could be a promising feedstock to address the state's forest products industry while also creating a thriving renewable fuels industry. However, harvesting of forest resources for bioenergy production could reduce terrestrial C storage in the near term. Therefore, forest-based bioenergy LCA should incorporate forest carbon stock assessment over a long time frame to estimate the total greenhouse gas (GHG) emissions. In this study, the Carbon Budget Model of the Canadian Forest Sector (CBM-CFS3) is applied to simulate the C stock change in Michigan over a greater than 200 year time frame for aspen-dominated forests, where aspen is intensively harvested for biofuel and bioenergy production. Under intensive harvesting, additional aspen biomass could produce an extra of 18.46 billion gallons of ethanol to blend with gasoline for the transport sector over the next 250 years. Alternatively, the same biomass could be used to produce 32.25 billion gallons of bio-oil by fast pyrolysis process, which can be combusted to generate electricity or upgraded to renewable gasoline and diesel. Model results show that the biofuels (ethanol or pyrolysis oil) and bioenergy (electricity) result in higher GHG emissions than using fossil fuels and business-as-usual aspen harvesting due to the forest carbon loss in the short term. However, the intensive management and harvest of aspen helps accumulates more C in the ecosystem over the long term, which results in negative dLUC emissions (credit) eventually. The time required for this initial carbon debt to be repaid is approximately 60 years for both biofuels; thereafter, this aspen harvesting and management approach accrues GHG savings compared to use of fossil fuels. This study also shows that forest carbon dynamics and dLUC emissions are highly sensitive to the assumed post-harvest growth curve, which can change the time required to overcome the initial C debt.

(575b) Light-Limited Continuous Culture of Micro Algae in a Taylor Vortex Reactor

Vigil, R. D., Iowa State University
Kong, B., Iowa State University

Originally presented on: 11/6/2013 15:37:00 - 15:59:00

Recently, it has been demonstrated that Taylor vortices – hydrodynamic structures that arise in the annular region between two concentric cylinders when the inner cylinder rotates – can substantially improve the growth rate of algal biomass in a batch photobioreactor by inducing the flashing light effect. In order to assess the potential for using Taylor vortex flow to continuously culture algae, experiments were carried out in a continuous flow Taylor vortex algal photobioreactor using various microalgae. Two important operating parameters were varied: the dilution rate and the inner cylinder rotation speed. For a fixed inner cylinder rotation speed, biomass productivity decreased at low dilution rates but was constant at higher dilution rates. Biomass productivity was found to increase slightly with increasing inner cylinder rotation speed for a fixed dilution rate, and this weak dependence on rotation speed may indicate that the flashing light effect was nearly maximized even at the lowest cylinder rotation speeds employed. Overall, it is demonstrated that a continuous flow Taylor vortex algal photobioreactor can be used to produce and sustain high biomass production and carbon dioxide capture rates when operated in continuous flow mode.

(575c) Algae Derived Bio-Oil Separation By Supercritical Fluid Fractionation

Ginosar, D. M., Idaho National Laboratory
Agblevor, F., Utah State University
Petkovic, L. M., Idaho National Laboratory
Newby, D. T., Idaho National Laboratory
Moses, J. M., CF Technologies, Inc.
Originally presented on: 11/6/2013 15:59:00 - 16:21:00

Bio-oils produced by thermal processes, such as fast pyrolysis, catalytic pyrolysis or hydrothermal liquefaction, are a promising source of sustainable, low greenhouse gas (GHG) alternative fuels. These thermal treatment processes are also well suited to decentralized energy production due to low capital and operating costs. However, pyrolysis bio-oils from algae have complex chemical compositions, including large concentrations of water, carboxylic acids, carbohydrates, and protein-derived substances. The oils are viscous, reactive, and thermally unstable. Due to the occurrence of decomposition products together with water and an abundance of compounds of many classes, range of polarities, and the large fraction of oxygenated compounds the mixture is difficult to fractionate by distillation, chromatography, or membrane based processes.

Several methods have been explored to separate bio-oils into stable fractions. By far the most effective separation and stabilization method has been solvent extraction. Although effective, solvent extraction poses two main obstacles to commercialization: (1) the significant amounts of energy required to remove the solvent from the product and (2) the propensity of the solvent to be contaminated with minerals from the char or ash. Separation of thermally produced bio-oils using supercritical fluids offers the advantages of liquid solvent extraction while drastically reducing energy demands and the predisposition to carry over solids into the extracted phase.

Scenedesmus dimorphus was pyrolized at 350, 400, and 450°C with and without a catalyst. Near critical and supercritical carbon dioxide and propane were evaluated for the extraction and fractionation of algae pyrolysis oils into stable fractions. Extractions were carried out at 25, 40, 65 and 110°C at a reduced pressure of 2.0. The effect of pyrolysis conditions and extraction conditions on product composition and stability will be discussed.

(575d) Enhanced Separation of Algal Biomass By Vibratory Shear Microfiltration for the Sustainable Production of Biofuels

Slater, C. S., Rowan University
Savelski, M. J., Rowan University
Kostetskyy, P., Rowan University
Originally presented on: 11/6/2013 16:21:00 - 16:43:00

The dewatering step in the production of algal biofuels is energy intensive resulting in a significant carbon footprint. Membrane separations are one of the processes being investigated to increase efficiency of the operation, but have limitations due to concentration polarization and fouling. This is a concern that limits the scale-up potential of this process for commercial biorefineries. To overcome issues with flux degradation, we have investigated shear-enhanced cross-flow membrane filtration processes. High membrane shear rates were achieved by high-frequency torsional oscillations of the membrane unit. Operated at the resonant frequency, these systems can provide enhanced separation performance at minimal power inputs. Several microfiltration membranes have been evaluated for the separation of dissolved *Chlorella Vulgaris*, a potential feedstock for biodiesel production. This process provides a greener alternative to conventional systems, as high values of flux and separation efficiency can be maintained and the biomass can be effectively concentrated. Studies to evaluate the effect of algae feed concentration, membrane pore size, trans-membrane pressure, and vibrational amplitude on permeate flux were performed. The integration of vibratory field membrane separation into an overall design for the dewatering of algal biomass can result in a more sustainable, greener system.

(575e) Multi-Objective Optimization of Biobutanol Production: Continuous Fermentation Coupled With Vacuum Separation Unit

Sharif Rohani, A., University of Ottawa
Thibault, J., University of Ottawa
Mehrani, P., University of Ottawa
Originally presented on: 11/6/2013 16:43:00 - 17:05:00

Butanol produced via acetone-butanol-ethanol (ABE) fermentation of biomass, is a liquid biofuel that offers many advantages over other biofuels such as bioethanol and is targeted to partly replace gasoline. In ABE fermentation process, a strain of bacteria such as *Clostridium acetobutylicum*consumes sugar to produce ABE solvents. Due to the inhibition of butanol, ABE fermentation is known for its low solvent concentrations in the broth. Thus, to produce butanol more economically, it is required to operate the fermentation process under optimal conditions. Multi-objective optimization has been used to determine Pareto-optimal solutions where optimal decisions are made in the presence of two or more conflicting objectives. Based on an available kinetic model and a set of related mass balances for ABE fermentation, a continuous fermentation system coupled with a vacuum separation unit was simulated and a large number of Pareto-optimal solutions were obtained based on 4 decision variables (feed sugar concentration, dilution rate, evaporation rate and biomass concentration), and 3 objective functions (specific productivity, butanol concentration and sugar conversion) in which were maximized.

The optimal solutions were then ranked using the Net Flow Method and were compared to the continuous fermentation without the vacuum separation unit. It was found that for the best solution, the selective removal of solvents via the vacuum separation unit enhanced the specific productivity of the ABE fermenter by 51% for the same substrate conversion. The final average butanol concentration in the vacuum fermentation was 14.8 g/L which represents an increase of 70% compared to that in the standard continuous fermentation. Moreover, for both fermentation cases, it was shown that the sugar loss is extremely high which has an adverse effect on profitability of the process. In addition to sugar loss, butanol yields over sugar consumption for the best points in standard continuous system and continuous system coupled with vacuum were 10.1% and 13.9%, respectively.

(575f) Growth, Harvesting, and Modeling of Lipomyces Starkeyi Lipids On Waste Sweet Potatoes

Mason, M.,
Dufreche, S., University of Louisiana at Lafayette
Zappi, M. E., University of Louisiana at Lafayette
Subramanyam, R.,
Bajpai, R., University of Louisiana at Lafayette
Originally presented on: 11/6/2013 17:05:00 - 17:27:00

The production of fuels from renewable resources has traditionally faced many barriers, with the most common being limited feed stocks for utilization. The growing need for cost effective renewable resources is a driving force in the development of potential feed stocks. The sweet potato has shown in previous research to have a very high starch content. In addition, up to 20% of the annual sweet potato harvest is discarded due to unsuitable use as a food item. The yeast *Lipomyces Starkeyi* had previously shown the ability to produce lipids from pure starch, but not raw extract from whole sweet potatoes. The soluble starch obtained from this waste was used for the microbial conversion of starch to lipids utilizing a cheap feedstock. The soluble starch was evaluated for methods that would optimize production of the starch for conversion to lipids. Techniques were developed for preparing the sweet potatoes and maximizing extractable soluble starch. The starch obtained was then used as a feedstock for *Lipomyces starkeyi* to evaluate lipid production. Results to be presented include analysis of yeast growth and lipid accumulation, extraction efficiencies, and an economic analysis of full-scale lipid production from sweet potatoes.

(575g) A New Approach to Study Selected Aspects of Biomass Oxidation in Supercritical Water

Mohammad, J.,
Hegde, U. G.,
Hicks, M.,
Kozinski, J. A., University of Saskatchewan
Originally presented on: 11/6/2013 17:27:00 - 17:49:00

A variety of reactor systems are used in supercritical water oxidation (SCWO) including stirred reactors, transpiring wall reactors, flow through reactors and centrifugal reactors. Chemical reactions using SCWO technology take place in the thermodynamically favorable supercritical regime where high conversion rates and are not noticeably limited by heat transfer and mass transfer rates.

One of the advantages of using SCWO for biomass conversion is that supercritical water acts as an environmentally green solvent. Because biomass has an initial water content of 10-25 wt.%, SCWO technology is advantageous because the supercritical water serves as a solvent to breakdown the organic material prior to oxidation without the need for a drying step. However, conventional thermo-chemical techniques, such as pyrolysis and gasification utilized for the production of green fuels and chemicals, require the biomass to at least be partially dried.

During this study a portable SCWO reactor system, the SCWO Test Facility developed by NASA, has been utilized to study the conversion of beetle-infested wood into valuable fuels and chemicals. The SCWO Test Facility is originally designed for studying the use of SCWO for processing waste streams comprising plastics, papers and bio-wastes that would be generated during extended space missions. However, this unique system can also be used for processing aqueous waste streams containing small organic molecules such as methanol and acetic acid that would otherwise be difficult to be removed by conventional methods. A detailed study involving both biomass model compounds and the raw biomass has been undertaken to evaluate the system's overall conversion efficiency.

One of the salient features of the NASA SCWO Test Facility is that it can be operated as a constant volume batch reactor or a constant pressure continuous reactor. A piston accumulator is installed downstream to regulate the reactor gas pressure. Real-time in-situ measurements of the reactant and product concentrations are performed using the high temperature optical fiber and laser-assisted probe connected to a Raman spectrometer.

During this study the SCWO reactor temperatures and pressures have been varied in the range of 400-500 ^{0}C and 3400-4400 psi by maintaining a constant feed/water ratio. The primary objective of this green technology is to recover synthetic gas (H_2–CO) as a gaseous fuel while maximizing the production of value-added chemicals such as levoglucosan, furfural, hydroxyacetaldehyde, acetic acid, and methanol. Both gas and liquid product yields are investigated, analyzed (GC-MS, NMR and FTIR) and presented.

(592a) Catalytic Fast Pyrolysis of Forest Thinnings in a Bubbling Fluidized Bed Reactor With ZSM-5 Catalysts

Paasikallio, V., VTT Technical Research Centre of Finland
Agblevor, F., Utah State University
Originally presented on: 11/7/2013 8:30:00 - 8:48:00

Catalytic pyrolysis is a technique which has been receiving increasing attention in the recent years due to its ability to produce partially deoxygenated pyrolysis oil with improved fuel properties. This change in the chemical composition of pyrolysis oil can be seen as a shift from its typical polar nature more towards conventional petroleum-derived fuels. This transformation in turn opens up possibilities for the utilization of pyrolysis oil in new applications such as co-processing in conventional oil refineries. While a multitude of biomass types have been subjected to catalytic pyrolysis, a low-cost and abundantly available feedstock is one of the main prerequisites for a full-scale industrial process. In countries with strong forest industry backgrounds, forest residues and forest thinnings are an attractive source of biomass with otherwise low economic value.

In this work, conventional and catalytic fast pyrolysis of Finnish forest thinnings was carried out in a 200 g/h bench-scale bubbling fluidized bed reactor using silica sand and two ZSM-5 catalysts as the fluidized bed material. The first catalyst was a ZSM-5 based fluid catalytic cracking additive, whereas the second catalyst was a typical ZSM-5 zeolite supplied as extrudates. The resulting pyrolysis oils, which were collected both at the middle and at the end of the catalytic two hour experiments, were characterized for their moisture content, elemental composition, heating value, viscosity and pH. Chemical composition of the pyrolysis oils was analysed using ^{13}C nuclear magnetic resonance spectroscopy.

When compared to the silica sand at a reaction temperature of 475°C, the ZSM-5 based fluid catalytic cracking additive did not induce major changes in the physical properties or the chemical composition of the pyrolysis oil. The second ZSM-5 zeolite proved to be more effective in converting the pyrolysis vapours, and produced pyrolysis oil with oxygen content as low as 20 wt-%. The change in elemental composition was accompanied by an increased heating value and lower viscosity. Elimination of carbohydrate degradation products resulted in an increased proportion of aromatic compounds in the pyrolysis oil. The second catalyst was selected for further experiments where the effect of varying the reaction temperature was investigated. Decreasing the reaction temperature to 400°C caused a clear increase in the yield of organic liquids, which was accompanied by an increase in the oxygen content of the pyrolysis oil. Compared to 475°C, the pyrolysis oil produced at 400°C exhibited lower viscosity, which cannot therefore be explained by elimination of oxygenated molecules. On the other hand, increasing the reaction temperature to 550°C caused an increase in the viscosity of the pyrolysis oil. Other physical properties were not affected to such a large extent by the temperature increase.

(592b) Effects of Reactor Configurations and Catalyst Properties for Vapor Upgrading in HZSM5 Catalyzed Pyrolysis Vapor Upgrading

Wan, S., University of Oklahoma
Waters, C., University of Oklahoma
Gumidyala, A., University of Oklahoma
Jentoft, R., University of Oklahoma
Lobban, L., University of Oklahoma
Crossley, S., University of Oklahoma
Resasco, D. E., University of Oklahoma
Mallinson, R. G., University of Oklahoma
Originally presented on: 11/7/2013 8:48:00 - 9:06:00

Effects of Reactor Configurations and Catalyst Properties on HZSM5 Catalyzed Upgrading of Biomass Pyrolysis Vapors

Shaolong Wan, Christopher Waters, Abhishek Gumidyala, Rolf Jentoft, Lance Lobban, Steven Crossley, Daniel Resasco, and Richard Mallinson

Since the mid 1980s, HZSM-5 has been shown to be a superior catalyst for upgrading the vapors from biomass pyrolysis to produce aromatic hydrocarbons.Â However, this comes with relatively low liquid yields, significant coke and gas formation and rapid deactivation.Â We have studied HZSM-5 catalysis of biomass pyrolysis vapors in a custom CDS pyroprobe with a separate reactor in order to understand how to increase liquid yield and improve catalyst stability.Â Compared with pyrolyzing the catalyst mixed with the biomass, different product selectivities are observed with the separate reactor due primarily to differences in residence time.Â The pyroprobe pyrolysis of a biomass-catalyst mixture lasts for seconds and results in high conversion of a primary cellulose decomposition product, levoglucosan, as well as light oxygenates including acetic acid.Â When configured in a separate reactor with much shorter residence time, high conversion of the light oxygenates is obtained, but low conversion of levoglucosan is observed.Â In this separate configuration, multiple pulses of pyrolysis vapors may be passed over the catalyst, thus allowing the deactivation characteristics to be studied.Â For example, at a constant number of acid sites in the catalyst bed, the deactivation rate was found to be much faster with a catalyst containing a higher acid site density.Â Different HZSM-5 materials have been tested, with different separate reactor temperatures.Â These have included variation in the Si/Al ratio from 40 to 11.5 and the use of core-shell materials with a core of silicalite, an outer layer of HZSM-5, and also a material with a further outer layer of silicalite.Â Additionally, the effect of the passivation of surface acidity has been examined.

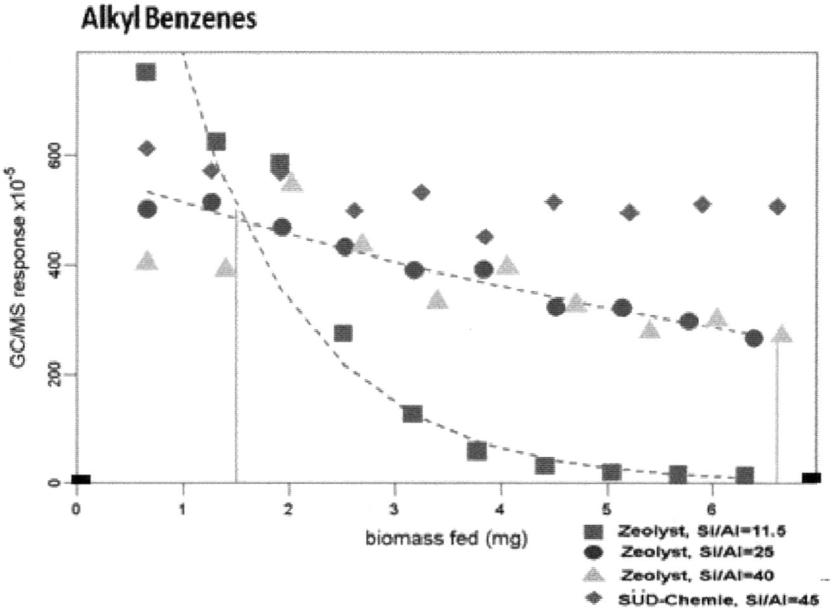

(592c) CFD Modeling of the Catalytic Pyrolysis of Biomass With Zeolite in Fluidized Bed

Sun, D.,
Papadikis, K.,
Gu, S.,

Originally presented on: 11/7/2013 9:06:00 - 9:24:00

Fast pyrolysis of biomass using zeolites as the catalyst material is a direct method to improve the quality of bio-oil in the pyrolysis reactor. In this paper, catalytic pyrolysis of sawdust is simulated in a fluidized bed. Two different cases, one with zeolite catalyst and one without catalyst are modeled and compared.

The flow behavior in the fluidized bed is modeled by means of gas-solid two-fluid model with kinetic theory of granular flow for the granular phase. Two kinds of granular are used in the reactor, i.e. the catalyst and biomass particles. The momentum transfer between the solid phases can be modeled by the drag coefficient simplified for the kinetic theory by Syamlal(1987). Kinetic scheme for catalytic pyrolysis is based on the kinetic research by Atutxa et al.(2005). The kinetic experiment has been carried out to study the catalytic pyrolysis of sawdust with zerlite HZSM-5 in a conical spouted bed reactor at the temperature 400°C. The effect of space time is also incorporated in the kinetic mechanism.

The predicted product yield distribution from catalytic pyrolysis is in good agreement with the experiment data of Atutxa et al. (2005). The gas yields, due to the cracking of tar, increase with an increasing amount of catalyst. Gas, tar and char products vary along the height of the reactor, which give a good prediction of the residence time of biomass into the bed. Along the height, concentration of tar increases from bottom of the riser result from the pyrolysis reaction of biomass and then reduced by the catalytic reaction to gas. Three regions can be divided to indicate the biomass heating, pyrolyzing and catalyzing process. The simulation also shows the velocity of two solid phases in the riser. The velocity of catalyst is higher than the sawdust because of the larger diameter. This velocity difference between them can enhance the collision occasions and promote the turbulence intensity and solids distribution in the riser, hence benefiting chemical reaction and consequently the final pyrolysis yields.

(592d) Deoxygenation of Bio-Oil Intermediates: Examination of Methane As a Direct Reducing Agent

Oberg, A. A., Pacific Northwest National Laboratory
Miller, D. J., Michigan State University
Originally presented on: 11/7/2013 9:24:00 - 9:42:00

Thermochemical conversion of biomass into bio-oil is a promising way to create renewable liquid transportation fuels. However, untreated bio-oil is not suitable for use a fuel, because it is thermally unstable, corrosive, has low heating value, and is often immiscible with hydrocarbon fuels. Upgrading of bio-oil to remove oxygen, saturate double bonds, and increase energy density is therefore required; such upgrading currently is done via catalytic hydrodeoxygenation with molecular hydrogen. Another possible source of reducing equivalents for upgrading is methane, available either from anaerobic digestion of biomass or from abundant fossil reserves. Each mole of methane is capable of providing up to four reducing equivalents; however, methane is the most stable hydrocarbon having strong C-H bonds (104 kcal/mol), stable tetrahedral geometry, and no reactive functional groups to facilitate reaction.

The present work examines methane activation over supported metal catalysts as a first step in using methane directly for bio-oil upgrading. In addition to methane as a source of reducing equivalents, the possibility of methane incorporation into bio-oil species, which would further increase fuel value because of methane's high energy content (~50 MJ/kg), is explored. Initial studies of methane activation over Pt and Ru catalysts at 150°-250°C showed that methane readily undergoes H-D exchange between CH_4 and D_2, and between CH_4 and CD_4 to give mixed CH_xD_{4-x} species. The $CH_4 - CD_4$ exchange takes place without H_2 or D_2 present.

Direct incorporation of methane into benzaldehyde and other species as model bio-oil compounds over Pt and Pt/Cu catalysts was not observed. However, it was shown that over $Pt/Cu/SiO_2$ benzaldehdye was quantitatively converted to toluene in the presence of hydrogen, and into benzene in the presence of methane or nitrogen. Copper was confirmed as the active catalyst species for deoxygenation.

(592e) Fast-Hydropyrolysis and Catalytic Hydrodeoxygenation for Conversion of Biomass to Liquid Fuel

Venkatakrishnan, V. K., Purdue University
Degenstein, J., University of North Dakota
Delgass, W. N., Purdue University
Agrawal, R., Purdue University
Ribeiro, F. H., Purdue University
Originally presented on: 11/7/2013 9:42:00 - 10:00:00

The production of high volumetric energy density fuels from a sustainable carbon source, such as lignocellulosic biomass, to supply the liquid fuel needs of the transportation sector is a grand challenge to be addressed. As one of the plausible solutions, a novel process, *H_2Bioil*[1], based on biomass fast-hydropyrolysis and catalytic hydrodeoxygenation (HDO), has been proposed for conversion of lignocellulosic biomass to liquid fuel with high estimated energy and carbon efficiencies. The *H_2Bioil* process integrates fast-pyrolysis thermochemical conversion, involving rapid heating of biomass, in a high-pressure (up to 50 bar) hydrogen environment, to produce fast-hydropyrolysis vapors that are subsequently catalytically upgraded in the vapor phase and quenched to form a deoxygenated liquid product with high energy density.

This presentation summarizes our work on *H_2Bioil* process development, using laboratory-scale high-pressure fast-hydropyrolysis and catalytic hydrodeoxygenation with cellulose, as a model biomass feedstock, and lignocellulosic feedstocks. To identify and quantitatively compare major chemical species present in the liquid product, we have developed and utilized a novel liquid chromatography-mass spectrometry (LC-MS) analytical method which accounts for up to 95% by mass of the cellulose hydropyrolysis liquid products. We have also used gas chromatography-mass spectrometry (GC-MS) and other analytical techniques such as elemental (C, H, O) and Karl Fisher water content analysis to characterize the liquid product. The effects of process conditions such as pyrolysis temperature (480°C-580°C), hydrogen partial pressure (25 bar, 50 bar) were studied with a cellulose model feedstock. The major product of high-pressure fast-pyrolysis and fast-hydropyrolysis of cellulose is the monomer (levoglucosan), which accounts for ~35-40% of the starting mass of cellulose pyrolyzed. As the pyrolysis temperature is increased, the product distribution shifts towards the lower molecular weight compounds, like glycolaldehyde and formic acid, which is attributed to increased thermal cracking of pyrolysis vapors. The first stage cellulose fast-hydropyrolysis does not lead to significant deoxygenation, even up to 50 bar hydrogen partial pressure, but presence of HDO catalysts leads to improvements in deoxygenation in the liquid product. The results from testing of candidate Pt-based and Ru-based catalysts for on-stream vapor-phase catalytic hydrodeoxygenation suggests that a balance of metal and acid catalytic functions is necessary for improving the extent of deoxygenation along with high carbon recovery in the liquid product and avoiding coking on the catalyst.

[1]Agrawal R, Singh NR. Synergistic Routes to Liquid Fuel for a Petroleum-Deprived Future, *AIChE J.* 2009; 55: 1898-1905

(592f) Low-Oxygen Bio-Crude From Hydropyrolysis in a Pressurized Fluid-Bed Reactor

Carpenter, J. R., RTI International
Hlebak, J., RTI International
Farmer, J., RTI International
Dayton, D. C., RTI International
Originally presented on: 11/7/2013 10:00:00 - 10:18:00

Pyrolysis oil prepared by conventional (non-catalytic) fast pyrolysis is not suitable for integrating into the current fuel infrastructure or into a petroleum refinery because it is : a) thermally unstable with a high fouling tendency, b) corrosive due to high carboxylic acid content (pH 2.2 – 2.4 typically), and c) not miscible with refinery feedstocks. Fast pyrolysis bio-oil tends to have high oxygen content similar to the solid biomass feedstock used as the starting material. The thermal instability and high water content require specialized hydroprocessing to lower the oxygen content of the bio-oil before use. Hydropyrolysis is a biomass thermal liquefaction process option that combines biomass, hydrogen, and a catalyst to integrate pyrolysis and a degree of hydroprocessing into a single unit operation.

The National Advanced Biofuels Consortium (NABC) is working to develop sustainable, cost-effective processes to produce advanced "drop-in" biofuels that are compatible with today's transportation infrastructure. As part of the NABC, the hydropyrolysis of loblolly pine has been explored to produce low-oxygen content bio-crude. A pressurized fluid-bed reactor was designed and fabricated to test catalysts and reaction conditions (T, P, and pH_2). Testing was performed with a nominal loblolly pine feed rate of 200 g/hr at temperatures between 350°C to 500°C and pressure up to 300 psig with varying hydrogen concentrations. Full mass closures are obtained by measuring all liquid, solid, and gas products. Bio-crudes are characterized by ultimate analyses, Karl-Fischer titration, total acid number (TAN), pH, and GC/MS. In addition to parametric testing, several catalysts were screened and performance with a corn stover was evaluated.

A selected catalyst was tested over an extended trial of 32h on stream with over 8kg of biomass. Consistent activity was observed based on product yields and composition when catalyst regeneration occurred after conversion of 1 kg of biomass. The liquid product separated into an aqueous layer that was 97wt% water and a bio-oil that was less than 1wt% water and had an oxygen content of 2.5wt%. The bio-crude had high levels of aliphatic and aromatic compounds with reduced concentrations of acids, sugars, and ketones in comparison with conventional fast pyrolysis bio-oil. Hydropyrolysis has the potential to produce bio-oils with improved properties and suitability for integration into refinery processes.

(592g) Catalytic Fast Pyrolysis of Isotopically Labeled Biomass and Plastic Blends

Dorado, C., Eastern Regional Research Center, Agricultural Research Service, USDA
Mullen, C., USDA
Boateng, A. A., USDA Agricultural Research Service
Originally presented on: 11/7/2013 10:18:00 - 10:36:00

This study aims to address two problems important to agriculture, production of drop-in fuels from biomass and disposal of agricultural plastics via co-pyrolysis of both materials. Mixtures of biomass (cellulose and switchgrass) and polypropylene (PP) or polyethylene terephthalate (PET) were subjected to catalytic fast pyrolysis (CFP) at 650°C in the presence of H-ZSM5 using a micro pyrolyzer coupled with GC/MS (py-GC/MS). Yields of the aromatic hydrocarbons, toluene, naphthalene and p-xylene, were enhanced for the CFP of mixtures of cellulose and plastic versus the CFP of biomass or plastic alone, suggesting a synergistic effect that enhanced the carbon efficiency of the CFP process. To further probe this effect, CFP of uniformly labeled ^{13}C biomass with non-labeled (^{12}C) PP and PET in the presence of H-ZSM5 were co-pyrolyzed. The resultant aromatic compounds had varying $^{13}C_x{}^{12}C_y$ compositions, proving that each reactant provides carbon to a pool from which aromatics are formed. Further analysis allowed for a more detailed picture of the source of carbon for each individual product. For example, it was found that for CFP of ^{13}C biomass and PP pyrolysis, 69% of the toluene product was composed of at least four carbon atoms from PP. In this presentation the results of the catalytic co-pyrolysis of biomass and plastics and the isotopic labeling experiments will be discussed in detail in terms of yields of individual products, product selectivities and implications for the operative CFP mechanisms.

(592h) Mild Pyrolysis of P3HB/Switchgrass Blends for the Production of Bio-Oil Enriched With Crotonic Acid

Mullen, C. A., USDA Agricultural Research Service
Boateng, A. A., USDA Agricultural Research Service
Dorado, C., Eastern Regional Research Center, Agricultural Research Service, USDA
Schweitzer, D., Metabolix Inc.
Sparks, K., Metabolix
Snell, K., Metabolix, Inc.
Originally presented on: 11/7/2013 10:36:00 - 10:54:00

Poly-3-hydroxybutyrate (P3HB) is a naturally occurring polyester that is produced by some microorganisms as a method of carbon and energy storage. In addition to being utilized as a renewable, biodegradable substitute for plastics produced from petroleum, it can be thermally degraded *in situ* to crotonic acid in high yield. Crotonic acid can be used as a platform chemical enabling an alternative access to high-valued industrial and fine chemicals. For example, it can be reduced to *n*-butanol for chemical or fuel applications or oxidized to maleic anhydride. Other targets available through simple transformations include propylene and acrylic acid. In the work presented in this presentation, the mild pyrolysis of switchgrass/P3HB blends that mimic P3HB-producing switchgrass lines was studied in a pilot scale fluidized bed reactor with the goal of simultaneously producing crotonic acid and switchgrass-based pyrolysis oils to be further converted to biofuels. While pure P3HB decomposes under distillation conditions to provide a near quantitative yield of crotonic acid, pyrolysis of P3HB-containing switchgrass requires an additional crotonic acid purification step. Factors such as pyrolysis temperature, residence time, flow rates and particle size of the P3HB were studied for their effects on the recovery of crotonic acid as a component of the pyrolysis oil produced from the mixture. Effects of the presence of P3HB and crotonic acid on the switchgrass bio-oil in terms of chemical composition and properties will also be examined. Furthermore, efforts to isolate pure crotonic acid from the switchgrass bio-oil will be discussed. Finally, an example of the utility of crotonic acid as a chemical feedstock will be briefly discussed. Hydrogenation of the *n*-butyl ester of crotonic acid was found to produce *n*-butanol in excess of 90 % yield over commercial Cu-based catalysts under the optimized conditions.

(641a) Algal Biorefinery: Production of Biocrude Oil & Byproducts

Reddy, H. K., New Mexico State University
Selvaratnam, T., New Mexico State University
Muppaneni, T., New Mexico State University
Nirmalakhandan, N., New Mexico State University
Schaub, T., New Mexico State University
Dungen, B., New Mexico State university
Appuhamilage, N. S.,
Lammers, P., New Mexico State University
Voorhies, W., New Mexico State university
Deng, S., New Mexico State University
Originally presented on: 11/7/2013 8:30:00 - 8:52:00

We have developed a strategy needed for the production of sustainable algal biofuels with hydrothermal liquefaction of three different algal species. Hydrothermal liquefaction experiments with 10% solid loading of *Chlorella sp., Nannochloropsis sp.*, or *Galdieria sp.* were carried out between 180°C-300°C and 30 min. of reaction time to investigate the effects of HTL conditions on the important liquefaction reactions and final products. The liquid biocrude products were analyzed with GC-MS and FT-ICR for their chemical compositions, and micro bomb calorimeter for their high heating values. The composition of residual algal biomass was analyzed after each experimental run to identify the biochemical composition of the biomass. The residual water collected after liquefaction was also analyzed to quantify the nutrients and dissolved organic compounds in the water phase. This water phase produced in the HTL process was recycled for cultivation of *galdieria sp.* at different levels of dilution. An integrated processing strategy was developed for individual algal species based on the values of three products (bio-crudes, solid residues and water phase) obtained in the HTL process.

(641b) Biodiesel Production Through Microwave Assisted Transesterification of Microbial Cells

Cui, Y., Southern Illinois University

Originally presented on: 11/7/2013 8:52:00 - 9:14:00

One strain of oleaginous yeasts, Cryptococcus curvatus (ATCC 20509) has been studied to grow on several substrates including biodiesel production byproduct crude glycerol and sweet sorghum juice. After cultivation, yeast cells were heated under microwave radiation to extract lipid and produce biodiesel through in-situ transesterification.

Firstly, the yeast growth with crude glycerol was studied.On the second step, an optimization of yeast fermentation with crude glycerol was conducted. The resulting parameters of the response surface method optimization were used in a fed-batch fermentation where crude glycerol was automatically pumped in responding to exhausted oxygen levels in the fermentor.At the end of 12 days, the biomass density and lipid content were 44.53 g/L and 49%. Compared with our previous fed-batch experiment which was conducted under un-optimized condition, the yield of biomass and lipid increased 35.26% and 25.29%.

Considering its capability to utilize a broad range of substrates, C. curvatus was tested on its lipid productivity on sweet sorghum juice. When cultured in a fed batch process for 3 days, the final biomass density was 23.6 g/L with a lipid content of 51%. To extract lipids from cells in an effective and fast fashion, a domestic microwave oven was used with different solvents. With only methanol, a lipid yield of 33.2% of yeast cells was obtained in 4 min.

In the end, to convert yeast lipids to biodiesel directly without the step of lipid extraction,the in-situ transesterification method used microwave irradiation on the simultaneous extractionand transesterification of wet yeast biomass to biodiesel. Based on the experimental results and RSM analysis, the optimal conditions for this process were determined as: methanol to yeast biomass (v:w) ratio of around 50:1, catalyst concentration about 5 wt.%, and reaction time of 2 min. The biodiesel samples were analyzed with GC and the FAME content in biodiesel is about 50%. Compare the with other researches, this process proved to be a fast and easy method to produce biodiesel from wet yeast biomass.

(641c) Design Optimization By Response Surface Methodology for Continuous Fermentative Production of 1,3 Propanediol From Waste Glycerol By Product of Biodiesel Processes

Kanjilal, B., University of Connecticut
Noshadi, I., University of Connecticut
Intoci, N., University of Connecticut
Dowding, M.,
Hale, W.,
Brendel, B., University of Connecticut
Srivastava, R., University of Connecticut
Parnas, R., University of Connecticut
Originally presented on: 11/7/2013 9:14:00 - 9:36:00

The production of ~10 gallons of crude glycerol for every 100 gallons of biodiesel has led to a current glycerol excess of ~2.52Bn pounds per annum by the US biodiesel industry and non biodiesel sources. The bio-fermentation of crude glycerol to 1, 3-propanediol, an emerging intermediate platform in biorenewable chemicals with a nascent market of $575 million worldwide, as a value added product gives a recourse to capitalizing on the current over capacity of glycerol. This work presents the design optimization for a green and efficient process for the batch and continuous production of 1,3 propanediol from industrial waste glycerol using soil based bacterial inoculums. The optimal process parameters for the best possible yield of 1,3 propanediol were determined using a design of experiment by varying paramaters such as pH, glycerol concentration and yeast concentration. Response surface methodology (RSM), an important branch of experimental design, was used to find the relationship between experimental parameters and the yield and biomass growth in order to find the optimal conditions for the process. The optimization of 1,3 propandeiol production, in this way, with several inputs, helps to enhance the conversion in an economic manner. pH was seen to have the most profound effect on the process both in terms of yeild as well as bacterial growth. The optimised conditions were used to run a continous fermentation and arrive at a steady state for a given dilution rate, while the inhibitory solvent effects were mitigated by cycling between acetate and butyrate laced batches. While the addition of acetate enhanced biomass growth, butyrate addition enhanced 1,3 propanediol production with distinct inhibitory effects on the growth of bacteria. The effects were utilized to systematically maximize the survival of the mixed inoculum in a continuous run.

(641d) Review of Bio-Jet Fuel Conversion Pathways

Wang, W. C., National Renewable Energy Laboratory
Tao, L., National Renewable Energy Laboratory
Originally presented on: 11/7/2013 9:36:00 - 9:58:00

Bio-derived jet fuel has become a key subject in the aviation industry strategy to address the challenges from reducing operating cost, environmental performance and impacts to the global ecosystem. For producing a long-term renewable aviation fuel, economic and environmental sustainability are necessary. Researchers from oil refining industry, commercial aviation industry, government, biofuel companies, agriculture organizations and academic systems are looking forward to developing a commercial viable process that has not only low production cost but low GHG emissions. Additionally, the fuel produced from the process has to meet the ASTM specifications and is intended to be 100% drop-in for the replacement of current petroleum jet fuel. This study reviewed the current major technologies for converting bio-based materials into renewable aviation fuels. All the conversion pathways are classified to four categories: (1) syngas-to-jet, includes Fischer-Tropsch(F-T) synthesis and gas fermentation; (2) renewable-oil-to-jet, includes hydroprocessed renewable jet (HRJ) via plant oil and algal oil, catalytic hydrothermolysis (CH) and hydrotreated depolymerized cellulosic jet (HDCJ, also named pyrolysis) via pyrolysis oil; (3) alcohol-to-jet, includes the conversion of ethanol, butanol and methanol to jet fuel from lignocellulosic biomass feedstocks; (4) sugar-to-jet, includes fermentation of sugars to hydrocarbon or hydrocarbon precursors as well as catalytic conversion of sugars or sugar intermediate to fuels. Discussions for each pathway are primarily focused on the available feedstocks, conceptual process design, process economics, life-cycle assessments and commercial readiness, based on the literature studies. With the help of this comprehensive review on bio-derived jet fuel conversion technologies in recent literatures, it is believed that, whereas there are significant issues related to feedstock price and availability and energy consumption of the process, bio-derived jet fuels could progressively replace a great portion of conventional jet fuel required to meet the commercial and military demand.

(641e) Pellets From Lignocellulosic Materials: Briquetting Process Evaluation

Gonzalez, Y., Universidad Libre
Rincon, S., Universidad Libre
Camargo, G., Universidad Libre
Sierra, F., Universidad Nacional
Forero N, C. A., Universidad Nacional
Originally presented on: 11/7/2013 9:58:00 - 10:20:00

Colombia is a palm oil producer country and the implementation of biodiesel industry has generated huge quantities of the residues. On another hand, there are several regions in Colombia without a sufficient energy supply for a better and faster development. The purpose of this research is investigate the use of agro industrial residues coming from palm oil for obtain briquettes that can be used as fuel in remotes region in Colombia. Different parts of the agricultural waste they be employed in the manufacture of briquettes, the fibers and the seed. The biomass was characterized by ultimate analysis and in the densification process two different levels of pressure was employed. Finally densities, mechanical strength, heat of combustion were evaluated for the briquettes. The initial results shown that the waste from the seed materials presents a better performance for the densification process than the fiber materials and the briquettes elaborated a higher pressure exhibit more elevated calorific power

(641f) Pyrolytic Liquid Yields From Bamboo Species (Bambusa Oldhamii and Bamboo Angustifolia)

Pliego-Bravo, Y. S., Instituto Tecnologico de Orizaba
Cervantes- Pesado, L., INSTITUTO TECNOLOGICO DE ORIZABA
Bolanos-Reynoso, E., Instituto Tecnologico de Orizaba
Gonzalez-Garcia, P.,
Originally presented on: 11/7/2013 10:20:00 - 10:42:00

PYROLYTIC LIQUID YIELDS FROM BAMBOO SPECIES (*Bambusa Oldhamii and Bamboo Angustifolia*)

Abstract

A characterization of the species of bamboo (*Bambusa Oldhamii and Guadua Angustifolia*) with respect to compositional analysis, elemental analysis and thermochemical calorific using as a carrier gas, nitrogen, to identify desirable traits for subsequent use in a thermal conversion process such as pyrolysis, this energetic process provides greater value to convert biomass into a liquid fuel. Moisture content lignin degradation temperatures and ash content are indicators to determine the ideal raw material for conversion to a renewable process.

Introduction

Among the new alternatives that have been implemented for reducing the amount of emissions of CO, SOx, NOx, there being a source of biomass energy that partially replaces fossil fuels and decreases the concentration of these gases emitted into the atmosphere. (Zhang *et al*, 2007). Immediate analysis, elemental and calorific thermochemical analysis are useful in the understanding of the properties of the species to consider a material as an energy source, allowing to know the effective thermochemical process for producing a biofuel (Bridgwater, 2001 and Brown 2011). In this paper we characterize the species of bamboo (GA and BO) to propose an alternative to using craft or promote the efficient use of waste generated by the industry based on the characterizations mentioned above in order to obtain a liquid fuel

Methodology

The proximate analysis, elemental, calorific power and thermogravimetric were performed these are described below.

Materials

In lignocellulosic analysis were used: Ethanol analytical reagent grade (99.5% purity), analytical reagent grade sulfuric acid (98.2% purity), analytical reagent grade toluene, analytical reagent grade nitric acid (69% purity) all of J. T. Baker and distilled water. In thermochemical analysis was used 99% pure nitrogen and air.

Immediate analysis

Establishes the relative amounts of moisture (ASTM D 2867-99), volatile matter (ASTM D 5832-98), fixed carbon, ash (ASTM D 2866-99). Get the amount of carbon fixed subtracting the amounts of ash and volatile matter of a total of 100%.

Elemental analysis and calorific power

The amounts of C, H, N and S of the constituents of bamboo species were estimated experimentally using elemental analyzer (Vario EL III) based on moisture and ash free (LHC), the oxygen obtained by difference. The calorific value was estimated by the equation Reed and Das (1998) expressed in MJ / Kg.

Thermochemical analysis

Thermogravimetric analysis were performed with a non-isothermal model held in the TG Mettler Toledo 851e with software version Star. Samples between 6-10 mg were analyzed, studies were carried out using nitrogen and air, and before each experiment was performed a nitrogen purge for 10 min to all samples with a flow of 100 mL / min.

Getting pyroligneous liquids

For testing proceeded with an experimental 3^2 where the factors were time (15, 30 and 60 min) and the temperature (350, 500 and 600 ° C) for each of the species studied. These were carried out on a laboratory scale reactor with stainless steel fixed bed capacitors and nitrogen supply 200 mL / min.

Results and discussion

Moistures were between 16-17%, whereby the raw material was suitable for a thermochemical process. The high ash contents cause slag deposits on equipment deteriorating and decreasing the life thereof. The species studied showed percentages of 4.92 of ash for BO and 3.33 for GA making it attractive for use in these applications. Low fixed carbon content was related to the age of the material, in young bamboo lignin was removed easily due to low contents of total polysaccharides and cellulose, whereas an older kind more lignin is harder removing (Garcia *et al*, 2009).

The calorific value was obtained by correlations of Reed and Das (1998). In thermochemical analysis TGA curves showed that the decomposition of the species begins about 300 ° C, the displacement of the curves to the right with increasing temperature. For Bambusa oldhamii degradation stabilized 550 ° C to 600 ° C heating speeds of 15 and 25 ° C / min, while at 5 ° C / min stabilized at 500 ° C. In the case of Guadua angustifolia behavior was observed for more stable speeds of 5 and 15 ° C / min stabilizing at approximately 480 ° C whereas at 25 ° C / min from around 540 ° C. For older BO liquid yields were achieved at 600 ° C with 46 ± 0.5% in the times of 30 and 60 min, for the species GA best performance was obtained at 600 ° C for a time of 60 min of 43 ± 2%.

Referencias

[1] Bridgwater, A., V. (2001). *Progress in Thermochemical Biomass Conversion*. Backwell Science. Volume 2 IEA bioenergy.

[2] Brown R. C., *Thermochemical Process of Biomass*. Wiley, U.S.A. (2011).

[3] García, L, Marquez, F, Aguiar, L, Arauso, J, Carballo, L, Orea, U Zanzi, R (2009). *Rendimiento de los productos de la descomposición tÃ©rmica de la madera Eucalyptus saligna Smith a diferentes alturas del fuste comercial.* Revista Chapingo. Serie ciencias forestales y del ambiente, Vol. 15, Núm. 2. Julio_diciembre. pp. 147-154

[4] Higman C. y van der Burgt M., *Gasification*. Segunda edición. Elsevier. U.S.A. (2008).

[5] Reed T. B., y Das A., *Handbook of biomass downdraft gasifier engine system*. Solar Energy Research Institute, U.S. department of energy. U.S.A. (1998).

[6] Zhang Qi, Chang Jie , Wang Tiejun , Xu Ying (2007). Review of biomass pyrolysis oil properties and upgrading research. Energy Conversion and Management 48 (2007) 87–92.

(641g) The Impact of Ethanol and Iso-Butanol Blends on Regulated Emissions, Air Toxics, and Particle Emissions from SIDI Vehicles

Karavalakis, G., UC Riverside
Short, D., UC Riverside
Vu, D., UC Riverside
Durbin, T., UC Riverside
Asa-Awuku, A., UC Riverside
Originally presented on: 11/7/2013 10:42:00 - 11:04:00

Spark-ignition gasoline direct injection (SIDI) vehicles have been the subject of several studies, with much of this work has been done outside the U.S., as this is where GDI vehicles have been more prominent. The penetration of SIDI vehicles in the U.S. market is rapidly growing with most domestic automotive manufacturers offering direct injection versions in their existing models. The SIDI technology has emerged from the need to improve the thermodynamic efficiency (and thus reduce fuel consumption) and power output of spark ignited engines. The improved fuel efficiency compared to the conventional port fuel injection (PFI) vehicles could potentially provide the manufacturers the means to meet the target set in the U.S. on the fleet-average carbon dioxide (CO_2) emissions of future vehicles. It is foreseen that this category of vehicles will dominate the gasoline market, eventually replacing the conventional and less efficient PFI vehicles. However, SIDI engines are known to emit more particulate matter (PM) emissions than homogeneous-charge PFI spark-ignited gasoline engines and diesel engines equipped with diesel particle filters (DPFs) in the exhaust system. In particular, it is likely that SIDI vehicles will be higher than 3 mg/mile CARB LEVIII PM Standard through 2021, whereas PFI vehicles generally emit below 1 mg/mile, and often below 0.5 mg/mile, providing the potential to more than double PM emissions from new light-duty vehicles. The PM characteristics of direct injected spark ignition engines in comparison to conventional PFI engines can be mainly attributed to the injection characteristics. There is a growing consensus amongst the health experts that particles in the ultrafine range (smaller than 100 nm) are potentially more toxic and have greatest adverse health effects on human health.

Limited studies have investigated the influence of alternative fuels on the exhaust emissions, especially particulate emissions, from SIDI vehicles. The fuels employed in these studies were mainly low and mid-level ethanol blends. Currently, ethanol is the most widely used renewable fuel for transportation in the U.S. and with the push to use increasingly higher levels of renewable fuels, there has been an accompanying push to further increase the ethanol level in gasoline. The EPA Renewable Fuel Standard version 2 (EPA-RFS2) and the California Low Carbon Fuel Standard (CA-LCFS) are driving the U.S. market for the development of biofuels. The EPA-RFS2 requires that 36 billion gallons of renewable fuel are available in the U.S. market by 2022, with ethanol expected to make up the majority of this requirement.

Another pathway that could be used to reach the congressionally mandated biofuel volumes is the use of biobutanol. Analogous to ethanol, butanol can be produced both by petrochemical and fermentative processes. Biomass-derived butanol can be produced by alcoholic fermentation of biomass and agricultural feedstocks, such as corn, wheat, sugar beet, sugarcane, etc. Butanol offers a number of advantages over ethanol for transportation use. Butanol has considerably higher energy content than ethanol. Butanol is less hydroscopic and less corrosive than ethanol, therefore, it is more suitable for distribution through pipelines, whereas ethanol must be transported via rail or truck. Moreover, in blends with gasoline, butanol is less likely to separate from the base fuel than ethanol if the fuel is contaminated with water. Butanol has a lower vapor pressure and a higher flash point than ethanol. The heat of vaporization of butanol is also less than half of that of ethanol.

The literature is particularly poor with respect to studies on butanol/gasoline blends on SIDI engines. In addition the effect of oxygenated fuels especially on particulate emissions from SIDI engines is not well understood. The purpose of this investigation is to elucidate the effects of fuel type and blend concentration for ethanol and iso-butanol on the exhaust emissions from two different modern technology stoichiometric SIDI wall-guided light-duty vehicles fitted with three-way catalysts (TWC). The vehicles were tested over the Federal Test Procedure (FTP) and the Unified Cycle (UC) on light-duty chassis dynamometer. The fuel blends included E10, E15, E20, Bu16, and a E10/Bu8 blend. A total of seven fuels were employed in this study, including an E10 fuel, which served as the baseline fuel, and two more ethanol blends, namely E15 and E20. For this study, iso-butanol isomer was blended with gasoline at proportions of 16, 24, and 32% by volumes, which are the E10, E15, and E20, respectively, equivalent based on the oxygen content. The seventh fuel was an alcohol mixture consisting of 10% ethanol and 8% iso-butanol. This mixed alcohol formulation was equivalent to E15 based on the oxygen content. Each vehicle was tested on each fuel over three FTP and three UC tests. Measurements of regulated emissions were made for nitrogen oxides (NO_x), carbon monoxide (CO), total hydrocarbons (THC), nonmethane hydrocarbons (NMHC), methane (CH_4), and CO_2. Special attention was paid to gaseous air toxics, including carbonyl compounds (aldehydes and ketones) and volatile light aromatic hydrocarbons (benzene, toluene, ethylbenzene, and xylenes), and 1,3-butadiene. Particle emissions were characterized by defining PM mass, total particle number, particle size distributions, and black carbon concentrations. The results of this study will enhance our understanding of alcohol fuel formulations effects on gaseous and particle emissions from emerging vehicle technology.

(652a) Development of a Heterogeneous Guanidine Base Catalyst for the Conversion of Lipids to a Sustainable Biofuel

Mollaeian, K., Lamar University
Islam, M. R., Lamar University
Dickerson, B.,
Benson, T. J., Lamar University
Originally presented on: 11/7/2013 12:30:00 - 12:50:00

The production of biodiesel, a mixture of fatty acid methyl esters, has grown over the last two decades. However, due to limited feedstock supplies, excessive amounts of process wash water, and a low-value byproduct (i.e. glycerol), the biodiesel industry, and the renewable fuels industry as a whole, may see a decline in the coming years. This research focuses on the production of a biofuel, which is similar to biodiesel, except it not only produces fatty acid methyl esters but also fatty acid glycerol carbonates (FAGCs) by replacing methanol with dimethyl carbonate (DMC). The envisioned process would use no water washing steps, could convert triglycerides and phospholipids, and would have no unwanted, or low-valued, byproducts.

From early experiments, it was discovered that changing the methylating agent from methanol to DMC would require a change in catalyst from the typical sodium methoxide used in biodiesel manufacturing. The criteria for a novel catalyst are 1) readily able to catalyze the DMC to form the methoxy anion, 2) requires no neutralization step, and 3) able to also catalyze phosopholipid-containing feeds.

One of the drawbacks to the current biodiesel industry is that the process is limited to refined and degummed plant oils. If a conversion process could be developed that would also convert phospholipids, then lower value microbial-based feedstocks could be used in place of expensive plant-derived oils. Therefore, a series of hybrid organic-inorganic composite materials was synthesized by grafting the guanidine base 1,5,7-triazabicyclo[4.4.0]dec-5-ene (TBD) on Mg–Al layered double hydroxide (Mg/Al LDH with a Mg/Al molar ratio of 3 to 5) intercalated with sodium dodecylsulfate (SDS). The basal spacing of SDS-LDH was controlled simply by tuning the pH of the aqueous solutions and the Mg/Al molar ratio during the catalyst preparation. In our approach, TBD is attached with a coupling agent (3-Glycidyloxypropyl) trimethoxysilane (3GPS), and *N*-cetyl-*N,N,N*-trimethylammonium bromide (CTAB) molecules added afterwards, in the SDS-LDH gallery. The catalyst materials were characterized by XRD, SEM, EDX, FTIR, and Raman spectroscopy. Based on these results, a schematic representation for the functionalization of Mg/Al LDH surface with immobilized TBD will be presented. Results have indicated that this heterogeneous catalyst would indeed fit the initial criteria. Development of the catalyst and a proposed mechanism for converting both triglycerides and phospholipids will be presented.

(652b) A Highly Active and Stable Silica-Carbon Supported Palladium Catalyst for Decarboxylation of Oleic Acid to Produce Green Diesel

Sari, E., Wayne State University
Kim, M., NextCat Inc
Salley, S. O., Wayne State University
Ng, K. Y. S., Wayne State University
Originally presented on: 11/7/2013 12:50:00 - 13:10:00

A large amount of waste cooking oil and grease (495 million gallon/year) is produced in the U.S. every year. This waste oil/grease which is a non-food competing, inexpensive and high energy value containing feedstock can be exploited for liquid biofuel generation. In particular, brown grease, which contains mainly oleic acid, can be a potential source to obtain straight chain hydrocarbons in the diesel fuel range (green diesel). Hydrodeoxygenation (HDO) is the currently available process for green diesel production from oil/grease. However, this technique requires high pressure and excess H_2, therefore leading to high production cost. On the other hand, in a decarboxylation pathway, oxygen is eliminated as carbon dioxide without the need of additional H_2 in its stochiometry.

There has been considerable study of the decarboxylation of free fatty acids (FFA) to hydrocarbons. However, there is not yet an active and selective catalyst that can be used for decarboxylation of unsaturated FFAs to hydrocarbons in the absence of H_2. Due to the competitive adsorption and reaction of active C=C bonds on the catalyst surface, the decarboxylation yield decreases while yield of the side reactions increases over Pd/C catalyst. Although Pd/C exhibit high activity and selectivity for FFA decarboxylation, it readily deactivates due to coking and adsorbance of unsaturated products and reactants.

In this study a new catalyst that can improve the selectivity to green diesel in the absence of additional H_2 was explored for conversion of oleic acid as a model unsaturated FFA molecule. Also the correlation of catalyst structure and acidity on the decarboxylation activity was investigated. Additionally, a procedure to maintain high catalytic activity was developed.

5% Pd on silica-carbon supports with various Si:C weight ratios were prepared and oleic acid decarboxylation activity tests were conducted in a 100 ml batch reactor in the absence of H_2 at 300 °C and 1.5 MPa. A significant correlation between the Pd metal particle size and the catalytic activity was observed. Pd/Si-C-4 catalyst exhibited the highest selectivity to n- heptadecane. High catalytic activity of Pd/Si-C-4 was attributed to the accessible, small and well-distributed metallic Pd nanoparticles inside the hybrid mesopores of this catalyst. It was also observed that the Pd on Si-C supported catalyst favors a direct decarboxylation of oleic acid unlike a Pd/C catalyst which catalyzes C=C double bond hydrogenation prior to the oleic acid decarboxylation. Furthermore, the novel catalyst was highly stable for oleic acid conversion selectively for green diesel production in the absence of H_2.

(652c) Catalytic Deoxygenation of Cyanobacteria-Derived Fatty Acids to Hydrocarbons

Schulz, T., North Carolina State University
Oelschlager, M., Arizona State University
Nielsen, D. R., Arizona State University
Vermaas, W., Arizona State University
Lamb, H. H., North Carolina State University
Originally presented on: 11/7/2013 13:10:00 - 13:30:00

A phototrophic cyanobacterium, *Synechocystis* sp. PCC6803, was genetically modified to secrete fatty acids (FAs), predominantly lauric acid (LA) into the growth medium.[1] Previous research demonstrated that LA can be deoxygenated using a Pd/C catalyst with high selectivity to *n*-undecane and CO_2.[2] In contrast, cyanobacteria-derived fatty acids (CBFAs) recovered from the growth medium by adsorption on a hydrophobic resin were deoxygenated much less efficiency and with lower CO_2 selectivity. Herein, we demonstrate that the decarboxylation activity of the catalyst is related inversely to the sulfur concentration in the CBFA sample. Since this is likely due to contaminants such as sulfolipids, β-hydroxy FAs and proteins, effective removal of sulfur-containing compounds from the FA feedstock is essential to achieving high alkane yield and CO_2 selectivity. Activated carbon decolorization, C18 adsorption chromatography, acid hydrolysis, and alkaline hydrolysis were each employed as purification methods for the raw CBFA samples. Samples were analyzed for sulfur and phosphorus by inductively coupled plasma-optical emission spectroscopy (ICP-OES). CBFA samples were deoxygenated in a semi-batch reactor at 300°C under 5% H_2 using a 5 wt.% Pd/C catalyst. Deoxygenation yield and CO_2 selectivity were found to be highly correlated to sulfur concentration. A CBFA sulfur concentration of <20 ppm was required to achieve results equivalent to those obtained previously with LA.

CBFAs also contain β-hydroxy fatty acids that are constituents of an endotoxin produced by gram-negative bacteria such as *Synechocystis* sp. PCC6803.[3, 4] Deoxygenation experiments with 10% β-hydroxy myristic acid (BHMA) and 90% LA under standard semi-batch conditions evidenced that the presence of BHMA does not significantly decrease the final alkane yield or CO_2 selectivity of LA deoxygenation; however, it does inhibit the deoxygenation kinetics. Deoxygenation of BHMA proceeded with high CO_2+CO yield (92%), liquid product yield (86%) and CO_2 selectivity (85%). The reaction also produced a strong initial H_2 evolution peak. Major liquid-phase products were *n*-tridecane, 2-tridecanone and 2-tridecanol. Both 2-tridecanone and 2-tridecanol also were reacted over 5 wt.% Pd/C. Conversion of 2-tridecanol proceeded mainly via dehydrogenation to 2-tridecanone, which explains the H_2 evolution observed during BHMA deoxygenation. Neither 2-tridecanol nor 2-tridecanone was deoxygenated to *n*-tridecane in significant yield. BHMA deoxygenation to *n*-tridecane most likely proceeds via a myristic acid intermediate with 2-tridecanol and 2-tridecanone as byproducts.

References

1. X. Liu, J. Sheng and R. Curtiss, III, *PNAS* **108** (17), 6899-6904 (2011).

2. J. P. Ford, J. G. Immer and H. H. Lamb, *Top. Catal.* **55** (3-4), 175-184 (2012).

3. B. Szponar, L. Krasnik, T. Hryniewiecki, A. Gamian and L. Larsson, *Clin. Chem.* **49** (7), 1149-1153 (2003).

4. M. Asayama, *Appl. Microbiol. Biotechnol.* **95** (3), 683-695 (2012).

(652d) Co-Solvent Enhanced Production of Fuel Precursors From Lignocellulosic Biomass

Cai, C. M., University of California, Riverside
Nagane, N., University of California, Riverside
Zhang, T., Center for Environmental Research and Technology, University of California
Kumar, R., University of California, Riverside
Wyman, C. E., University of California, Riverside
Originally presented on: 11/7/2013 13:30:00 - 13:50:00

Lignocellulosic biomass is the most abundant source of organic carbon on Earth and presents the only option with the potential to economically and sustainably replace fossil resources for large-scale production of liquid transportation fuels. Furfural, 5-hydroxymethylfurfural (5-HMF), and levulinic acid (LA) are promising platform chemicals made by dehydration of biomass sugars and can be catalytically hydrogenated to gasoline, jet fuel, diesel-range alkanes, and fuel blending components compatible with the existing fuel infrastructure (so-called "drop-in" fuels). However, the natural resistance of lignocellulose to breakdown to fuel precursors including monomeric sugars and their dehydration products is the major obstacle to low cost fuels and must be overcome for biomass fuels to be economically competitive. Here, we demonstrate that tetrahydrofuran (THF) is uniquely suited as a water-miscible co-solvent to promote biomass deconstruction, sugar dehydration, and lignin extraction simultaneously and achieve much higher selectivity to fuel precursors than previously reported. We propose a novel biomass-to-fuels platform in which THF is both a recoverable co-solvent and final co-product and compare the performance of this system using different raw substrates and various acid catalysts to achieve the targeted production of fuel precursors without the need for additional pretreatment. We also discuss the advantages of co-solvent strategies for the direct processing of raw feedstock in comparison to two-phase and steam-stripping operations.

(652e) Direct Conversion of Oxygenated Fuels to Power Using Solid Oxide Fuel Cells

Lancaster, B. R., University of Colorado at Boulder
Linic, S., University of Michigan
Originally presented on: 11/7/2013 13:50:00 - 14:10:00

Over the last several decades, intermediate temperature (600-800°C) solid oxide fuel cells (ITSOFCs) have been developed that can generate electricity with very high efficiencies. Although the high temperature of these fuel cells allows for power generation directly from carbonaceous fuels, current commercial nickel SOFCs cannot be run directly on carbonaceous fuels because the carbon quickly degrades the catalyst on the anode, or fuel side of the cell. Recently, surface alloys have been made from nickel and tin that show improvements in the carbon tolerance of ITSOFCs on methane and iso-octane. Concurrently, biofuels including ethanol and butanol have been developed from renewable plants and other biological sources which have a lower carbon footprint than conventional fossil fuels. SOFCs run on these renewable fuels would create a highly efficient process to generate electricity with a small carbon footprint.

In this work, fuel cells with monometallic nickel and tin/nickel surface alloy anodes are made and tested with oxygenated fuels such as ethanol and butanol. The performance and resistances across the fuel cells are monitored over time using chronoamperometry, linear sweep voltammetry, and electrochemical impedance spectroscopy. The degradation of the fuel cell anodes is evaluated using scanning electron microscopy. When run on ethanol, the nickel catalysts show changes in performance over time whereas the tin/nickel surface alloy catalysts have very stable performance. The nickel catalysts also show more degradation over time than the alloy catalysts under the same conditions.

(652f) Kinetic Characterization of Catalytic Wet Air Oxidation of Polymers for Waste Management

Tillie, C. F., Cleveland State University
Adhikari, S., Cleveland State University
Cmar, J. C., Cleveland State University
Reeves, S. A., Cleveland State University
Gatica, J. E., Cleveland State University
Originally presented on: 11/7/2013 14:10:00 - 14:30:00

Increasing energy demands and sustainability have recently renewed attention on waste-to-energy routes as sustainable waste management strategies. Wet thermal catalytic oxidation (WTCO) processes are considered in this study as low-temperature waste management or reduction.

Gasification processes promote low-temperature conversion of long polymeric chains into synthetic or "supply" gas (syngas). As the activation energy has proven to be reduced significantly in the presence of a catalyst, WTCO is considered an attractive option to more traditional, high-temperature, gasification techniques.

This research examines the catalytic oxidation of low and mid-fidelity space exploration trash simulants (single polymers and combinations of polymers) for different reactor configurations: (1) catalyst particles combined in a slurry of water and waste polymers, and (2) catalyst deposited on solid substrates.

Data collected in a research-grade laboratory reactor are characterized by gas chromatography and used for the kinetic characterization of catalytic gasification processes. Experiments are complemented by thermal characterization of simulants and catalytic slurries.

Kinetic analysis is used to formulate and characterize a reaction mechanism of the gasification process. The kinetics of this gasification process are combined with detailed thermodynamics in a process simulator. Scaled-up flow processes for continuous waste management are investigated with this model for yield and selectivity optimization.

(652g) Kinetics of Self-Buffering Hydrolysis of Pectin in Batch Reactor

P Asher, H.,

Rivas, R. C., Texas A&M University- Kingsville

Mills, P. L., Texas A&M University-Kingsville

Originally presented on: 11/7/2013 14:30:00 - 14:50:00

Kinetic study for a self-buffering hydrolysis of pectin in batch reactor

Heman P. Asher, Raul Rivas-Cantu, Patrick L. Mills

Processing of citrus waste generates large amount of wastes. From 2008-2011, the USA generated almost 10.5 million tons of Citrus Processing Waste (CPW) from processing of grapefruits and oranges. The CPW is dried and then used as a low cost feed for livestock. Due to low nutritional values of feedstock and high energy consumption during drying, this operation is not economical and is used as a remedial process for waste disposal. The unused CPW is sent to landfills, which incurs environmental hazards and high economical costs. Grapefruit Processing Waste (GPW) is the primary source of CPW in Texas. GPW can be used as a renewable source of pectin-rich biomass, which on hydrolysis gives Galacturonic Acid (GA) and other value added products for biorefinery applications. GA on catalytic reduction may produce hexoses (glucose, galactose), which can be used to produce bio-ethanol.

The objective of this work was to obtain kinetic parameters for self-buffered enzymatic hydrolysis of pectin in a batch reactor to determine the reaction kinetics with an emphasis on its conversion into galacturonic acid (GA). Lab scale experiments were performed in bottom- baffled Erlenmeyer flasks submerged in water bath shaker at pH 4.8, temperature 45 °C and 175 rpm.

The first step was to evaluate the experimental results of total hydrolysis of various concentrations of pectin solutions in 75 mM sodium acetate (NaOAc) buffer solution and in a self-buffering solution of sodium galacturonate (NaGA) produced during enzymatic hydrolysis and formed due to addition of sodium hydroxide (NaOH) to raise the initial pH to 4.8. Experiments for determination of kinetic parameters were performed for three different enzyme loadings for 0.5 wt% pectin solutions monitored at periodic time intervals for 5 hours. Commercially available enzyme complexes, such as Pectinex^R Ultra-SPL and Accelerase^R XY, were used as biocatalyst. The initial results showed that the self-buffering system behaved similar to externally buffered system. Subsequently, the kinetic parameters for a self-buffered system were determined based on experimental data using the Michaelis-Menten model. Additional experiments at pH 6.0 are being evaluated for improving enzymatic activity and to increase the reaction yield. The determined kinetic parameters are shown in table 1. Fig. 1 shows the behavior of GA concentration with time during the enzymatic hydrolysis of pectin and Fig. 2 shows the normalized M-M plot for enzymatic hydrolysis.

These and other results, including the utility of the kinetic modeling for reaction scale-up by a volume factor of more than 150, will also be discussed

Table 1 ? Determined kinetic parameters for various enzyme loading

Kinetic Parameters	**25 µL Pectinex, 10 µL Accelerase XY**			**50 µL Pectinex, 20 µL Accelerase XY**			**75 µL Pectinex, 30 µL Accelerase XY**		
	Lineweaver	Eadie-Hofstee	Hanes-Woolf	Lineweaver	Eadie-Hofstee	Hanes-Woolf	Lineweaver	Eadie-Hofstee	Hanes-Woolf
Michaelis-Menten Constant (Km, mmol/L)	13.47	13.47	13.47	14.27	13.76	13.76	14.26	14.16	14.16
Max. forward velocity (Vm, mmol/L-min)	0.093	0.093	0.093	0.082	0.079	0.079	0.16	0.16	0.16

Fig.1 ? Concentration of GA *vs.* time during enzymatic hydrolysis of pectin at different enzyme loading

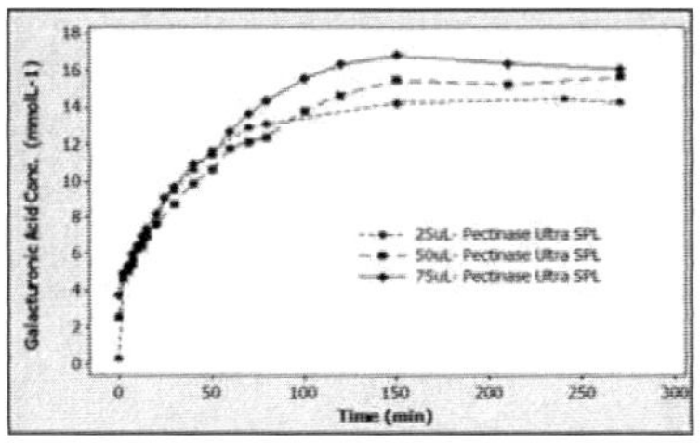

Fig.2 ? Normalized M-M plot for conversion of GA during enzymatic hydrolysis reaction of pectin at 75 µL Pectinex^R- Ultra-SPL and 30 µL Accelerase^R XY loading

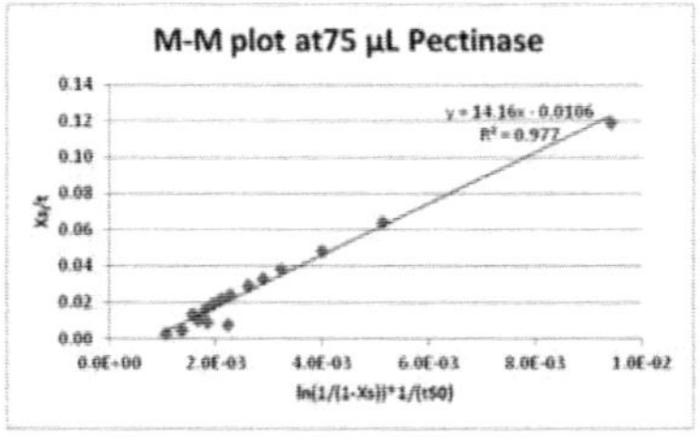

(653b) Compositional Analysis and Advanced Distillation Curves for Mixed Alcohols Produced Via Syngas On a K-Comosx Catalyst

Hensley, J., National Renewable Energy Laboratory
Lovestead, T., National Institute of Standards and Technology
Christensen, E.,
Dutta, A., National Renewable Energy Laboratory
Bruno, T.,
McCormick, R., National Renewable Energy Laboratory
Originally presented on: 11/7/2013 12:45:00 - 13:00:00

The distillation behavior of mixed alcohols was studied by use of the Advanced Distillation Curve (ADC) methodology. Crude mixed alcohols (oxygenates) were generated from syngas over a potassium-promoted cobalt-molybdenum-sulfide catalyst and assayed for major and minor products. Distillation (boiling) curves were generated for the crude mixed oxygenate products and composition channel data were collected. The crude mixed alcohols consisted primarily of methanol with significant quantities of ethanol, 1-propanol, 1-butanol, methyl acetate, and ethyl acetate. These 6 species constitute 93.7 % – 95.8 % (mass/mass) of the total product. Ester, ether, and aldehyde impurities were identified, as well as thiols and organic sulfides. Considering just the alcohol products without impurities, these can be blended into gasoline at 8.5 % (vol/vol) and meet the requirements of the Octamix waiver if an appropriate corrosion inhibitor were also included (blend would contain 3.0 % – 3.4 % methanol, more than 2.5 % higher alcohols (vol/vol) and a total oxygen content of 3.7 % (mass/mass)). Distillation targeted at 50% methanol removal increased the volume of product that could be blended to over 9 % (vol/vol). Methanol, aldehydes, and dimethyl sulfide were the first to vaporize from the mixture, and all C_{4+} alcohols remained within the last 20 % of the distilled volume. Other products, including ethanol, propanols, esters, and organic sulfur species distilled over a range of boiling temperatures. ADCs suggest the presence of one or more azeotropes in the distillate, consistent with a large number of known binary azeotropes between components found in the mixed oxygenate product. Enthalpies of combustion were calculated for multiple distilled fractions and ranged from 890 $kJ \cdot mol^{-1}$ in the first drop of distillate to 1150 $kJ \cdot mol^{-1}$ in the first drop collected after distilling 80 % of the original liquid volume. This energy density is low compared to 91 octane gasoline at 3700 and 4940 $kJ \cdot mol^{-1}$ in the first drop and at 80 %, respectively. Comparisons of fractional distillation of the mixed oxygenate products showed directional agreement between experiment and simulation with Aspen Plus. This study provides useful insights into mixed oxygenate products derived from a sulfided catalyst, including considerations for process recycle, product constituents and their blending, and the applicability of distillation information from process simulators.

(653c) Production and Characterization of Straight-Run Thermal Deoxygenation (TDO) Fuels

Eaton, S., University of Maine
Wheeler, M. C., University of maine
Beis, S. H., University of Maine
Karunarathne, S. A., University of Maine
van Heiningen, A., University of maine
Originally presented on: 11/7/2013 13:00:00 - 13:15:00

Thermal DeOxygenation (TDO) refers to a simple, non-catalytic, process which produces highly deoxygenated crude oils from levulinic and formic acids formed by acid hydrolysis of biomass feedstocks. The TDO reaction is a single-step decomposition reaction which converts the neutralized acids to crude hydrocarbons, chars, CO_2, water and volatiles at 450°C and at ambient pressure. The TDO crude oils have long-term shelf-stability and very low water miscibility compared to fast pyrolysis oils. Successful TDO scale-up to 50L semi-batch operation has produced crude oils with broad boiling point distribution, between 75° and 585°C, and oxygen content of 1-8 wt.%.

Choice of basic cation is important in TDO reactions as it controls reaction temperatures and product mass distributions. In this work, calcium and magnesium are compared to understand their influence on crude oil composition. Straight-run crude oils obtained using each cation underwent fractionation and material and fuel property characterizations to explore suitability in gasoline, diesel and residual fuel applications. The products showed high aromaticity, low H/C ratios at about 1.3/1 and higher heating values of 38-40 MJ/kg. Crude oil oxygen content, total acid number and final boiling points were all higher for the magnesium case. In the distillate region, fuel density and cetane index were equivalent. Diesel combustion properties were characterized by comparing a 100% straight-run TDO distillate cut to specification ultra-low sulfur diesel in a single-cylinder research engine. Combustion was maintained over the entire operating window under intake air heating with measured increases in hydrocarbon and CO emissions.

(653d) Oxy-Combustion of Low-Volatility Fuel With High Water Content

Yi, F., Washington University in St. Louis
Axelbaum, R. L., Washington University in St.Louis
Originally presented on: 11/7/2013 13:15:00 - 13:30:00

High water content fuels, such as coal fine, microalgae and bioethanol, are abundant in nature. Fundamental study is necessary to investigate the novel idea of direct combustion of high water content fuels, which avoids the energy intensive dewatering process. Our previous research investigated the flame stability for high-volatility fuels, including ethanol, 1-propanol and t-butanol. In this study, we continue the study and focus on low-volatility fuel and glycerol is chosen as the candidate. It is found that self-sustained flame can only be obtained for glycerol solution with concentration higher than 50 wt%, when it is burned in pure O_2. However, the flame is lifted far away from the nozzle, which is not typical for a swirl stabilized burner. Knowing ethanol having strong preferential vaporization over water from previous study, ethanol was added as additive to enhance the flame stability. Experiment showed that anchored flame can be obtained by burning the mixture of 10 wt% ethanol, 30 wt% glycerol and 60 wt% water under oxy-fired condition. The flame stability for this mixture was characterized under 100% and 85% swirl flow conditions. It is found in our previous research that t-butanol shows even stronger preferential vaporization behavior than ethanol. The flame stability for the mixture of 8.3 wt% t-butanol, 30 wt% glycerol and 61.7 wt% water, which has the same energy content as the mixture of 10 wt% ethanol, 30 wt% glycerol and 60 wt% water, was characterized under the same flow conditions as ethanol case. Under 100% swirl flow condition, the blow-off limits are approximately the same for both cases. Under 85% swirl flow condition, the blow-off limits for t-butanol case are much lower in the low flow rate region. Additionally, the lift-off limits for t-butanol case are shifted to the left compared to the ethanol case, which means the flame stability for t-butanol case is better. Further, flame structure was characterized. Contours of temperature across the chamber's centerline were obtained for flame burning 10 wt% ethanol, 30 wt% glycerol and 60 wt% water under three different running conditions, including 100% O_2 and 100% swirl, 100% O_2 and 85% swirl, 80% O_2 and 100% swirl. It is found the anchored flame becomes narrower as the swirl intensity decreases and the peak temperature zone shifts downstream as the O_2 concentration decreases.

(653e) Hierarchical Pore Structure Zeolites for the Catalytic Upgrading of Tars

Akkalkotkar, A.,

Du, S., University of Connecticut

Bollas, G. M., University of Connecticut

Valla, J., University of Connecticut

Originally presented on: 11/7/2013 13:30:00 - 13:45:00

Hierarchical pore structure zeolites

for the catalytic upgrading of tars

Ameya Akkalkotkar, Shoucheng Du, George M. Bollas, Julia Valla

Introduction

Thermochemical conversion of biomass via gasification has attracted increasing scientific interest lately. However, one of the most important technical barriers towards the successful commercialization of biomass gasification is the efficient removal of impurities present in the synthesis gas (syngas), like tars. Tars are a complex mixture of light and heavy molecular weight Poly-Aromatic Hydrocarbons (PAH). PAHs deactivate catalysts, cause blockages in transfer lines, damage downstream units and lower the overall process efficiency. Moreover, due to their carcinogenic nature, the disposal of tars is impossible.

Modifying the gasification operation parameters can affect and potentially reduce the tars content in the syngas, while newly developed entrained flow gasifiers produce almost no tars, but require very flexible designs and are associated with high capital cost. Downstream thermal cracking is another way to reduce tars. However, it involves heating the tars to very high temperature (1000°C). Catalytic cracking of tars appears to be one of the most promising options for tars reduction, because it has the potential to convert tars into lighter hydrocarbons, which can later be reformed to valuable syngas.

Zeolites are microporous materials that have the potential to catalytically eliminate tars from syngas. Zeolites exhibit a unique network of properties like acidity, microporosity and shape selectivity. For this reason they are extensively used during the Fluid Catalytic Cracking (FCC) Unit, where heavy hydrocarbons are cracked into lighter valuable products, like gasoline and diesel. Notwithstanding the positive effect of the presence of micropores with respect to shape selectivity, the zeolite micropore structure may often induce a negative effect due to the low diffusion rates of heavy molecules into the zeolite crystals, as well as unwanted adsorption effects of the reactants and/or products that can lead to the formation of coke.

The aim of our research is twofold: a) to test the effectiveness of zeolites for tars cracking, and b) to modify the zeolites in order to create hierarchical pore structure network that will reduce the diffusion limitations of the heave PAHs in and out of the zeolite pores, hence improving their cracking reactions.

Experimental Section

Model Compounds

Model compounds, representative of the tars mixture, have been used during this study including: benzene, naphthalene, fluoranthrene and coronene, as they are the most abundant compounds in the groups of light

PHA and heavy PHA. The structure of the above model compounds is shown in Figure (1).

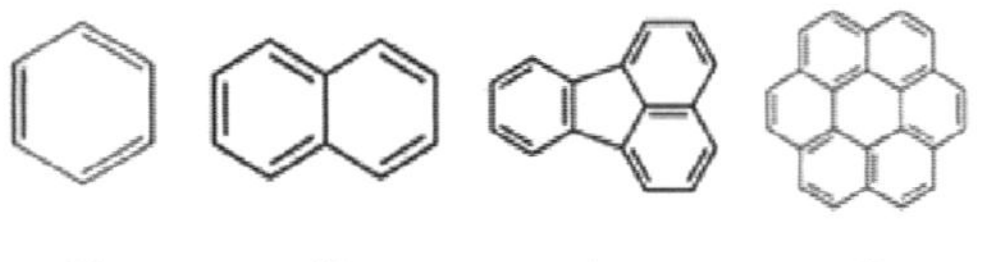

Figure 1. Tars model compounds used in this study: (a) benzene, (b) naphthalene, (c) fluoranthrene and (d) coronene

Experimental Set-up

In Figure (2) the experimental set up for the tars cracking tests is shown. 5 gm of each model compound is placed in a stainless steel vessel and heated up just below its boiling point. Nitrogen is bubbled through the liquid compounds and picks up the model compound. This mixture then passes over the catalyst bed (0.5 gm), which is held in a quartz tube reactor (7mm ID). The reactor is placed in a furnace for the catalytic cracking experiments. Two different temperatures have been tested (625°C and 700°C). Stainless steel piping is used to withstand the high system temperature. The piping is heat taped to avoid tars condensation inside the system. The products are passed through a train of impingers, which contain isopropanol as a solvent. The solvent dissolves the heavy tars, which are then analyzed using a GC equipped with Simulated Distillation and GC Mass Spectroscopy. The gas is analyzed using GC equipped with Thermal Conductivity Detector (TCD).

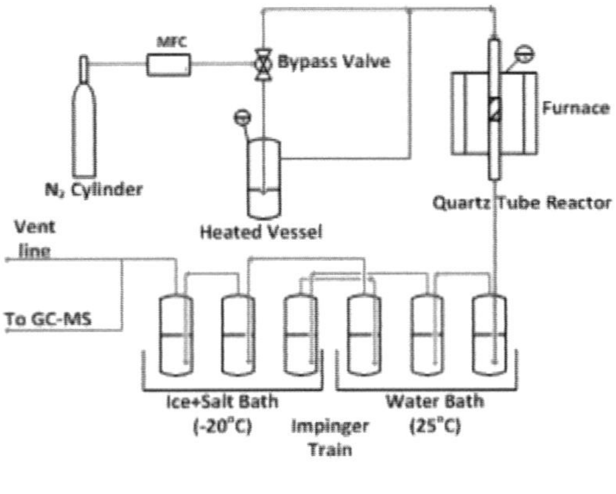

Figure 2. Experimental set-up for the tars model compounds cracking reactions

Materials Preparation and Characterization

In this study ZSM-5 and Y zeolites obtained from Zeolyst International were used. Hierarchical mesoporous zeolites ZSM-5 and Y zeolites were prepared by sequential alkaline-acidic treatment [1]. During the treatment, the ammoniated form of the zeolite is calcined at 550°C to obtain the protonated form. This zeolite is then subjected to treatment with NaOH (30 ml/g of zeolite) at 65°C for 30 min. The suspension is quenched, filtered, washed till neutral pH and dried. The powder is then treated with HCl (100 ml/g of zeolite) at 65°C for 6 hrs. It is again filtered, washed and dried. The product is then ion-exchanged 3 times with 0.1N NH_4NO_3 for 12 hrs each. The resultant suspension is again filtered, washed and dried and then calcined at 550°C. The materials were characterized using X-ray Diffraction (XRD) and N_2 adsorption. The coke deposited on the catalyst after the cracking experiments was analyzed using Thermogravimentric

Analysis (TGA).

Results and Discussion

N_2 adsorption of the parent Y and ZSM-5 zeolites, as well as the various mesoporous USY and mesoporous ZSM-5 are shown in Figure 3. There are two important observations: 1) the concentration of the alkaline media (NaOH) plays a key role on the extent of the desilication of the zeolites and consequently on the introduction of mesoporous and the destruction of microporous structure of the zeolites, and 2) the acid treatment after the alkaline treatment (desilication) is a necessary step that helps to the unblocking of the microporous due to the formation of extra framework aluminium [2]. Based on the N_2 adsorption results, it has been revealed that the optimum conditions for maximum mesoporosity incorporation and microporosity retention is 0.1M NaOH/0.02M HCl and 0.2M NaOH/0.02MHCl for the USY and the ZSM-5, respectively.

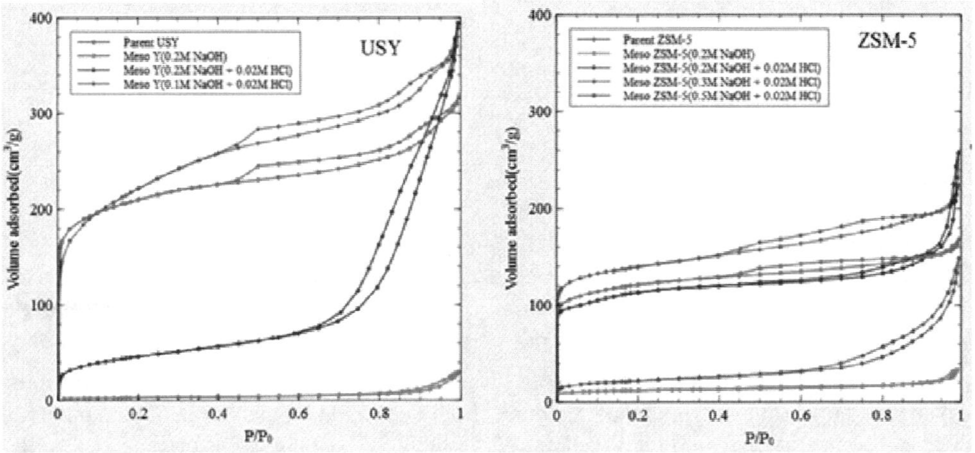

Figure 3. N_2 adsorption/desorption curves for the parent and the mesoporous USY and ZSM-5 zeolites

Catalytic cracking experiments have been conducted using both benzene and naphthalene. Benzene is the smallest aromatic hydrocarbon, while naphthalene is the smallest PAH. In Figure 4 the benzene conversion to gas, liquids and coke at two temperatures (625°C and 700°C) using parent ZSM-5 (a), mesoporous ZSM-5 (b), parent Y (c) and mesoporous Y (d) is presented.

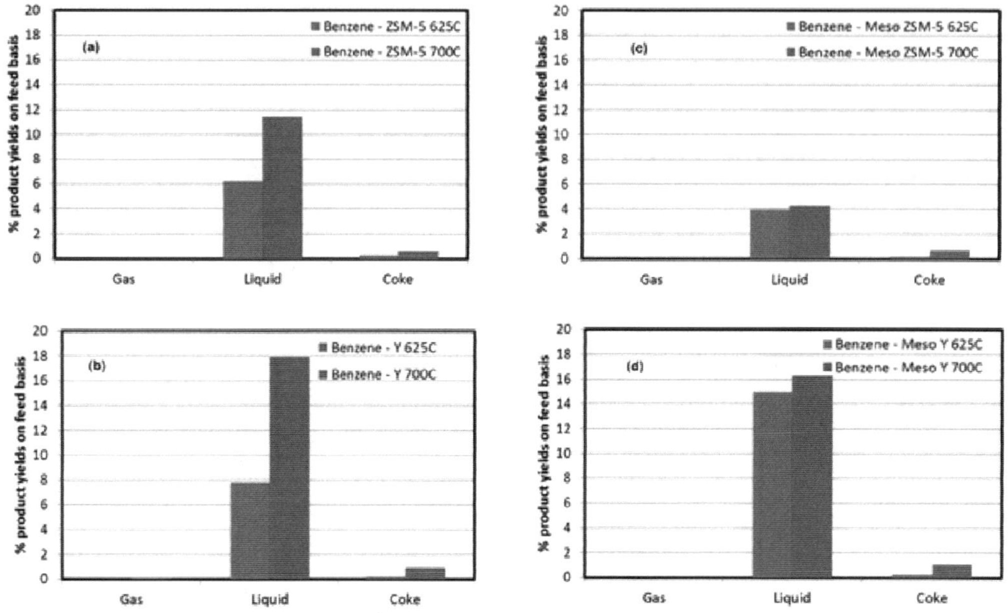

Figure 4. Benzene conversion to gas, liquids and coke using ZSM-5 (a), Mesoporous ZSM-5 (b), Y (c) and Mesoporous Y (d).

Firstly, parent Y zeolite (Figure 4(b)) leads to slightly higher benzene conversion than ZSM-5 (Figure 4(a). This was expected since the parent Y zeolite has higher both microporous and mesoporous area. Secondly, for both parent ZSM-5 and Y zeolites (Figure 4(a) and Figure 4(b)), increasing the reaction temperature from 625°C to 700°C favors the overall benzene conversion. With respect to the mesoporous zeolites, mesoporous ZSM-5 (Figure 4(c)) does not favor the conversion of benzene, compared to the parent ZSM-5. The above results lead us to the conclusion that the reaction of the benzene to the ZSM-5 is not diffusion limited. On the contrary, benzene conversion is favored by the active sites in the ZSM-5, which are partially destroyed during the introduction of mesoporosity. In the case of Y zeolites, the results are more complicated. It appears that at low temperature (625°C) mesoporous Y favors the conversion of benzene compared to the parent Y. However, at high temperature (700°C) no difference in benzene conversion was observed between the parent and the mesoporous Y. The above results as well as the results of the catalytic cracking of naphthalene over Y, ZSM-5 and the corresponding mesoporous zeolites will be discussed.

Conclusion

A laboratory experimental set up has been designed and built for the evaluation of zeolites and hierarchical pores structure zeolites on the catalytic cracking of PAHs derived from biomass gasification. Hierarchical pore structure ZSM-5 and Y zeolites have been prepared using the desilication method. Preliminary cracking tests using benzene and naphthalene as model compounds at two different reaction temperatures and using the parent and the corresponding hierarchical zeolites have been conducted and the results will be presented and discussed.

Acknowledgement. The study is funded by the National Science Foundation Award CBET-1236738.

References

[1] Radwan A.M.; Kyotani T.; Tomita A., Appl. Catal. A., **2000,** *192*, 43.

[2] Groen J.C.; Moulijn J. A.; and J. P¨|rez-Ram¨ªrez, Microporous Mesoporous Mater., **2005,** *87*(2), 153.

(653f) Production of Fuels From Natural Gas Through a Hybrid Thermochemical/Biochemical Gas Fermentation Process

Griffin, D. W., LanzaTech Limited

Originally presented on: 11/7/2013 13:45:00 - 14:00:00

Natural gas is currently an abundant, low-cost feedstock in the United States that can be used to produce fuels and chemicals through a variety of thermochemical conversion routes. Over recent decades, there has been an ever-increasing focus on developing alternative technologies to produce fuels through biological pathways either by using biomass as a feedstock or utilizing a biological process. LanzaTech has developed a hybrid thermochemical/biochemical gas fermentation process which produces fuels or fuel intermediates from natural gas.

Gas fermentation is a biological process in which microbes convert climate-active gases, such as methane, carbon monoxide, or carbon dioxide, into low-carbon fuels and chemicals. LanzaTech currently operates a gas fermentation process that utilizes biological renewable resources to convert carbon monoxide to ethanol and other fuel derivatives. The carbon monoxide can come from a variety of sources including industrial waste gases, such as steel mill off-gas, or any CO-containing syngas. Natural gas can be readily reformed to syngas through a variety of conventional thermochemical technologies, which can be converted to fuel through gas fermentation. The gas fermentation process has a number of distinct advantages over conventional thermochemical gas-to-liquid (GTL) routes, including syngas composition flexibility, reduced economic operating scale, and reduced gas clean-up requirements, providing cost and emissions benefits.

The integrated thermochemical/biochemical LanzaTech gas fermentation process for the production of fuels from natural gas is described including process economics and life cycle analyses; there is also active research on the direct fermentation of natural gas to fuels which will also be described. There are many natural gas assets that are currently uneconomical to exploit due to relatively low available volumes and/or high costs of gas transportation, CO_2 removal, or gas clean-up. Gas fermentation presents an enabling technology for small-scale deployment and accessing sources that are currently uneconomical to process and may be ultimately flared, vented, or emitted.

(653g) JP-10 Combustion Studies Using Automatic Reaction Mechanism Generation

Gao, C., Massachusetts Institute of Technology
Yee, N., Massachusetts Institute of Technology
Magoon, G. R., Aerodyne Research Inc.
Edwards, R., Aerodyne Research Inc.
Wong, H. W.,
Vandewiele, N., Ghent University
Green, W. H. Jr., Massachusetts Institute of Technology
Originally presented on: 11/7/2013 14:00:00 - 14:15:00

The synthetic jet fuel JP-10 has important applications in propulsion technologies and has 11% higher volumetric energy density than conventional jet fuel JP-8. JP-10 is a single-component fuel consisting of the highly strained tricyclic molecule (exo-tricyclo(5.2.1.0(2,6))decane), which undergoes complex combustion chemistry at high temperatures. Here, the first detailed, comprehensive model of JP-10 combustion and pyrolysis kinetics was constructed using Reaction Mechanim Generator, an open-source automatic mechanism generation software package. Its predictions agree well with published experimental ignition delay data. A series of shock tube experiments were performed for multiple oxidation and pyrolysis conditions over a temperature range of 1000-1400 K at pressures of 5-8 bar, with products quantitatively analyzed using GC/MS. Comparisons of the model against shock tube data show that the model captures qualitative and quantitative trends in the product distribution over most experimental conditions. The combined experimental and modeling efforts demonstrate that the initial JP-10 decomposition pathways and branching ratios play an important role in product speciation. Analysis of the discrepancies between the model predictions and the experiments highlights where future work is needed.

(653h) Fischer-Tropsch Synthesis: Comparison of Process Based On Fossil and Biomass Feedstock

Davis, B. H., Center for Applied Energy Research, University of Kentucky

Originally presented on: 11/7/2013 14:15:00 - 14:30:00

The conversion of synthesis gas to fuels is independent of the feedstock once the proper ratio of hydrogen/carbon monoxide has been obtained. In one approach to obtain the proper ratio for the syngas, the water-gas shift reaction is used and this can be done in a separate reactor or in the Fischer-Tropsch reactor. The feedstock will determine the amount of adjustment that is required to attain the proper hydrogen/carbon monoxide ratio; in general, the amount of adjustment will be greatest for biomass than for coal than for natural gas. Approaches to conduct this adjustment will be described as well as the limitations that they place on the process and on lthe catalyst that is used.

(653i) Conversion of Dimethylether Into Target Hydrocarbons Over Zeolite Catalysts

Gonzales, A. C., Ohio University
Tschannen, R. E., Ohio University
Lee, S., Ohio University
Originally presented on: 11/7/2013 14:30:00 - 14:45:00

The continued depletion of easily extractable petroleum reserves leading to elevated transportation fuel prices has furthered the desire to produce transportation fuels from non-petroleum sources. Natural gas has tremendous potential for conversion to transportation fuels due to the abundance and low cost caused by the shale booms in the eastern and northwestern parts of the country. One possible route to liquid transportation fuels is reformation of the natural gas into syngas, and syngas conversion to dimethylether or DME. DME can then be used directly as a transportation fuel or converted to more traditional transportation range hydrocarbons such as gasoline or diesel. The Dimethylether-to-Hydrocarbon process converts DME to higher order hydrocarbons by reacting DME over the acid sites of zeolite catalysts. This work examines the effects of temperature, weighted hourly space velocity (WHSV), and varying the silica to alumina ratio of the catalysts on the selectivity and yield of the hydrocarbon products.

(653j) Characterization of Heavy Petroleum Fractions and Its Application in Design and Operation of Upgrading Related Processes

Riazi, M. R., Kuwait University

Originally presented on: 11/7/2013 14:45:00 - 15:00:00

As needs for use of heavy and unconventional oils are on the rise, conversion of these oils into light and middle distillates is becoming increasingly important. The nature and constituents of heavy oils and heavy residues are significantly different from those of light and conventional oils [1]. In this presentation first the differences between light and heavy oils especially from compositional and elemental points of views are discussed. Heavy oils are rich in elements such as sulfur, nitrogen and metals. Methods used for characterization of light and conventional oils cannot be directly used for heavy oils and their fractions. Traditional methods used for estimation of thermophysical properties of light fractions are not applicable to those of heavy fractions. In the second phase of this presentation we discuss appropriate characterization methods for heavy fractions and residues. Properties and specification which can be measured in laboratory to characterize heavy oils and fractions are discussed. It will be shown how the amounts of hydrogen, nitrogen, sulfur and metals in a heavy fractions can be estimated from those of its original crude oil. Data for properties of very heavy residues are used to demonstrate accuracy of several characterization methods in representing amounts of various compounds in the mixture. In addition method of experimental measurement on the rate of dissolution of light gases into heavy oils for upgrading purposes under different conditions of temperature and pressure is presented.

In the final stage latest processes developed for upgrading heavy oils and heavy residues, advantages and disadvantages of each process are reviewed [2]. A set of data measured on conversion of a heavy petroleum fraction into lighter products at different operating conditions are used to demonstrate how product specifications and conversions can be estimated using feed specifications and appropriate characterization method. In summary several characterization schemes utilizing measurable properties for heavy oils and residues will be presented which can be used to properly estimate thermodynamic and physical properties needed for optimum design and operation of upgrading units.

[1] Riazi, M. R., "Characterization and Properties of Petroleum Fractions," ASTM International, 2005. 430p. (2nd Print: 2007)

Web: http://www.astm.org/BOOKSTORE/PUBS/MNL50.htm

[2] Riazi, M. R., Eser, S., Agrawal, S. S., Pena Diez, J. L., "Petroleum Refining and Natural Gas Processing" ASTM International, Conshohocken, PA, USA, 2013. 811p.

Web: http://www.astm.org/BOOKSTORE/PUBS/MNL58.htm

(712a) Nature-Inspired Fractal Gas Distributor for Low-Temperature PEM Fuel Cell Cathodes

Marquis, J., Rensselaer Polytechnic Institute
Coppens, M. O., Rensselaer Polytechnic Institute
Originally presented on: 11/7/2013 15:15:00 - 15:30:00

This work examines a novel approach to the design of a fuel cell gas distributor by mimicking the structure and functionality of the human lung[1]. A fractal flow field geometry is proposed to address the fundamental shortcomings of conventional flow field designs, namely the non-uniform distribution of reactant gases over the surface of the catalytically active area. The latter lead to highly non-uniform transport and reaction conditions throughout the cathode catalyst layer.

The proposed design is inspired by ability of the fractal geometry of human lungs to uniformly distribute oxygen throughout the lung from a single inlet (trachea) through many outlets (alveoli). Our novel flow field design utilizes a similar fractal branching structure to provide nearly uniform coverage of reactant oxygen gas across the cathode gas diffusion layer. This results in an almost constant oxygen concentration boundary condition across the entire gas diffusion layer - cathode catalyst layer interface. This eliminates concentration depletion along the flow paths, seen in conventional flow geometries[2].

The fractal geometry for the gas inlet is a repeated "H" shape that has one inlet and 4^n outlets, where n is the number of branching generations. Fractal geometry has been applied to fuel cell components (flow field / current collector) in the past; however, this design is fundamentally different, as it is a truly three-dimensional branching structure as compared with previously reported two-dimensional designs[3,4].

Simulations of this gas distributor geometry have been conducted using COMSOL Multiphysics 3.5a. The simulation results provide insight into the necessary spacing between adjacent outlets to deliver a uniform concentration of oxygen across the catalyst layer interface.

Three gas distributor prototypes were fabricated, each representing an increased number of generations (more tightly packed outlets). The prototypes were fabricated via sterolithography (one resin, two in stainless steel) and tested against a standard single serpentine channel. Results indicate that the higher generation fractal prototype outperforms the serpentine flow field when operating at elevated current densities (above 0.90 A/cm^2) with air as the cathode gas. Enhanced performance is attributed to uniform distribution of reactant gases in the cathode catalyst layer.

References:

1) S. Kjelstrup, M.-O. Coppens, J. G. Pharoah, P. Pfeifer, *Energy & Fuels,* 24, 5097 (2010)

2) A. Su, Y. C. Chiu, F. B. Weng, Int. J. Energy Res., 29, 409-425 (2005)

3) K. Tuber, A. Oedegaard, M. Hermann, C. Hebling, *J. Power Sources, 131, 175-181 (2004)*

4) J.-Y. Chang, Y.D. Kuan, S.-M. Lee, S.-R. Lee, *J. Power Sources, 184,180-190 (2008)*

(712b) Optimizing Rotary Reactor Parameters to Achieve Higher Process Yield in Ex-Situ Oil Shale Processing

Golpour, H., Missouri University of Science and Technology
Smith, J. D., Idaho National Laboratory
Originally presented on: 11/7/2013 15:30:00 - 15:45:00

Abstract

To meet the increasing demand for liquid transportation fuels, development of unconventional oil and gas reservoirs has gained tremendous momentum in recent years. America's oil shale resources are highly concentrated in major deposits found in Colorado, Wyoming and Utah. Different *ex-situ* & *in-situ* methods have been proposed to extract oil from these unconventional resources. An In-direct heated rotating pyrolytic reactor is an *ex-situ* process that is the main focus of the work reported in this paper.

In-direct heating of White River basin oil shale uses gas-burners firing on a horizontal rotary reactor with steam feed to accomplish high extraction efficiency with recovery and distillation at the end to produce syn-crude. The present work focuses on optimizing operational parameters including bed temperature profile and rotation speed to increase syn-crude yield and reduce energy requirements. Test results identified the critical bed temperature below which process yields significantly decreased (below 80%) and above which mineral components in the shale decomposed to produce excessive CO2 emissions. Changing reactor rotation speed affected process residence time which was identified as a key factor in transferring heat into the solids bed through bed mixing and changing residence time. Although, increasing the external heat supply could compensate for reduced processing time, increased local bed temperature resulting in increased CO2 emissions limited this option. Optimization of bed mixing, external heat source and reactor rotation speed were the key factors examined in this work. Results include optimized values for the key operational parameters for this indirect heated process to convert oil-shale to syn-crude.

(712c) Impact of Internals Size and Configuration On Bubble Dynamics in Bubble Columns

Kagumba, M., Missuori S&T
Al Dahhan, M. H., Missouri University of Science and Technology
Originally presented on: 11/7/2013 15:45:00 - 16:00:00

Slurry bubble columns have been used in wide range of applications in industry such as the Fisher-Tropsch synthesis and Liquid phase methanol synthesis among others. The effects of dense internals with different diameters and covering the same cross-sectional area and hence configuration on bubble properties are lacking in the open literature. Therefore the focus of this study is to investigate the effects of dense internals with different diameters and covering the same cross-sectional area and hence configuration on the bubble dynamics such as; specific interfacial area, local and overall gas hold-up, axial bubble velocity, bubble passage frequency and bubble chord lengths using four-points optical probe. The experimental work was carried out in 0.14 m diameter Plexiglas bubble column for an air-water system. The superficial gas velocities applied based on both total cross-sectional area as well as free cross-sectional area available for the flow were in the range of 0.03 to 0.45 m/s covering the bubbly flow regime through churn-turbulent flow regime. The internals used were both Plexiglas rods of 0.5-inch diameter and 1-inch diameter occupying 25 % of column cross- section area. The experimental data obtained suggest that 0.5-inch internals gave consistently higher overall and local gas holdup, specific interfacial area, and bubble passage frequency than the 1-inch internals or empty column particularly at the column center. The variation was insignificant for the gas holdup, particularly in the churn turbulent flow regime. While lower axial bubble velocity was obtained with the smaller diameter internals. Details of the results will be discussed during the presentation.

(712d) A New Closed-Loop Relay System for Oil Sand Extraction

Hong, A., University of Utah
Zhao, X., University of Utah
Originally presented on: 11/7/2013 16:00:00 - 16:15:00

Hot water extraction of bitumen from oil sands with variations in operation has been used in Canada for decades. Water-based oil sand extraction is prone to several disadvantages, such as high demand of water, large volume of tailing that are difficult to dewater and disposed of, low extraction efficiency on low-grade oil sands, etc. Solvent-based oil sand extraction can potentially eliminate these issues yet achieving good bitumen recovery, though some with its own drawbacks. We have developed a new, closed loop method for oil sand extraction. Laboratory experiments show that the closed loop oil sand extraction relay (CLOSER) technology uses a solvent system that inputs only oil sand and outputs only bitumen and clean spent sand. The technology uses no water, resulting in no water to treat or dispose of while the solvents are completely recycled and reused in the process. At the conference presentation, we will report the operation conditions and extraction results, and discuss the principle that enables this new technique. The technology provides a new, efficient way to develop unconventional fuel, particularly for Utah's oil sands that present a great challenge for existing methods; the new technique promises a more environmentally sound and sustainable practice of energy development.

(712e) Gas Production From the Scoop

Long, R. Jr., NMSU

Originally presented on: 11/7/2013 16:15:00 - 16:30:00

Gas Production from the SCOOP(South Central Oklahoma Oil Play) is of recent interest.The opportunity to provide CAPEX to the development is sometimes beneficial. And new technology can be useful in improved recovery. Horizontal drilling and hydraulic fracking are important areas of technology in the SCOOP. The methodology is now established.

(712f) Steam Gasification of Glycerin Using a Entrained Flow Gasifier

Pearson, L., Mississippi State University
Toghiani, H., Mississippi State University
Originally presented on: 11/7/2013 16:30:00 - 16:45:00

Glycerin was used as a feedstock in an existing entrained flow gasifier located at the Institute for Clean Energy Technology at Mississippi State University. This gasifier had previously been used to gasify conventional feedstocks such as sawdust and non-conventional feedstocks such as chicken litter. Preheated glycerin was introduced into the gasifier through a spray nozzle, which discharged the glycerin into the incoming superheated steam flow. The pilot scale gasifer provides for processing of more than 100 lbs glycerin/hr to produce synthesis gas. The synthesis gas composition can be adjusted based on the steam/glycerin ratio and processing temperature. Experimental results will be presented.

(712g) Cascade Utilization of Fuel Gas Energy in Gtl Plant

Deng, S., Hatch
Hynes, R.,
Vatanakul, M., Hatch
Originally presented on: 11/7/2013 16:45:00 - 17:00:00

This paper investigates on a GTL plant with ATR syngas production and proposes a new process to use GT/WHRGSP to replace the fired heater. The new process features cascade utilization of fuel gas energy, as fuel gas is firstly used in a GT at very high temperature and then lower-temperature GT exhaust gas is further used for preheating. Large exergy loss of heat transfer in the fired heater is eliminated. The improved process has an equivalent power generation efficiency of 80% which is significantly higher than conventional technology. Economic analysis indicates 129.8 M$ revenue would be produced over the lifetime if the extra power from a 15,000 bbl/d GTL plant can be exported to the grid at the price of COE for a conventional NGCC plant. The new process proposed in this paper would be economically attractive when a GTL plant is based on ATR syngas production and exporting excess power is feasible.

(712h) Distribution and Diffusion of Sara Fractions in Bituminous Layer of Athabasca Oil Sands: Instrumental Measurements and Molecular Dynamic Simulation

He, L., Tianjin University
Wu, G.,
Sui, H., Tianjin University
Li, X., Tianjin University
Originally presented on: 11/7/2013 17:00:00 - 17:15:00

The bitumen liberation and recovery of oil sands have been well known affected, to some extent, by the bitumen physical properties, especially the surface properties, such as the interfacial tensions. Previous research showed that the interfacial tension of bitumen-water system was dependent on the composition of the bitumen (Mehrotra et al., 1985). Therefore, the understanding of the composition of the surface materials on the sand grains is important and necessary for the oil sands processing and technology improvement. In this study, a conceptual model was proposed to illustrate the composition and distribution of SARA (saturate, aromatic, resin and asphaltenes) fractions in the bituminous layer on the surface of Athabasca oil sands, which was verified by elemental analysis, X-ray photoelectron spectroscopy, field emission scanning electron microscopy with energy dispersive spectrometer, and Fourier-transformed infrared spectrometer. The contents of S, N, and the ratios of C/S and C/N were characterized as potential indicators for evaluating the distribution of SARA fractions in the bituminous layer. The results indicated that the light fractions such as saturates tend to deposit at the outer bituminous layer, while the asphaltenes and resins were inclined to distribute at the inner layer. It is also suggested that the distribution of SARA fractions was thermodynamic dependent and was susceptible to thermal treatment. These findings came to support our simulation results with molecular dynamic (MD) simulation (Wu et al., 2013). Results of simulation showed the sorption of saturates on quartz decreased by 31% when temperature increased from 298 to 398 K while opposite trend was observed for resins, but insignificant changes were found in asphaltenes and aromatics. It was found that the main contribution of interactions was van der Waals energy (>90%) irrespective of molecular components and temperatures. The diffusion coefficient of saturates was predicted as 10.8×10^{-10} m^2s^{-1} while that of the remaining fractions was about $4\times10^{-10}m^2s^{-1}$at 298 K. This work provides experimental evidences and simulation supports for validating the distribution patterns of SARA fractions on the surface of Athabasca oil sands. The proposed distribution model is expected to serve a basis for future studies on the liberation of bitumen from oil sands and the operation conditions for oil sands processing, especially for the solvent extraction and water-based extraction with chemical addition prior to the extraction. Also, it will provide insights to understand the oil remediation in oil fields or oil polluted fields.

(712i) Direct Cracking of Raw Oil Sands Bitumen - A New Approach

Ng, S. H., CanmetENERGY-Devon

Wang, J., CANMET Energy Technology Centre

Ding, F.,

Zheng, Y., University of New Brunswick

Wei, Q., LANSCE-LC

Yui, S. M., Edmonton R & D Centre

Originally presented on: 11/7/2013 17:15:00 - 17:30:00

Abstract

The objective of this study is to determine the viability of direct fluid catalytic cracking (FCC) of the resids-free portion of the raw Athabasca oil sands bitumen in the presence of the easily-cracked synthetic crude oil (SCO) after removing its light 343°C? fraction. A 343?525°C cut heavy gas oil (HGO) was prepared by distilling a vacuum topped bitumen from the vacuum distillation unit at Syncrude commercial upgrading plant[1] (Figure 1) in northern Alberta, Canada. It was then blended with various amounts of 343°C+ fraction of commercial SCO, called Syncrude Sweet Premium (SSP). Five HGO blends were prepared containing 100, 75, 64, 61, and 48v% of SCO HGO. The two blends with 64 and 48v% were referred to here the synsynbit and synbit HGOs, respectively. An earlier marketability study[2] on oil sands products in Asian countries predicted that (1) synsynbit, blended to approximate the quality of Middle East light sour crudes, would have the largest exporting volume and diversity of marketability, compared with those for the sweet SCO, and (2) Asian refineries would be able to accept synsynbit without any major modifications of the plant structure.

Each HGO blend was catalytically cracked at 500 and 520°C using a bench-scale Advanced Cracking Evaluation (ACE) unit. The results showed the acceptable performance of bitumen HGO when adequate amount of SCO HGO was added. The conversion, defined as the summation of dry gas (H2, H2S, C1?C2), liquefied petroleum gas (LPG, i.e., C3+C4), gasoline (C5?221°C), and coke, and selectivities of gasoline, light cycle oil (LCO, a diesel pool component, 221?343°C), and coke were either improved or little affected, in comparison with those of 100% SCO HGO used as a bench mark (Table 1). In-house analyses supported by those reported in the literature[3] showed that the bitumen HGO, which accounted for 40.6wt% vacuum topped bitumen, actually contained less nitrogen than its counterpart and had relatively low Ni+V and CCR. These imposed no serious threat to the cracking catalysts which were vulnerable to these poisons. Perhaps, the major challenge was the mild increase in the dry gas yield notably due to the augmentation of H2S, and higher sulfur levels in gasoline and diesel which might need desulfurization prior to being sent to gasoline and diesel pools. As sulfur species in bitumen HGO exist mostly as sulfides, mercaptans, and small thiophenes, they should be easily removed by mild hydrotreating or other low cost process.

In summary, the proposed route provided an interesting and promising option to directly crack the raw bitumen leading to a substantial saving in energy and reduction in green house gases by eliminating the costly upgrading of the bitumen HGO.

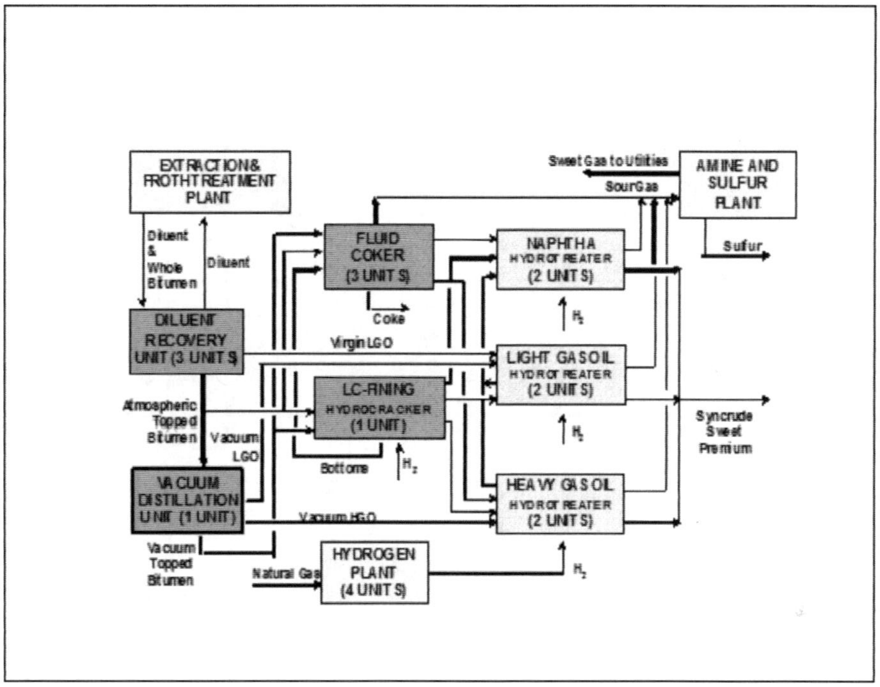

Figure 1. Syncrude upgrading process flow

Table 1. Product distributions at 72 wt% conversion for HGO blends

	v% SSP in Blend	Yield, wt%						Delta Coke[b] wt%
		Dry Gas	LPG	Gasoline	LCO	HCO[a]	Coke	
500	100	1.7	15.6	48.2	16.6	11.4	6.4	0.63
520		2.2	16.7	47.3	16.0	12.0	5.9	0.78
500	75	2.2	14.7	48.7	17.2	10.8	6.3	0.70
520		2.7	15.8	48.2	16.5	11.5	5.8	0.82
500	64 (Syn synbit)	2.6	15.4	46.0	17.6	10.4	7.8	0.73-
520		3.1	16.4	45.3	17.2	10.8	7.3	0.80
500	61	2.7	15.4	45.3	17.3	10.7	8.9	0.81-
520		3.3	16.3	43.1	17.2	11.1	8.5	0.80-
500	48 (Synbit)	2.9	16.0	45.3	18.1	10.4	8.9	0.78-
520		3.4	17.1	43.4	17.5	10.7	8.8	0.81-

[a] heavy cycle oil, sometimes used as heavy fuel oil

[b] the amount of coke on catalyst **References**

[1] Yui, S., Chung, K. ***?Syncrude Upgrader Revamp Improves Product Quality?***, Oil & Gas Journal, December 17, 2007, pp. 52?59.

[2] du Plessis, D., Wu, N. ***?Marketability of Oil Sands Products in Asian Countries?***, presentation to HUTF, June 20, 2007.

[3] Chung, K.H., Xu, C.M., Hu, Y., Wang, R. ***?Supercritical Fluid Extraction Reveals Resid Properties?***, Oil & Gas Journal, January 20, 1999, pp. 66?69.

(712j) Modeling of Plant Simulation of MRC Plant Under Variable Working Conditions and Operation Optimization

Sun, H., China University of Petroleum (Beijing)
Yan, F., China University of Petroleum (Beijing)
Liu, F., China University of Petroleum (Beijing)
Originally presented on: 11/7/2013 17:30:00 - 17:45:00

A LNG plant with MRC cycle would not work under the designed condition all the time. It is significant to know the performance of a fixed MRC plant working at changed working condition. The problem involves the problem of plant simulation, which is somewhat similar to but actually different with the process simulation which is used in the design of a LNG plant. A model of MRC plant simulation was established and brief comparison between the two kinds of simulations was made. Here, the performance of main devices at different working conditions is simulated using fitting method and the rest is regarded as constant. The calculation of devices in the MRC process is connected using the HYSYS soft so as the entire plant simulation at different conditions could be carried out.

Two kinds of MRC plant of LNG are studied. One kind is the LNG plant with SMR cycle which generally used in small-scale plant. The other is the plant with C3-MR process which often used in large base-load plant is studied. The follow cases, including ambient temperature variations, feed gas pressure variations and load variations, are simulated and evaluated. The adaptabilities of the two LNG processes are analysis and compared. The results show the SMR cycle has good adaptability under changed working condition. Since C3-MR process is more sophisticated and has more devices, the effect of changed working conditions on the performance of the LNG plant is larger. The results could be referred in the selection of liquefaction process in medium-scale LNG plant.

To reduce the energy consumption, it is necessary for LNG plant operation to adapt the mixture refrigerant composition and other process parameters to the changed working conditions for LNG plant. The operation optimization is a reasonable way to obtain the object. The operation optimization is also rather different with the design optimization. The operation optimization should be conducted on the basis of plant simulations. Here, the object function is to minimize the operation power consumption and corresponding constraint condition of operation should be made. The BOX method is used to solve the optimization problem. The operation optimizations of SMR plant and C3-MR plant are done under the same condition of changed ambient temperature. The results show that the obvious energy saving could be achieved for the both kinds of plants, although the total energy consumptions is still higher than that of design optimizations. The SMR plant also shows better performance at changed working conditions than that of C3-MR plant. The work could be referred for actual operation of a MRC plant for variable working conditions.

(722a) Water-Gas Shift Catalysis Over Supported Platinum Nanoparticles

Shekhar, M., Dow Chemical
Lu, J., Argonne National Laboratory
Akatay, M. C., Purdue University
Elam, J., Argonne National Laboratory
Miller, J. T., Argonne National Laboratory
Delgass, W. N., Purdue University
Ribeiro, F. H., Purdue University
Originally presented on: 11/7/2013 15:15:00 - 15:35:00

The water-gas shift (WGS) reaction ($CO + H_2O \rightarrow CO_2 + H_2$) is an important industrial chemical process for hydrogen production. Supported Pt catalysts have turnover rates (TOR's) comparable to the industrial $Cu/ZnO/Al_2O_3$ catalysts but are more robust.

The WGS TOR's, normalized by surface Pt atoms, for Pt/Al_2O_3, Pt/SiO_2, Pt/ZrO_2, and Pt/TiO_2 catalysts are independent of average Pt particle size. The apparent reaction orders for Pt catalysts do not vary with particle size. Thus, all surface Pt atoms exhibit the same rate.

The WGS TOR and H_2O order (in parenthesis) vary as Pt/Al_2O_3 (0.93) ~ Pt/SiO_2 (0.84) < Pt/TiO_2 ~ Pt/ZrO_2 ~ Pt/CeO_2 (0.66-0.72) at 300 °C. The CO, CO_2 and H_2 orders do not vary with the support. Further, the TORs for Pt/ZrO_2 and Pt/TiO_2 catalysts decrease by 125 and 10 times their original TOR, respectively, upon addition of 70 atomic layer deposition (ALD) cycles, ~ 75 wt.% Pt. This is due to excess Pt coverage that limits the number of support sites, shown from their transmission electron microscopy (TEM) images, available for H_2O activation. We interpret these data to show that the support plays a direct role in activating H_2O.

Lastly, the WGS TOR for 70 ALD cycles 76.4wt.%Pt/np.ZrO_2 catalyst is 0.004 s^{-1} at 300 °C, 10 times less than that for 1.6wt.%Pt/Al_2O_3 catalysts. Thus, the WGS TOR for the unsupported Pt surface is at least 10 times less than Pt/Al_2O_3, implying that the assumption that Pt/Al_2O_3 can be modeled as just unsupported Pt surface may not be true.

(722b) A Kinetic Study On the Structural and Functional Roles of Lanthana in Iron-Based Catalysts for High-Temperature Water-Gas Shift Reaction

Hallac, B., Brigham Young University
Argyle, M. D., Brigham Young University
Brown, J. C., Brigham Young University
Originally presented on: 11/7/2013 15:35:00 - 15:55:00

Hydrogen is an invaluable highly-reactive chemical that is used in numerous industrial processes, including ammonia synthesis, fuel cells, and hydrocarbon cracking. This study focuses on the effects of lanthana addition to iron-based catalysts (8 wt% Cr_2O_3, 4 wt% CuO, (88-x) wt% Fe_2O_3, x wt% La_2O_3) used to catalyze the high-temperature water-gas shift reaction for a better hydrogen yield. Lanthana is well known for its thermal stability and carries +3 charge in its more stable oxide form. Therefore, La^{3+} can integrate into the iron-chromium spinel crystal structure as it replaces some Fe^{3+} cations to enhance the structural and functional properties of the catalyst.

Different batches of catalysts were prepared *via* co-precipitation with varying wt% of lanthana (0, 0.5, 1, 2, and 5) added at the expense of the Fe_2O_3 phase. They were tested for water-gas shift activity at four different reaction temperatures (350, 375, 400, and 425°C). The results show that addition of 0.5 wt% lanthana increased the water-gas shift activity from 27 to 31 mmol of CO/ g_{cat}·min and reduced the deactivation rate of the catalyst from 23% to 11% at 400°C after 10 days on-stream when compared to the catalyst with no lanthana in it. Compared to literature, this catalyst performed more actively at 425°C than chromia-free catalysts operating at 550°C.Further additions of lanthana (>0.5 wt%) appeared to impose a negative effect on the activity and stability of the catalysts which becomes worse at higher temperatures. According to characterization by X-ray diffraction, the enhanced catalytic stability with the addition of 0.5 wt% lanthana is due to the stabilization of the iron-chromium spinel structure. Further loadings of lanthana (> 0.5 wt%) apparently disrupt the spinel structure due to the larger size of La^{3+} compared to Fe^{3+}. BET surface area measurements show that the most stable spent catalyst with 0.5 wt% lanthana had the highest surface area with 67 m^2/g_{cat} compared to 40 m^2/g for the catalyst with no lanthana. Scanning electron micrographs of the spent catalysts confirm that this catalyst is the most physically stable with the smoothest surface and least cracks. Temperature-programmed reduction with hydrogen shows that 0.5 wt% lanthana facilitates the reduction of Fe_2O_3 to the active iron oxide phase, Fe_3O_4, by lowering the reduction temperature 15°C.

The kinetic data were fit to three rate law models (power-law, Langmuir-Hinshelwood, and redox) to acquire a mechanistic insight that reasons how the addition of lanthana affects the functionality of the catalysts and makes the catalyst with 0.5 wt% lanthana the most active catalyst. Statistical analysis results showed that the power-law model provides the best fit and that the reaction follows an absorptive mechanism that is well explained by a Langmuir-Hinshelwood rate model with CO adsorption being the rate-determining step. 95% joint confidence regions were obtained for the fitting parameters (adsorption equilibrium constants for water and CO) to show that the adsorption equilibrium constant for CO is largest for the catalyst with 0.5 wt% lanthana (940 L/mol), indicating that the adsorption of CO is facilitated by this minor addition of lanthana. Furthermore, the modeling showed that water adsorbs more strongly on the catalyst surface than CO and therefore inhibits the reaction from taking place.

In summary, 0.5 wt% addition of lanthana increases the activity and stability of iron-based catalysts for high-temperature water-gas shift reaction by stabilizing the iron-chromium spinel structure and facilitating the adsorption of CO on the catalyst surface.

(722c) Flame Spray Pyrolysis Synthesized Cu-CeO2 Catalysts for Low Temperature WGS Reaction

Gunugunuri, K. R., University of Cincinnati
Inturi, S. N. R., University of Cincinnati
Smirniotis, P., University of Cincinnati
Originally presented on: 11/7/2013 15:55:00 - 16:15:00

Recently, the water-gas shift (WGS) reaction has attracted renewed attention because of production of high purity H_2 in conjunction with fuel cell power generation. Cu is the active metal for the low temperature WGS reaction. The main problem with the current catalyst formulations is that they are sensitive to even small amounts of sulfur. So far several wet synthesis methods has been reported for the preparation of Cu based catalysts for the low temperature WGS reaction. Recently, the synthesis of catalysts using flame spray pyrolysis method become attractive since it is a rather inexpensive route for large-scale production of thermally stable, high-surface area nano-materials. The FSP technique converts precursor droplets into solid nano-particles in flames and has been widely employed for the large-scale production of carbon black, while vapor-fed flames have been used for synthesis of several catalysts. In this way, high-purity materials with novel meta-stable phases are made that are not accessible by conventional wet chemical and solid-state processes. Particularly, FSP allows for the controlled synthesis of nano-particles with high specific surface area (SSA), which resulted from the small size of nano-particles, and well-defined chemical composition, as recently demonstrated on both a laboratory and pilot scale. The present study mainly aimed at synthesis of sulfur tolerant Cu-CeO_2 catalysts for the low temperature WGS reaction using Flame Spray pyrolysis synthesis method. Various ratios Cu-CeO_2 solid solutions have been synthesized to find out the optimum ratio between Cu and CeO_2. These catalysts were characterized using XRD, TPR, BET, N_2O chemisorption, TEM and XPS techniques. The catalysts were evaluated for low temperature WGS reaction in the presence and absence of H_2S. The TPR analysis showed the presence of Cu promoted the reduction of CeO_2 to much lower temperatures indicating strong interaction between Cu and Ce. Catalytic activity measurements suggest that our FSP synthesized catalyst exhibits better activity than co-precipitated synthesized catalyst. These interesting findings will be discussed in the presentation.

(722d) On the Importance of Metal-Oxide Interface Sites for the Water-Gas Shift Reaction Over Pt/CeO2 Catalysts

Heyden, A., University of South Carolina
Aranifard, S., University of South Carolina
Ammal, S. C., University of South Carolina
Originally presented on: 11/7/2013 16:15:00 - 16:35:00

The objective of this work is to understand the unique catalysis and importance of the three-phase boundary (TPB) of CeO_2 supported Pt catalysts for the water-gas shift (WGS) reaction. The mechanism of the WGS at the Pt/CeO_2 interface has been investigated using plane wave DFT calculations by considering the following pathways: (i) redox pathway, (ii) associative carboxyl pathway, and (iii) associative carboxyl pathway with redox regeneration. Analysis of a microkinetic model containing all these pathways indicates that the CO promoted carboxyl pathway with redox regeneration is the most favorable pathway with the lowest activation barrier and highest rate. The calculated apparent activation barrier and reaction orders for this pathway are in excellent agreement with experimental values from three independent research groups. In general, this study reveals that the Pt sites away from the TPB likely act primarily as a CO reservoir and the oxide support plays an essential role in O-H bond breaking and H-atom transfer to the metal surface. Overall, catalysis seems to be dominated by interface sites.

(722e) Theoretical Analysis of Formic Acid Decomposition On Transition-Metal Catalysts

Yoo, J. S., Stanford University
Studt, F., SLAC National Accelerator Laboratory
Abild-Pedersen, F., Technical University of Denmark
Norskov, J. K., Technical University of Denmark
Originally presented on: 11/7/2013 16:35:00 - 16:55:00

Hydrogen is a promising energy carrier that can be used to generate electricity in fuel cells without pollution. To implement this fuel cell based hydrogen economy, however, practical solutions to controlled storage and release of hydrogen are essential. Previously, hydrogen storage has been achieved by compressing the gas at high pressures, liquefying it at low temperatures, and storing it in the form of chemical/metal hydrides. All these methods, however, suffer from loss of hydrogen, safety issues, and low energy densities.

Recently, formic acid has been suggested as a sustainable material for hydrogen storage. It contains 4.4 wt.% of hydrogen with a volumetric capacity of 53.4 g/l at STP. Even though its hydrogen content falls short of the milestones (5.5 wt.%) set by the US Department of Energy for 2010, its volumetric capacity surpasses that of most other storage materials used today. Hydrogen stored in formic acid can be released on demand by decomposing formic acid into H_2 and CO_2. If we can produce formic acid again by CO_2 hydrogenation, a carbon-neutral hydrogen storage cycle would be established on the globe. To develop formic acid as a hydrogen storage material, however, we first need to find catalyst materials that can actively decompose formic acid into H_2 and CO_2.

Previous studies have shown that some elements of noble metals were highly active for formic acid decomposition. For example, Solymosi and co-workers investigated formic acid decomposition over a range of monometallic noble metal catalysts (Ir, Pd, Pt, Ru or Rh supported on carbon) at 423 K. All their catalysts showed promising activity with turnover numbers of 1280-4120 hydrogen molecules per site per second. When the performance of different catalysts was compared, the catalytic activity trends of Ir ≈ Pt ≈ Pd > Ru > Rh were observed. However, these trends may not reflect the intrinsic catalytic properties of the noble metals because the metal dispersion of the catalysts was different ranging from 5.6% to 26.1%.

For direct applications in fuel cells, formic acid needs to be decomposed via dehydrogenation (HCOOH --> $H_2 + CO_2$) rather than via dehydration (HCOOH --> $H_2O + CO$) such that virtually CO-free hydrogen is produced. Many researchers have tried to achieve this by adding a secondary metal to some noble metals. The improvements in hydrogen selectivity as well as catalytic activity were particularly significant on Pd-Au and Pd-Ag catalysts that produced CO below 100 ppm under ambient conditions. However, these catalysts are heavily based on expensive rare-earth elements. We need to find alternative materials that are not only active and selective for formic acid dehydrogenation but also cost-effective.

First principles based analysis was used to investigate formic acid as a sustainable material for hydrogen storage. The reaction mechanism and the energetics of formic acid decomposition were obtained based on the adsorption energies of the reaction species calculated over five (211) surfaces: Ag(211), Cu(211), Pd(211), Pt(211) and Rh(211). Formic acid decomposition was found to proceed via dehydrogenation rather than via dehydration on the surfaces we investigated, although the minimum energy pathway varied depending on the surface. The DFT-calculated adsorption energies decreased significantly for some species adsorbed on Cu(211) when the BEEF-vdW functional was used instead of the RPBE functional. This was interpreted as vdW forces playing an important role in the interaction of the species with the metal surface.

We then extended the study to other transition-metal surfaces by scaling the vdW-included adsorption energies of the reaction species with CO and OH adsorption energies (E_{CO} and E_{OH}). The scaling relations for individual species were independent of the surface structure, allowing us to estimate the adsorption energies of the reaction species adsorbed on other surfaces by calculating two parameters per surface: E_{CO} and E_{OH}. A micro-kinetic model was used to construct a theoretical activity volcano that mapped out the turnover frequencies for hydrogen production from formic acid as a function of E_{CO} and E_{OH}. Hydrogen and CO selectivity maps were also established as a function of E_{CO} and E_{OH}. The catalytic activity and selectivity trends obtained from our first-principles analysis were consistent with many experimental observations reported in the literature. The interpolation concept was used to quickly search for binary alloys that can actively and selectively decompose formic acid into H_2 and CO_2. As a result, Cu_3Ni alloy was newly identified to be interesting. We hope to find even more interesting catalyst materials when we add more data points to the volcanoes presented in this study.

(722f) TiO2/Au Nanocomposites for Photocatalytic Water Splitting: Understanding the Mechanisms By Which Au Impacts Activity

Przybylek, M., University of Michigan
Linic, S., University of Michigan
Originally presented on: 11/7/2013 16:55:00 - 17:15:00

We aim to understand the mechanisms by which Au impacts the water-splitting activity of TiO_2. Although TiO_2 is one of the most widely studied materials for photocatalytic water splitting, it has two major drawbacks— slow oxygen evolution kinetics and the inability to absorb visible light. The addition of Au to TiO_2 may create a composite catalyst with increased activity compared to TiO_2 alone. This is because Au has strong absorbance in the visible range from localized surface plasmon resonance (LSPR). Recent studies have shown that LSPR can spatially localize the creation of electron holes as well as impart visible-light activity when paired with a photocatalyst that is unable to absorb visible light. By spatially localizing the creation of holes, the Au may facilitate the oxygen evolution reaction by increasing the chances that an adsorbate encounters a hole and proceeds down the reaction pathway. To investigate which effects derive from LSPR and which are catalytic effects of the Au, we run the water-splitting reaction under different light conditions (UV-only, visible-only, and UV-visible). Since Au only exhibits LSPR in the visible region, we determine the effect that adding Au nanoparticles has on TiO_2's absorbance and kinetics resulting from both LSPR and non-LSPR mechanisms.

(722g) Dendritic Au/TiO2 Nanorod Arrays for Visible-Light Driven Photoeletrochemical Water Splitting

Gong, J., Tianjin University
Su, F.,
Zhang, J.,
Wang, T., Tianjin University
Zhang, P.,
Originally presented on: 11/7/2013 17:15:00 - 17:35:00

In this talk, I will describe the rational design and synthesis of TiO_2 branched nanorod arrays (TiO_2 BNRs) with plasmonic Au nanoparticles attached on the surface. Such Au/TiO_2 BNR composites exhibit high photocatalytic activity in photoelectrochemical (PEC) water splitting. The unique structure of Au/TiO_2 BNRs shows enhanced activity with a photocurrent of 0.125 mA/cm^2 under visible light (≥420 nm) and 2.25 mA/cm^2 under AM 1.5 G illumination (100 mW/cm^2), which is significantly higher and comparable to the highest value ever reported. Furthermore, the Au/TiO_2 BNRs achieve the highest efficiency of ~1.27% at a low bias of 0.42 V vs RHE, indicating elevated charge separation and transportation efficiencies. The superior PEC performance is mainly due to the plasmonic effect of Au nanoparticles, which enhances the visible light absorption, together with the large surface area, efficient charge separation and high carrier mobility of the TiO_2 BNRs. The carrier density of Au/TiO_2 BNRs is nearly 6 times higher than the pristine TiO_2 BNRs as calculated by the Mott-Schokkty plot. Based on the analysis of UV-Vis spectroscopy, electrochemical impedance spectroscopy, and photoluminescence, a mechanism was proposed to explain the high activity of Au/TiO_2 BNRs in PEC water splitting. The capability of synthesizing highly visible light active Au/TiO_2 BNR based photocatalyst are useful for solar conversion applications, such as PEC water splitting, dye-sensitized solar cells and photovoltaic devices.

(768a) Hot Water Flowthrough Pretreatment of Cellulosic Biomass

Archambault-Leger, V., Dartmouth College
Shao, X., Dartmouth College
Lynd, L. R., Dartmouth College

Originally presented on: 11/8/2013 8:30:00 - 8:45:00

Transportation is a key energy use sector which will likely need liquid fuels for a long time if not indefinitely. Cellulosic biomass is a promising source of liquid fuels for very large scale sustainable energy production. Agricultural residues, particularly sugarcane bagasse, are an important low hanging fruit source of cellulosic biomass worldwide. The main obstacle to achieve commercial, cost competitive biological conversion of cellulose to fuel is biomass recalcitrance, referring to the incomplete accessibility of attack by microbes and their saccharolytic enzymes. As employed in the biological conversion field, pretreatment refers to the key limiting process of preparing the biomass to overcome biomass recalcitrance. There are a wide variety of pretreatment processes, generally involving exposing biomass to a solvent at elevated temperature and pressure. Pretreatment decisively improves the subsequent yield of sugar hydrolysis, but accounts for a substantial fraction of the cost of overall conversion.

The conundrum of conventional pretreatment is that conditions severe enough to yield reactive fiber are accompanied by sugar degradation. We report that a compressed hot water flowthrough (FT) pretreatment allows 85% sugar recoveries compared to 40% with batch pretreatment at conditions severe enough to provide 90% glucan conversion. FT solubilizes all of the hemicelluloses and 75% of the lignin, while reducing the capital cost and use of chemicals compared with other pretreatment methods [1]. The improved hemicelluloses and lignin removal in a FT configuration is a result of liquid being removed from the reactor, allowing less time for solubilized sugars to degrade and minimizing recondensation of solubilized lignin and xylan on cellulose fibers upon cooling [2] . However, FT pretreatment is challenging to implement because it involves liquid flowing through solids at controlled flow rates, elevated pressure (250 PSI) and high solids concentration (~25-40% solids). The purpose of this study is to propose a hot water pretreatment configuration for sugarcane bagasse that realizes the advantages of FT pretreatment at scale, addresses the mechanical complexities of arranging a bed of biomass for FT configuration and avoids unacceptably high energy requirements and sugar dilution. We evaluated the time and temperature effect on key performance metrics and measured key physical properties—pressure drop, viscous compaction and water retention. We carefully analyze the fluid dynamics of fluid flow through biomass and we use kinetic and process modeling to investigate several configurations—batch, co-, counter- and cross-flow.

Compared to batch pretreatment, FT pretreatment consistently resulted in higher hemicelluloses recovery, higher removal of non-carbohydrate carbon and higher glucan solubilization by simultaneous saccharification and fermentation (SSF). Hemicelluloses removal trends were similar between feedstocks whereas glucan conversion trends were significantly different, suggesting that factors in addition to hemicelluloses removal impact amenability of glucan to enzymatic attack. Corn stover exhibited higher hydrolysis yields than bagasse or poplar, which could be due to higher removal of non-carbohydrate carbon. Hemicelluloses were more easily degraded in bagasse than in corn stover and poplar. The observed pressure drop using biomass knife-milled to pass through a 2 mm sieve was highest for bagasse, intermediate for switchgrass, and lowest for poplar. The highest pressure drop correlated with the presence of more fine particles, greater viscous compaction of the biomass and the degree of water absorption. Grasses and agricultural residues are mechanically more challenging than wood, absorbing more water, compressing more easily and containing smaller particle sizes. The process and kinetic model indicate what solids concentration and configurations will yield acceptable energy consumption and sugar dilution.

1. Mosier N, Wyman CE, Dale BE, Elander R, Lee YY, Holtzapple M, Ladisch M: **Features of promising technologies for pretreatment of lignocellulosic biomass.** *Bioresource Technology* 2005, **96:**673-686.

2. Liu C, Wyman CE: **The effect of flow rate of very dilute sulfuric acid on xylan, lignin, and total mass removal from corn stover**. *Ind Eng Chem Res* 2004, **43**:2781-2788.

(768b) Visualization Studies of the Evaporation and Combustion of Alternative Liquid Fuels At Elevated Temperatures

Venkatasubramanian, V., Auburn University
Duan, J., Auburn University
Giles, S. A., Auburn University
Duke, S. R., Auburn University
Originally presented on: 11/8/2013 8:45:00 - 9:00:00

A number of liquids offer potential as alternative energy sources for high temperature energy conversion. Glycerin and certain waste oils were chosen as liquid fuels for this study to determine their viability as supplements for the energy needs of a cement plant. Visualization studies were carried out in a laboratory scale drop tube furnace system and an infrared heater to analyze the behavior, degradation, and combustion of these alternative liquid fuels. Techniques using high speed cameras and high magnification lenses are discussed for studies with suspended and free-falling droplets in the environments where the temperature and the gas composition are precisely controlled. Suspended droplets studies were conducted using micron sized graphite capillary tubes and thin wire thermocouples. Free-falling micron-sized droplets were generated using a custom piezoelectric drop-on-demand generator. High resolution visualization images and video showing the interaction of alternative liquid fuel droplets at elevated temperatures are presented. Results are used to estimates rates and mechanisms for evaporation and combustion of the liquids at specified gas atmospheres and high temperatures, particularly those that simulate energy generation in cement production. The goal is to use the results to evaluate co-firing options for coal and other traditional fossil fuels.

(768c) Continuous Vacuum Drying of Wood Chips for Residential/Commercial/Industrial Heating Applications

Marquis, J., Rensselaer Polytechnic Institute
Bequette, B. W., Rensselaer Polytechnic Institute
Originally presented on: 11/8/2013 9:00:00 - 9:15:00

AIChE Annual Meeting – Minneapolis 2013

Title: Continuous Vacuum Drying of Wood Chips for Residential/Commercial/Industrial Heating Applications

Authors: Jeffrey A. Marquis, B. Wayne Bequette, John E. Okonski

Abstract: Wood Chips are an abundant, renewable and affordable fuel source throughout the northeast United States. Decreasing the moisture content from approximately 50 wt-% to 20 wt-% nearly doubles the heating value per unit mass of final product[1]. Dried woodchips can be burned in high efficiency woodchip boilers ranging in size from 15 kW to 500 kW[2]. This work examines an unconventional continuous screw conveyor vacuum drying system. The decreased boiling point of water at elevated vacuum levels is exploited to expedite the water removal rate.

Existing technologies for wood chip drying such as air drying, kiln drying and rotary drying are slow (taking up to months of processing) expensive, and do not scale well[2,3]. The processes examined allows for the fast and efficient removal of excess moisture from wood chips. A series of bench-scale batch experiments provided the necessary insight into required design parameters for a pilot-scale system: Wall temperature, T_{wall}, vacuum pressure, P_{vac}, and residence time, τ. The resulting design parameters were $T_{wall} = 90°C$, $P_{vac} = 28$ inHg, $\tau = 15$ minutes.

A 100 lb dry mass per hour pilot scale system has been designed and constructed based upon the insight gained from the bench-scale testing. It is expected that this pilot-scale system will allow for fast, efficient, and inexpensive drying of wood chips over competing technologies.

References:

1) T. M. Maker, Biomass Energy Resource Center, Montpelier VT (2011).

2) www.evo-world.com, May 2, 2013

3) G. Lianbai, Drying Technology, **25**, 463-469 (2007)

4) S. Danielsson, A. Rasmuson, Drying Technology, **20**, 1427-1444 (2002)

(768d) Effect of Storage Conditions On Extraction of Algal Carbohydrates and Oils for Biofuel Production

Hestekin, C. N., University of Arkansas
Jernigan, A. C., University of Arkansas
Potts, T., University of Arkansas
Beitle, R. R., University of Arkansas
Hestekin, J. A., University of Arkansas
Originally presented on: 11/8/2013 9:15:00 - 9:30:00

There are several environmentally sound reasons to use algae as a feedstock for biofuels. The fact that they are readily abundant, can absorb unwanted elements out of waste water, and can use carbon dioxide as a carbon source are just a few of the reasons that algae are being examined as one of the most likely medium to replace petroleum fuels. An issue of some concern is that microalgae have a high rate of growth in the summer and lower rates in the winter, which can cause a depletion of feedstock for a bioreactor. In order to facilitate localized bioreactors that can operate year-round, the extraction of lipids and carbohydrate from algae stored for approximately 3 months to 1 year. Samples were harvested from water at a wastewater treatment plant as part of a water remediation process. In our study, it was observed that samples could be stored for up to 200 days without appreciable decreases in carbohydrate content. In addition, processing variables including moisture content, concentration of algae, and concentration of sulfuric acid were examined in addition to the production of butanol.

(768e) Methane-Derived Biofuels: Choosing Among Options

Woolston, B., The Pennsylvania State University
Stephanopoulos, G., MIT
Originally presented on: 11/8/2013 9:30:00 - 9:45:00

It has been estimated that the proven natural gas reserves in the U.S. could provide enough energy to meet the country's transportation energy demand for the next 50 years [1]. This, coupled with the increasing price gap between natural gas and gasoline over recent years [2], makes natural-gas derived fuels an attractive option for use in the transportation sector. However, the infrastructure to use natural gas directly (either as CNG or LNG) has not been developed, and so natural gas must first be converted to a liquid fuel. Chemical processes for accomplishing this (e.g. GTL, based on the Fischer-Tropsch reaction) suffer from high capital costs and low yields, thus there is significant interest in developing alternative processes for the high-efficiency conversion of methane to a drop-in transportation fuel [3].

Various microorganisms have evolved enzyme systems capable of metabolizing methane, either consuming it (methanotrophs) or producing it (methanogens), thus it is conceivable that, with the application of metabolic engineering technologies, a strain capable of producing a liquid fuel from methane could be developed. Before embarking on such an undertaking, however, it is critical to establish the inherent limitations of the chosen pathway, both from a carbon yield and energy yield standpoint, to maximize the likelihood of selecting a commercially viable process, and not just a novel scientific investigation. To that end, we present here an analysis of the various biological routes for the conversion of methane to a liquid fuel. Each process is divided into two segments: the upstream conversion of methane to an activated intermediate (either aerobically, anaerobically, or catalytically), followed by the downstream synthesis of a fuel molecule using several alternative pathways (Serine cycle, RuMP pathway, Wood-Ljungdahl pathway). We develop stoichiometric models for each scenario that allow calculation of the maximum potential yields, and conclude with the implication of these results in selecting an organism and pathway for the conversion of methane to a liquid fuel.

[1] "America's Energy Future: Technology and Transformation" (National Academy of Sciences, 2009).

[2] U.S. Energy Information Administration, Annual Energy Outlook 2012 (EIA Publication, DOE/EIA-0383, 2012;

http://www.eia.gov/forecasts/archive/aeo12/index.cfm).

[3] "Reducing Emissions using Methanotrophic Organisms for Transportation Energy (REMOTE)" 2013. ARPA-E. March 2013. <https://arpa-e-foa.energy.gov/FileContent.aspx?FileID=261ed0b9-6695-4c51-a8a5-60f58b81e4f4>

(768f) Green Algae Photobioreactor

Machado, J. B. III, South Dakota School of mines and Tec.
Anderson, B., SDSM&T
Buse, T.,
Originally presented on: 11/8/2013 9:45:00 - 10:00:00

Current methods of mass algae production utilize large, open ponds that produce harvestable biomass slowly and take up vast amounts of land area, making it an unpractical alternative to petroleum. The purpose of this project is to design a mass algae production method that most efficiently uses land area while maximizing the production of biodiesel. By collecting and distributing sunlight into a vertical, flat panel photobioreactor, this goal can be obtained. With controlled micro light-dark cycles through the use of static mixers, optimum algae biomass production can be achieved. Through manipulations of nutrients and heterotrophic cycles, the production of lipids can be greatly increased. By manipulating these variables, an economically feasible photobioreactor that minimizes ground space can be achieved.

(768g) A Parametric Study of Light-Duty Natural Gas Vehicle Competitiveness in the United States Through 2050

Peterson, M. B.,
Barter, G. E., Sandia National Laboratories
West, T. H.,
Manley, D. K.,
Originally presented on: 11/8/2013 10:00:00 - 10:15:00

We modeled and conducted a parametric analysis of the US light-duty vehicle park to examine the impact of natural gas vehicles as they compete with electric vehicles, hybrids, and conventional internal combustion engines. We find that low natural gas prices and sufficient public refueling infrastructure are the key drivers to natural gas vehicle adoption when matched with compressed natural gas powertrains from automakers. Due to the time and investment required for the build out of infrastructure and the introduction of vehicles by original equipment manufacturers, home natural gas compressor sales and bi-fuel natural gas vehicles serve as bridge technologies through 2030. By 2050, however, natural gas vehicles could comprise as much as 20% of annual vehicle sales and 10% of the LDV park fraction. We also find that natural gas vehicles may displace electric vehicles, rather than conventional internal combustion engines, as they both compete for consumers that drive enough miles such that fuel cost savings offset higher purchase costs. Due to this dynamic, natural gas vehicles in the LDV park offer little to no greenhouse gas emissions reduction as they displace lower emission powertrains.

(768h) Diesel Fuel Property Data and Predictions Using Highly-Saturated Bio-Renewable Feedstocks in Diesel Fuel Blends

Duncan, A. M., University of Kansas
Scurto, A. M., University of Kansas
Stagg-Williams, S. M., University of Kansas
Originally presented on: 11/8/2013 10:15:00 - 10:30:00

The use and production of renewable fuels continues to increase globally as a method to both lower net carbon emissions and to reduce consumption of petroleum derived sources. In the United States, biomass-based fuels will be required to have a production volume of 1.28 billion gallons in 2013. The majority of these gallons will be combusted as biomass blends.

Many of the renewable feedstocks that have been, and will be potentially blended with petroleum diesel fuel are very high in saturated fats. Saturated fats are present in high amounts for both biodiesel and renewable diesel feedstocks like animal fat and some vegetal sources. Saturated components of equivalent chain length present advantages over unsaturated compounds with regards to the flash point, storage stability and energy content of fuels. However, several properties important to engine performance can be significantly affected when large concentrations of long-chained saturated components are present, specifically cold flow and viscous properties. High-pressure viscosities up to 131 MPa, cloud point temperatures, and approximate freezing point pressures for blends of petroleum diesel with jatropha, palm, and tallow biodiesels are presented. Renewable diesel cold flow and viscous properties are predicted.

(768i) Property Prediction for Diesel Fuels Based Upon Surrogates Composed of Real Components

Reiter, A., Graz University of Technology
Wallek, T., Graz University of Technology
Pfennig, A., Graz University of Technology
Originally presented on: 11/8/2013 10:30:00 - 10:45:00

State-of-the-art correlations for prediction of fuel properties like density, viscosity or cloud point work well for atmospheric conditions when applied to non-polar fossil fuels. However, addition of biogenic energy carriers like fatty acid methyl esters (FAME) or ethanol will increase in future. For a successful simulation of such fuels and their mixtures. It is, therefore, necessary to establish reliable property prediction methods for mixtures of fossil and biogenic components over a wide range of pressure and temperature.

Surrogate fuels are a promising concept for fuel description. In this paper, a fuel is emulated by a surrogate composed of only few (up to 30) real components. The generation of such a tailor-made surrogate is based on the simultaneous fitting of a selection of significant fuel properties like boiling-curve, cetane number, liquid density or C/H-ratio, yielding the composition of the supplied components in the surrogate.

As the surrogate is exactly characterized in terms of its composition, rigorous thermodynamics like g^E-models or equations of state can be applied for property-data prediction. Properties predicted include Cetane number, PNA-analysis, flash point, heating value, different types of boiling-curves, cloud point, the mixture vapour pressure as well as pressure and temperature dependent properties like the liquid density and viscosity. Since the surrogate consists of real components these predictions can be compared to actual measurements.

Especially in the field of engine simulations reliable property prediction over a wide range of pressure and temperature is essential. One important property is the liquid density for which the Volume-Translated-Peng-Robinson (VTPR) equation of state is evaluated. In addition to the property prediction of surrogates containing only hydrocarbons, special attention is paid to blends of fossil fuel components and FAME representing Biodiesel.

(768j) Characterization Of Agro-Waste As An Alternative Fuel Source

Tahir, M. S., University of Gujrat
Danish, M., University of Gujrat
Ahmad, M. A.,
Saleem, M., Institute of Chemical Engineering and Technology, University of the Punjab, Lahore, Pakistan
Ahmad, A., East China University of Science and Technology
Feroze, N.,
Originally presented on: 11/8/2013 10:45:00 - 11:00:00

Scientific community is struggling to develop new processes for bio-fuels from last two decades. The present work is an attempt to utilize the agro-waste and to develop the regional energy resource in Pakistan. This work also presents an analysis of various agricultural wastes to determine the green energy potential in Pakistan.

Agro-waste like corn cob, wheat straw, rice husk, saw dust, cotton stalk, corn leave and bagass are selected for characterization. Investigation was conducted for measuring the calorific values of different agro-wastes with Bomb Calorimeter (Leco AC-500) and compared with empirically calculated values. Experimental calorific values were in good agreement with empirically calculated values.

Corn cob and saw dust shows higher calorific values than cotton stalks and corn leaves. Rice husks have least calorific value while bagasse and wheat straw have medium calorific values. Analysis (ultimate and proximate) of different wastes was also conducted to measure the constituents. Ultimate analysis shows that carbon contents were in the range of 39-47 per cent by weight, Hydrogen in the range of 4-6 per cent, Nitrogen in the range 0.3-0.95per cent, Sulphur 0.055-0.25 and Oxygen 38-48 per cent. Proximate analysis determined the moisture contents in the range of 7-12 per cent, Volatile matter 53-80 per cent, ash contents 2-15 per cent.

(771a) Maximizing Active Sites On Molybdenum Sulfide Nanomaterials: Hydrogen Evolution On Thiomolybdate [Mo3S13]2- Clusters

Kibsgaard, J., Stanford University
Jaramillo, T. F., Stanford University
Besenbacher, F., Aarhus University
Originally presented on: 11/8/2013 8:30:00 - 8:50:00

In order to develop improved catalysts, it is important to identify and understand the active sites responsible for reaction turnover in order to produce catalysts with a greater fraction of those sites and possibly even improve upon their turnover frequency. Oftentimes, the most active sites of a solid-state catalyst surface are those with special local structure and stoichiometry such as edges, corners, and kinks.[1]

MoS_2 is one example of such a catalyst: For the hydrogen evolution reaction (HER)[2] as well as for hydro-desulfurization (HDS)[3], MoS_2 edge sites are known to be catalytically active unlike the MoS_2 basal planes which are catalytically inert. In an effort to develop a scalable HER catalyst with an increased number of active sites, herein we report on a new type of Mo-S catalyst ? supported thiomolybdate $[Mo_3S_{13}]^{2-}$ nanoclusters[4] ? which are particularly interesting as most sulfur atoms in the cluster exhibit a similar structural motif to those found at MoS_2 edges, see Figure 1.

FIGURE 1

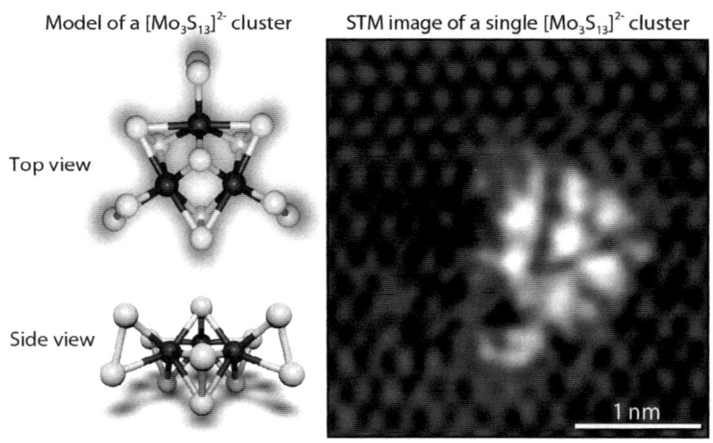

Moreover the thiomolybdate $[Mo_3S_{13}]^{2-}$ nano-clusters are synthesized by a facile, scalable route, and can be deposited onto a wide range of electrode surfaces by means of a simple drop-casting method using methanol as a solvent. The ability to deposit onto a wide range of supports in such a straightforward manner enables ready integration of these nanoclusters onto different device architectures and materials for electrochemical applications.

We evaluated the HER activity of the clusters on two types of substrates: (1) a high surface area graphite paper similar to that used in commercial electrochemical devices such as water electrolyzers and fuel cells, and (2) a highly orientated pyrolytic graphite (HOPG) substrate which allowed for fundamental studies on a sub-monolayer of nanoclusters by imaging them at the atomic-scale with scanning tunneling microscopy (STM), see Figure 1.

FIGURE 2

Cyclic voltammograms showing the HER activity of $[Mo_3S_{13}]^{2-}$ clusters.

Turnover frequency (TOF) of $[Mo_3S_{13}]^{2-}$ clusters together with several other molybdenum sulfide based HER catalysts.

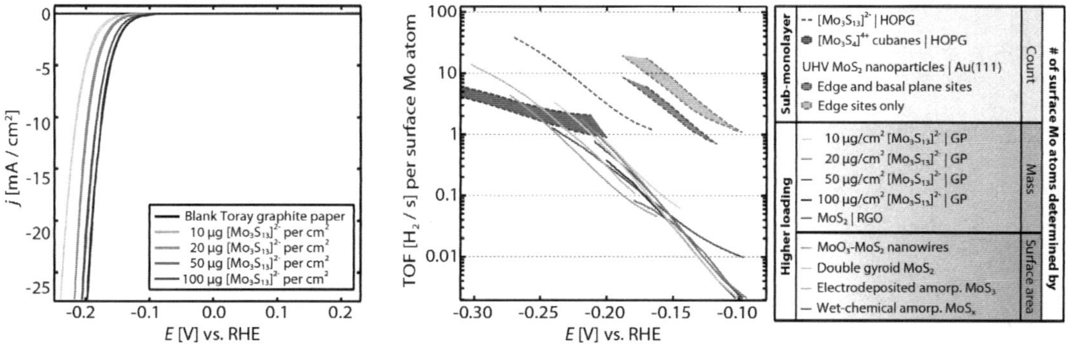

In a strong acid environment, these active and stable $[Mo_3S_{13}]^{2-}$ nanoclusters exhibit unprecedented turnover frequencies for the HER compared to all other molybdenum sulfide catalyst ever synthesized by non-vacuum methods, see Figure 2. We attribute this high activity to the fact that these small $[Mo_3S_{13}]^{2-}$ nanoclusters inherently expose a significant number of active edge sites.

REFERENCES

1. M. Boudart. *Chemical Reviews* **1995,** 95, (3), 661-666.

2. T. F. Jaramillo, K. P. Jorgensen, J. Bonde, J. H. Nielsen, S. Horch, I. Chorkendorff. *Science* **2007,** 317, (5834), 100-102.

3. J. V. Lauritsen, M. Nyberg, J. K. Norskov, B. S. Clausen, H. Topsoe, E. Laegsgaard, F. Besenbacher. *Journal of Catalysis* **2004,** 224, (1), 94-106.

4. J. Kibsgaard, T. F. Jaramillo, F. Besenbacher. *In preparation* **2013**.

(771b) Reforming of Model Biodiesel Over Ni-MoO2

Shah, S., Washington State University
Ha, S., Washington State University
Norton, M. G., Washington State University
Originally presented on: 11/8/2013 8:50:00 - 9:10:00

Currently, most of the hydrogen is produced by reforming of fossil fuels, which are a non-renewable source of energy and therefore a lot of research is going on in exploring different ways to produce hydrogen from renewable sources. One of the ways is to produce hydrogen from biofuels, which are not only a renewable source but also provide a CO2 neutral cycle, thus addressing the environmental concerns related to burning of fossil fuels. This paper explores the possibility of producing hydrogen via partial oxidation of biodiesel. Biodiesel contains almost no sulfur or aromatics and is an oxygenated fuel. It has a huge potential for hydrogen production via reforming, but there is a lack of studies to explore this field. Most of the catalysts used for reforming are either noble metal catalysts, which are not cost effective or nickel based catalysts, which face the problem of coking. Molybdenum dioxide (MoO2) is a non-noble metal catalyst, which has shown good catalytic activity for the oxidative reforming of various logistic fuels such as isooctane, dodecane and Jet-A fuel. One of reasons for having this high reforming activity for MoO2 is explained in terms of its high lattice oxygen mobility and metallic properties. Its highly mobile lattice oxygen allows its high coke resistance by reacting with the surface carbon, while its metallic property helps to efficiently reform carbon fragments. Therefore an initial activity test at O2/C=0.6 was conducted for reforming of model biodiesel, methyl oleate, over MoO2. MoO2 showed a good catalytic activity for reforming methyl oleate with nearly 100% conversion and average hydrogen yield of 78%. We further added nickel, to MoO2 via wet impregnation method with 10% loading by weight. As mentioned above, nickel is a cheap and cost effective material, which is known for its good catalytic activity for reforming except that it has problem of coking. But since MoO2 has high lattice oxygen mobility, its lattice oxygen could help prevent coking on nickel by scavenging the surface carbon deposited on it. We conducted a 24-hour activity test on the as synthesized Ni-MoO2 catalyst. It showed not only 100% conversion with no deactivation, but also a high hydrogen yield compared to MoO2, averaging around 92% over a period of 24 hours.

(771c) Oxidative Steam Reforming of Iso-Butanol Over Promoted Nickel Xerogel Catalysts

Mangalampalli, V. P. S., Hampton University
Akyurtlu, J., Hampton University
Akyurtlu, A., Hampton University
Originally presented on: 11/8/2013 9:10:00 - 9:30:00

Rapid development of alternative energy resources is the main challenge to meet the growing energy demand in the reality of depleting fossil fuels. Hydrogen energy is one of the clean alternatives. However main drawbacks involved in this process are direct distribution and storage; hence onsite production from various feed stocks is one of the choices. In this process, current route for H_2 production is the catalytic reforming of hydrocarbons and alcohols. Presently, there is a growing interest in the use of C2-C4 alcohols as the raw materials for reforming, since these compounds can be produced from biomass feedstock. However, formation of coke and other undesired by-products is the main obstacle in the process of steam reforming of these alcohols. Hence development of new catalysts for higher hydrogen production and less or no coke formation are of great interest. The oxidative steam reforming of alcohols is one of the best options to achieve these goals. It is a combination of steam reforming and partial oxidation. The addition of oxygen into the reaction prevents coke formation and results in a comparatively high hydrogen yield; it, also, requires less energy compared to the conventional steam reforming process. But thermal sintering of active and support materials make this process more challenging.

The present experimental study has been undertaken to investigate the effect of the operation variables on the oxidative steam reforming of isobutanol. In this present study, we have prepared Ni-based catalysts by the sol-gel method, which results in the retention of hydroxyl-rich surfaces, which exhibit unique textural and chemical properties compared with those prepared by other conventional methods, like the impregnation method. Alumina is used as the support material, Ni is the main active metal, and Ce-ZrO_2 and Ru are chosen as promoters. Catalysts were characterized by XRD, TPR, TPD of H_2, BET surface area and porosity measurements. The oxidative steam reforming of isobutanol was performed in a microreactor setup at different reaction temperatures, space velocity, oxygen/carbon ratio and steam/carbon ratio. The Ru-Ni-Ce-Zr-Al_2O_3 catalyst showed high catalytic activity, low carbon deposition, good resistance to sintering and prolonged stability compared to the other catalysts prepared in the project. The results from these experiments will be discussed during the presentation in terms of the effect of operation variables on the hydrogen yield and product gas composition.

Acknowledgement

The authors acknowledge financial support from ARO grant (W911NF-10-1-0514).

(771d) Fundamental Understanding of the Surface Structures of Bimetallic Catalysts for Aqueous-Phase Reforming of Glycerol

Wei, Z., Pacific Northwest National Lab
Karim, A., Pacific Northwest National Laboratory
King, D. L., Pacific Northwest National Laboratory
Wang, Y., Pacific Northwest National Lab
Originally presented on: 11/8/2013 9:30:00 - 9:50:00

Aqueous-phase processing of biomass and its derived chemicals has become a hot research topic for its application in fuel generation. Reforming of polyols in aqueous phase is one of the techniques developed within this scope and targeted at producing hydrogen as the renewable energy source. Much effort has been devoted in catalyst (including bimetallics) development for better performance. However, the surface changes of the catalyst encountered during the reaction are not very well understood. In this work, we studied the surface structure and chemistry of Pt/C and Pt-Re/C under a series of different environments that the catalysts may be exposed to for the process of aqueous-phase reforming of glycerol. Re was found to enhance the catalytic activity of Pt/C for this reaction and a previous explanation was the weakened CO interaction with catalyst surface due to PtRe alloy formation. However, our investigation indicates that with the PtRe alloy on the catalyst surface, CO adsorption can be stronger than that over Pt/C while giving rise to a much higher activity. With the *in situ* spectroscopic characterization tools, we were able to ascribe the enhanced activity to both the Pt surface modified by Re through charge transfer and the extra surface acidity introduced by the interaction between water and Pt-Re entities. Moreover, the selectivity was also greatly affected by the *in situ* generated surface acidity through a dehydration pathway.

(771e) New Insight Into Reaction Mechanisms of Ethanol Steam Reforming On Co/ZrO2

Sun, J., Washington State University
Karim, A., Pacific Northwest National Laboratory
Mei, D., Institute for Integrated Catalysis
Engelhard, M., Pacific Northwest National Laboratory
Wang, Y., Pacific Northwest National Laboratory
Originally presented on: 11/8/2013 9:50:00 - 10:10:00

New Insight into Reaction Mechanisms of Ethanol Steam Reforming on Co/ZrO_2

Junming Sun[†,‡], Ayman M. Karim[†], Donghai Mei[†], Mark Engelhard[†], Yong Wang[†,‡*]

[†] Institute for Interfacial Catalysis, Pacific Northwest National Laboratory, Richland, Washington 99352, United States

[‡] The Gene & Linda Voiland School of Chemical Engineering and Bioengineering, Washington State University, Pullman Washington 99164, United States

Due to its low carbon footprint and renewable source, steam reforming of biomass-derived ethanol (ESR) has been extensively investigated to produce hydrogen for hydrotreating of biomass derived oxygenates or the potential fuel-cell application.[1] Among the studied metals, cobalt based catalysts have been found to be promising in ESR due to its low-cost and high C-C cleavage activity.[2-4] Previous work has shown that acetone act as mainly intermediate toward coking.[5] In this work, a systematic investigation of ESR over Co/ZrO_2 has been studied using a combination of catalytic evaluation and characterizations such as X-ray diffraction, nitrogen sorption, *in situ* X-ray photoelectron spectroscopy and transmission electron microscopy, as well as DFT calculations. We show that acetone is a major reaction intermediate which is steam reformed to selectively form H_2 and CO_2 on metallic cobalt at 450 Â°C (Scheme 1, green highlighted path forward). In the newly discovered sequential reaction pathway, cobalt was found to play a bifunctional role in transforming ethanol to H_2 and CO_2 with high selectivity. For Co/ZrO_2 catalysts, non-reducible cobalt (cobalt species strongly interact with ZrO_2 support) passivates most of the strong acidic sites on ZrO_2, suppresses the undesired dehydration of ethanol on the support, and favors the dehydrogenation and condensation/ketonization reaction pathway, resulting in acetone formation. Metallic cobalt (reducible cobalt) on carbon filament formed during the ESR was found to be mainly responsible for the subsequent acetone steam reforming reactions.[6] The current study not only provides a fundamental new insight into the reaction mechanism of ESR on Co/ZrO_2, but also sheds a light on how to design high selective and durable cobalt catalysts for steam reforming of bio-mass derived small oxygenates (e.g., ethanol, acetone and acetic acid).

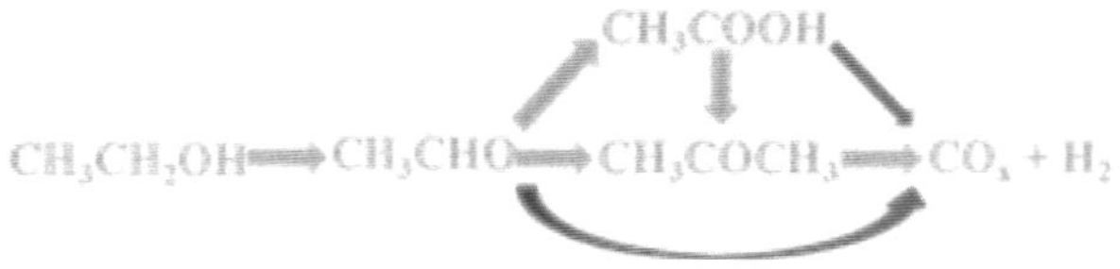

Scheme 1 Proposed main reaction pathway for ethanol steam reforming on Co/ZrO_2 catalysts

Key words: **Ethanol steam reforming**, **Acetone steam reforming**, **Cobalt, Hydrogen production**, **Reaction mechanism, Nanoparticles**

References

(1) Cortright, R. D.; Davda, R. R.; Dumesic, J. A. *Nature* **2002**, *418*, 964.

(2) Song, H.; Ozkan, U. S. *Journal of Catalysis* **2009**, *261*, 66.

(3) Lebarbier, V. M.; Karim, A. M.; Engelhard, M. H.; Wu, Y.; Xu, B. Q.; Petersen, E. J.; Datye, A. K.; Wang, Y. *ChemSusChem* **2011**, *4*, 1679.

(4) Karim, A. M.; Su, Y.; Sun, J. M.; Yang, C.; Strohm, J. J.; King, D. L.; Wang, Y. *Applied Catalysis B-Environmental* **2010**, *96*, 441.

(5) Mattos, L. V.; Jacobs, G.; Davis, B. H.; Noronha, F. B. *Chem. Rev.* **2012**, *112*, 4094.

(6) Sun, J.; Mei, D.; Karim, A. M.; Datye, A. K.; Wang, Y. *ChemCatChem* **2013**, 10.1002/cctc.201300041.

(771f) Modified Zirconia As Gold Catalyst Support for Low-Temperature Methanol Steam Reforming

Wang, C., Tufts University
Saltsburg, H. M., Tufts University
Flytzani-Stephanopoulos, M., Tufts University
Originally presented on: 11/8/2013 10:10:00 - 10:30:00

Modified Zirconia as Gold Catalyst Support for Low-Temperature Methanol Steam Reforming

Chongyang Wang, Howard Saltsburg, Maria Flytzani-Stephanopoulos*

Department of Chemical and Biological Engineering, Tufts University, 4 Colby Street, Medford

MA 02155, United States

Nanoscale zirconia (ZrO_2) is widely investigated for its catalytic relevance both as catalyst support and catalyst. Surface acidity/basicity of zirconia is proved to be a critical factor for developing strong interaction between catalyst support and metals. Pristine zirconia with strong Lewis acids catalyzes the methanol steam reforming reaction (SRM) through η^2-adsorption of the methoxy group, which produces CO as an undesired intermediate, requiring further processing (through the water-gas shift reaction) to maximize the yield of hydrogen. In this study, nanoscale zirconia was modified with zinc oxide (ZnO) at different amounts, and examined as support of gold catalysts for the SRM reaction. The gold precursor adsorbed more efficiently on the modified zirconia when using the anion adsorption synthesis method due to the higher point of zero charge (PZC) of the modified zirconia. Better dispersion of gold was also achieved on the ZnO-modified zirconia, which in turn improved the catalyst activity. Notably, the new method can be used to control the selectivity over both the low and moderately high temperatures (175°C to 375°C) by modulating the whole surface acidity/basicity of the catalyst. As with other gold catalysts, the η^1-adsorption of methoxy group is favored, producing CO_2 and H_2 as exclusive products of SRM.

To investigate the modification of zirconia, surface acidity titration has been conducted with probing molecules (isopropanol) both in temperature-programmed surface reaction (TPSR) and temperature-programmed desorption (TPD) modes. SRM-TPSR and CH_3OH-TPSR were performed on supports and gold-added catalysts to unravel the effects of surface acidity/basicity on the selectivity of methanol steam reforming reaction, and to understand the reaction mechanism of methanol steam reforming reaction on modified and pristine zirconia. Steady-state SRM reaction activity tests were also carried out on the bare and gold-loaded zirconia surfaces. In summary, the interaction between catalyst and support, and dispersion and stability of gold are enhanced by modification of the surface acidity of zirconia, and the overall catalyst selectivity to hydrogen is greatly improved.

(771g) Ni-Mg/Al2O3 and Ni-La/Al2O3 Catalysts for Steam Reforming of Biomass Tar

Garbarino, G., University of Genova
Finocchio, E., University of Genova
Valsamakis, I., Tufts University
Wang, C., Tufts University
Flytzani-Stephanopoulos, M., Tufts University
Busca, G., University of Genova
Originally presented on: 11/8/2013 10:30:00 - 10:50:00

Introduction

Hydrogen generation through a renewable route is highly desirable. The production of renewable hydrogen can be performed through biomass gasification, followed by treatment of the resulting syngas. The presence of heavier and lighter oxygenated organic compounds, commonly referred to as tars, causes several drawbacks in the syngas processing and thus the abatement of these species is necessary.

Catalytic steam reforming seems to be the most promising industrial strategy for tar abatement after biomass gasification. Ni-Al_2O_3 based catalysts are active for this reaction, but due to their deactivation by sulphur contaminants they have been reported as non-applicable below a reaction temperature limit. The sulphur content in syngas produced by gasification processes ranges between 20 and 200 ppm v_{H2S}/v. On the other hand, reaction temperature for the tar abatement may depend from the overall gasification plant configuration.

It is well known that alkali and alkali earth cations moderate the surface acidity of alumina and can also induce surface basicity, thus besides acting as structural stabilizers of the support, they might allow limiting catalyst coking. However, several studies suggest that these components, namely magnesium, may also increase the catalytic activity of Ni/Al_2O_3, e.g. for ethanol steam reforming and for methane reforming, Ni/$MgAl_2O_4$acting as excellent tar steam reforming catalysts.

In literature it is reported that Lanthanum is able to increase nickel dispersion and also has got a beneficial effect increasing sulphur resistance of the catalysts. In addition, the presence of lanthanide oxides enhances the stability of Ni-based catalysts, since these oxides are known promoters of carbon removal from metallic surfaces.

In this communication we present our results on the study of steam reforming of an ethanol-phenol mixture, used as a model for biomass tar, over Ni-Al_2O_3 based catalysts. The effect of catalysts doping with basic elements, such as lanthanum and magnesium, on the catalytic activity and the sulphur deactivation behaviour is also reported.

Experimental

Catalysts preparation

The nickel-based catalysts examined here, were synthesized via wet impregnation of the desired nickel weight amount onto commercial alumina (Siralox 5-170, Sasol, 170 m^2/g); Ni16 designates the catalyst containing 16% w/w of Ni. The magnesium containing catalyst (Ni16Mg, with 20% of MgO w_{MgO}/w_{Al2O3}) was prepared by sequential wet impregnation of Siralox 5/170 support with magnesium nitrate and nickel nitrate, both steps followed by drying and calcination. For these catalysts after each impregnation, drying at 363K for 8 hours and calcination at 973K for 5 hours have been performed.

Lanthanum containing catalysts were prepared through incipient wetness impregnation of commercial alumina, previously stabilized. Three different amounts of La_2O_3 were added: 5%, 20% and 80% w_{La2O3}/w_{Al2O3}. The samples were first calcined, and then nickel addition was done by the same impregnation technique. The final catalyst calcination was conducted from room temperature to 1023 K with a ramping of 2°C/min and a 5-hour hold at the final temperature. The NiO loading for all the catalyst prepared was 20% (w/w), corresponding to 16% Ni loading (w/w), because this amount has been shown from previous studies to give a good compromise between catalytic activity and sulphur deactivation resistance.

Characterization

The materials were characterized for BET, XRD, XPS, H2-TPR, IR, and IR with adsorption of probe molecules. The catalytic activity and a part of the material characterization on Ni-La/Al_2O_3catalysts it is now in progress while Ni16 and Mg-doped catalyst were fully characterized.

Catalytic reaction conditions

The total flow rate fed to the tubular quartz flow reactor was 40 Nml/min, with the following composition: 4.1% (v/v) of ethanol, 2% (v/v) of phenol, 54.6% (v/v) of water and 39.3% (v/v) of helium as carrier gas. The reactor was filled up for every test with 440 mg of sieved silica glass (60-70 mesh) and 44.1 mg of catalyst. Products analysis was performed with a gas-chromatograph Agilent 4890 equipped with a Varian capillary column "Molsieve 5A/Porabond Q Tandem" and TCD and FID detectors in series. Between them a Nickel Catalyst Tube was employed to reduce CO to CH_4. The sampling of the gases was made by injection, using a gas-tight with a nominal volume of $0.25*10^{-6}$ m^3. A sampling injection point was also available at the end of the preheating zone to analyse the reagents and determine the possible decomposition/change in the feed. Precise identification of the compounds in the products stream was also performed on GC/MS (ThermoScientific). Catalytic experiments were performed both in rising and in reducing reaction temperature (773K, 873 K, 973 K, 1023 K and reverse) in order to reveal if there a conditioning effect was affecting the catalytic performance.

Sulphur poisoning was examined adding tetrahydrotiophene (C_4H_8S, THT) in the feed. In one series of experiments, performed at 973 K, 210 ppm of THT were continuously fed with the reactants on stream for up to six hours. In a second series of experiments performed at the same temperature, two sequential pulses of THT (first 0.011 mol_S/mol_{Ni} and after 0.033 mol_S/mol_{Ni}) and a subsequent hold in the sulphur-free stream were performed.

Results

The experiments performed with sulphur free feed showed that pure alumina catalyses the conversion of ethanol to ethylene with traces of hydrogen exchange reaction to give ethane and acetaldehyde. Alkylation of phenol mainly to o-ethylphenol is observed. Experiments using increasing reaction temperature show that Ni containing catalysts catalyse the low temperature alkylation of phenol more than the pure alumina does. Complete steam reforming of both reactants to CO and CO_2 with traces of methane is obtained at 973 K for Ni5 and Ni16, while for Ni39 the steam reforming is almost complete at already 873 K. Experiments in decreasing reaction temperature revealed a catalyst conditioning effect that occured during high temperature operation, resulting in higher catalytic activity to steam reforming at lower temperature. Over Ni16 catalyst, in the increasing temperature experiment, ethanol shows an almost complete conversion already at 773 K while phenol is only barely converted. The main products are: alkylation products, ethylene, acetaldehyde and dialkylates. No activity in steam reforming was present. At 873 K phenol conversion remains around 25% while ethanol conversion is 86%. There is a little selectivity toward CO and CO_2 but the main products remain ethylene, mono-alkylate products, acetaldehyde and ethane. At 973 K and 1023 K we have complete activity to steam reforming, and CO and CO_2 as main products both in the increasing temperature experiment and in the decreasing one, accompanied by the complete conversion of both reactants and a hydrogen yield ranging around 80-82%.

For the magnesium containing catalyst (Ni16Mg), in the increasing temperature experiment, the main product at 773 K is acetaldehyde, followed by C8 alkylated product, ethylene, CO_2 and acetone. By increasing the temperature at 873 K ethanol conversion is complete while phenol one decreases to 28%. Acetaldehyde and ethylene together with CO_2, CO, mono-alkylates, methane, acetone and benzene are the detected products. At 973 K complete steam reforming of both reactants occurs to CO_2 (70% selectivity) and CO, with a hydrogen yield of 88%. By further increase of the reaction temperature to 1023K we have as expected complete steam reforming but with an increase of CO selectivity to 39% being CO_2 one 61%. The same activity and selectivity observed for the increasing temperature experiment is found at 1023 K and 973 K. A completely different behavior is observed at 873 K, where we have complete activity towards the steam reforming reaction and CO_2 is found as the main product with a selectivity of 74%, together with CO and methane. The hydrogen yield obtained in this case is 84%. At the lower temperature (773 K) almost complete conversion of ethanol and 44% conversion of phenol are observed. A different product distribution is found in this case: the main product is CO_2together with ethylene, monoalkylates, CO and acetaldehyde.

Conclusions

The tests performed on Ni16Mg show that this catalyst it is more active in steam reforming reaction than the catalyst without magnesium but it is also more prone to sulphur deactivation than Ni16. The performances of Lanthanum containing catalysts are now in progress, showing interesting preliminary results.

(800a) High-Titer and High-Yield n-Butanol Production From Lignocellulosic Biomass By Engineered Clostridium Tyrobutyricum

Du, Y., The Ohio State University
Yang, S. T., Ohio State University
Originally presented on: 11/8/2013 12:30:00 - 12:45:00

Fermentative n-butanol production by solventogenic clostridia usually suffers from low butanol titer, low butanol yield, and high substrate cost. In order to overcome these limitations, NADH driving force combined with adaptation and evolution mutagenesis strategy was used to achieve high-titer, high-yield, and cost-effective n-butanol production from lignocellulosic materials. In order to confirm that the *C. tyrobutyricum* mutant can consume glucose and xylose simultaneously and determine an optimal condition for n-butanol production from biomass hydrolysates, repeated-batch co-fermentation in a fibrous-bed bioreactor (FBB) using glucose-xylose mixture as a substrate was performed. Synchronized utilization of glucose and xylose with a high butanol titer (~ 20.0 g/L) and yield (> 0.30 g/g) was observed in the glucose-xylose co-fermentation. In addition, it is interesting to note that the consumption rate of xylose was dependent on glucose/xylose ratio. Then, the feasibility of using various cellulosic and lignocellulosic biomass hydrolysates, including Jerusalem artichoke, cassava bagasse, cotton stalk, sugarcane bagasse, soybean hull, and corn fiber, as alternative feedstocks for fermentative n-butanol production was investigated in both free-cell and immobilized-cell fermentation. The results showed that *C. tyrobutyricum* was capable of efficiently converting those biomass hydrolysates containing glucose and xylose to n-butanol, achieving a high titer of over 15 g/L and a high yield of over 0.30 g/g substrate. This study has demonstrated an economically competitive process for high-titer, high-yield, and cost-effective n-butanol production from abundant, renewable, and sustainable feedstocks.

(800b) Evaluation of Ascorbic Acid and Acetic Acid As Catalysts in the Transesterfication of Virgin Soybean Oil

Wenzel, J. E., Kettering University
Maricq, E., Kettering University
Stogsdill, M.,
Zand, A., Kettering University
Originally presented on: 11/8/2013 12:45:00 - 13:00:00

Biologically derived acids such as ascorbic acid and acetic acid may be used to catalyze the production of biodiesel in a supercritical fluid solvent. The use of binary mixtures of supercritical carbon dioxide and methanol with the aid of an acid catalyst can facilitate the transesterfication reaction. This may hold a potential advantage over non-catalytic supercritical methanol production of biodiesel due to methanol's higher critical temperature than that of the critical temperature of binary mixtures of carbon dioxide and methanol. The production of biodiesel using a ascorbic acid and acetic acid as a catalyst was evaluated in an agitated 500 mL Hastelloy C276 batch reactor. Comparison experiments between sub-critical methanol, supercritical methanol, and supercritical methanol with a carbon dioxide cosolvent were performed between 175 °C and 250 °C. Hold time was varied between 20 and 40 minutes and the methanol and soybean concentrations were varied. The effect of temperature and concentration upon conversion will be discussed as well as a comparison between catalysts and use of binary mixtures of supercritical carbon dioxide.

(800c) In-Situ Conversion of the FOG in Trap Grease Into Biodiesel

Tu, Q., University of Cincinnati
Chai, M., University of Cincinnati
Lu, M., University of Cincinnati

Originally presented on: 11/8/2013 13:00:00 - 13:15:00

Biodiesel is a type of alternative fuel made from oil-containing biomass and a variety of fats, oils, and greases (FOG). Biodiesel is a drop-in solution for the diesel engines to reduce the emissions and consequently the US biodiesel industry has been thriving in the recent years under the context of promoting clean fuels. With the production of biodiesel increasing sharply, a variety of feedstocks, especially non-food based oil sources such as algae and municipal sludge, have been introduced. However, one potential feedstock that has hardly been exploited is the trap grease, a mixture of water, solids, and FOG that accumulates in the grease interceptors and sewer pipelines. Trap grease has been a nuisance to the municipal system for years due to its FOG content which is the major cause of CSO and SSO in the US. Current management of trap grease is costly and ineffective. Considering its FOG content, significant yearly generation and low cost, utilizing trap grease as biodiesel feedstock can be a win-win solution for both environmental protection and biodiesel industry, particularly for the latter. Producing biodiesel from trap grease can significantly reduce the cost for the biodiesel producers and more importantly, improve the sustainability of biodiesel production by switching from oil crops (e.g. soybean oil) to waste materials and hence minimizing water and land footprints.

However, due to its highly heterogeneous nature, one issue related to the utilization of trap grease is that it takes significant amounts of efforts to extract FOG out of it before the transesterification reaction can be performed, which may not justify the application of trap grease as the feedstock in the industry. Therefore, direct transesterification, or *in-situ* conversion, is investigated in this study as an alternative to address this problem. *In-situ* conversion refers to the process where the reagent and catalyst directly contact the FOG content in the feedstock substrate without the extraction step being performed first. The elimination of the extraction step can benefit the biodiesel production by lowering cost and energy consumption, as well as the reduction in facility footstep. In this study, the *in-situ* conversion of FOG in the trap grease into biodiesel will be investigated and parameters such as methanol to trap grease ratio (v/m), catalyst concentration (wt %), temperature and time will be varied to assess their influences on the *in-situ* process and to obtain an optimum condition for the process. The yield of biodiesel from the *in-situ* approach will be compared with that of the conventional extraction-transesterification method to evaluate feasibility of its application in the industry.

(800d) Synthesis of Polyoxymethylene Dimethyl Ethers From Dimethoxymethane and Paraformaldehyde Catalyzed By Cation Exchange Resins

Zheng, Y., Tsinghua University
Tang, Q., Tsinghua University
Wang, T., Tsinghua University
Wang, J., Tsinghua University

Originally presented on: 11/8/2013 13:15:00 - 13:30:00

Synthesis of oxygenated compounds as diesel fuel additives from coal-based C1 chemicals is of great importance to solve the large surplus of C1 chemicals and the diesel supply shortage in China. When added to diesel, oxygenated compounds (methanol, dimethyl ether, dimethoxymethane, etc.) can significantly reduce smoke and engine emissions, which is closely correlated to the oxygen content of the compounds. However, methanol with low solubility in diesel, dimethyl ether with higher vapor pressure and lower viscosity than diesel and dimethoxymethane with a low cetane number are difficult for wide use as diesel fuel additives. As higher homologue of dimethoxymethane, polyoxymethylene dimethyl ethers($PODE_n$, $CH_3O(CH_2O)_nCH_3$) are new concerned oxygenated compounds as green diesel fuel additives. Among the series of $PODE_n$, $PODE_{3-5}$are ideal because of the proper physical properties for diesel.

Dimethoxymethane and paraformaldehyde are firstly reported to be used as raw materials to synthesize the $PODE_n$ compounds in this work, with cation exchange resins as catalysts. Compared with some reports on synthesis of $PODE_n$, no water is produced during this process, which is favorable for reducing side reactions and increasing the product yields. Seven different cation exchange resins were examined in this work. Among them, the NKC-9 resin was the most active catalyst. The characteristic results of scanning electron microscopy and nitrogen adsorption-desorption, together with the analysis of catalysts acid strength and the number of acid sites, were used to explain different catalytic performances. It was found that strong-acid cation exchange resin with large surface area, proper pore volume and pore diameter and high concentration of acid sites is favourable to give higher catalytic activity.

The effects of reaction temperature, DMM/CH_2O molar ratio, reaction time and catalyst loading on formaldehyde conversion and product distribution were studied. At optimized operating conditions, a formaldehyde conversion of 85.1% was obtained with the selectivity to $PODE_n$ near 100%, and the amount of $PODE_{3-5}$ in products was 36.6wt%. A possible mechanism, which includes decomposition of a paraformaldehyde molecule to formaldehyde molecules, protonation of formaldehyde molecule at the carbonyl group, formation of the intermediate and finally formation of a $PODE_{n+1}$ molecular, was proposed to give a step-by-step account of the bond reorganizations that take place in the course of the reactions. The catalyst characterizations and reaction data would lay a solid foundation for the further study on $PODE_n$.

(800e) Biodiesel Production With a Membrane Reactor Over Heterogeneous Alkali Catalyst

Xu, W., School of Chemistry and Chemical Engineering
Xiao, G., Southeast University

Originally presented on: 11/8/2013 13:30:00 - 13:45:00

Biodiesel, a mixture of fatty acid methyl ester(FAME), is a renewable and non-fossil fuel with less impact to the environment. However, the removal of residual oil from the product FAME is one of the mainly factors that plague biodiesel production costs, which largely limit its popularization and utilization. Membrane reactor could remove FAME from the system and simultaneously shifting the reaction balance to obtain relatively high oil conversion. Coupling heterogeneous alkali, a catalytic membrane reactor was applied to produce biodiesel from soybean oil, whose process combined the catalysis and separation in one step. Shaped KF/Ca-Mg-Al hydrotalcites catalyst was packed into the ceramic membrane tube. 30g soybean oil and 29.8g methanol (molar ratio around 1:24) were first mixed by strong stirring and heated to fixed temperature, and then charged into the preheated membrane reactor. Mixture of product FAME, glycerol and methanol was obtained from the permeate stream. And FAME could be obtained by layer separation after distilled excess methanol. The rest reactants as well as the distilled methanol were circulated. The central composite design experiments was applied to investigate the effects of reaction temperature, catalyst amount and circulation velocity on the biodiesel production. It was found that reaction temperature and catalyst amount had the positive effect on the FAME yielding rate while the circulation velocity had the negative effect on it. According to the experiment results, 70 °C, 0.531 g/cm^3 of catalyst amount and 3.16 mL/min of circulation velocity was proved to be the optimum condition, which achieved a 0.1820 g/min FAME yielding rate within 150 min.

(800f) Study On the Hydrogenation Technology of Vegetable Oil

Chong, P., Fushun research institute of petroleum and petrochemicals, SINOPEC
Xiangchen, F., Fushun research institute of petroleum and petrochemicals, SINOPEC
Ronghui, Z., Fushun research institute of petroleum and petrochemicals, SINOPEC
Originally presented on: 11/8/2013 13:45:00 - 14:00:00

Study on the Hydrogenation technology of Vegetable oil

ABSTRACT

Vegetable oil with low sulfur and nitrogen content, is a clean renewable energy. Studying on vegetable oil hydroprocessing for clean fuel production will help mankind reply less on non-renewable energy--oil, which can get significant benefit in environment and society.

This paper studied on the effects of catalysts, reaction temperature, reaction pressure and LHSV on vegetable oil HDO(hydrodeoxygenation) with soybean oil as feed in pilot plant. The results showed that different paths of vegetable oil happened on different catalysts. Direct HDO reaction mainly occurred on oxidated catalyst, other reactions were side reaction; whereas both direct HDO and decarboxylation happened on sulfided catalyst at the same time. Different reaction conditions had large effect on product performances. The suitable conditions for vegetable oil HDS were as followings: 16MPa of reaction pressure, 340~345°C of reaction temperature, 0.6~0.8h^{-1} of LHSV.

The results of the tests on hydroisodewaxing with vegetable oil demonstrated that hydroisodewaxing reaction was mainly occurred in low conversion, while in higher conversion, more paraffin hydrocracking reactions were happened. Jet fuel freezing point was influenced by cutting point and conversion. With the increasing of conversion, freezing point decreased; however, it increased remarkably with the increasing of cutting point. Therefore, when producing 3# jet fuel, only distillates with low FBP and narrow distillation could be cut. Hydroisodewaxing could help decrease diesel solidification point but caused loss on cetane number. However, diesel cetane number was more than 80 after hydroisodewaxing. After hydroisodewaxing, part of jet fuel blending components could be produced, and diesel solidification point was about +10°C, which was the superior diesel blending components. Using vegetable oil for HDO, then for hydroisodewaxing, high liquid yields could be attained, 3# jet fuel and diesel with low solification point were produced directly.

In order to broaden the feed sources for hydrotreating, the reaction conditions using gas oil blending with vegetable oil were investigated. The results indicated heavy distillates in vegetable oil had good conversion performance. During hydroprocessing, oxygen in vegetable oil was removed in the form of CO, CO_2 and H_2O. Though mutual competition between HDO and HDS, the dilution function of vegetable oil on feed showed more obviously than the inhibition of HDO on HDS. The diesel products performances after hydrotreated were largely improved, while basically there was no effect on unconverted oil properties.

(800g) Heterogeneous Catalysis for Biodiesel Production of Rapeseed Oil

Ramesar, N., CUNY City College of New York
Kiros, Y.,
Originally presented on: 11/8/2013 14:00:00 - 14:15:00

Heterogeneous Catalysis for Biodiesel Production of Rapeseed Oil

Naomi Ramesar[1] and Yohannes Kiros[2]

[1]Deparment of Chemical Engineering, CUNY City College, 160 Convent Avenue, New York, NY 10031

[2] Department of Chemical Engineering & Technology, KTH-Royal Institute of Technology, Teknikringen42, S-100 44 Stockholm, Sweden

AIChE Conference 2013: ***Abstract***

Biodiesel is an alternative fuel derived from vegetable oils, animal fats and used frying oils. It consists of long-chain fatty acid methyl esters (FAME), which is chemically produced through the transesterification reaction of a triglyceride with an alcohol (methanol or ethanol) in the presence of an alkaline or acidic catalyst. A new approach using heterogeneous catalysts has attracted much attention in recent years for the production of biodiesel because they are environmentally benign, recyclable, and cheaper than homogeneous catalysts. In our work, the two alkaline catalysts being used are calcium oxide (CaO) and lithium doped calcium oxide (Li-CaO). To observe the effect of both catalysts on biodiesel production, the reaction was carried out at 50 and 60°C with a weight percent of catalyst to oil of 3.5 and 2.5. Surface characterizations of both catalysts were determined by Powder X-ray Diffraction (XRD) and Brunauer, Emmett and Teller (BET) method. The percent yield of FAME was obtained via results provided from Gas Chromotography (GC).

(800h) Systematic Evaluation of Jatropha Curcas L.Oil Obtention By Mechanical and Solvent Extraction

Narvaez, P. C., Universidad Nacional de Colombia
Orjuela, A., Universidad Nacional de Colombia
Yate, A., Universidad Nacional de Colombia
Hernandez, A., Universidad Nacional de Colobia
Acevedo, H.,
Originally presented on: 11/8/2013 14:15:00 - 14:30:00

Systematic evaluation of *Jatropha curcas L.*oil obtention by mechanical and solvent extraction

Andrea Viviana Yate, Adriana Hernández Carrillo, Helmer Acevedo, Álvaro Orjuela, Paulo CÃ©sar Narváez

Abstract

Current biodiesel production relies mostly on edible oils (soybean, palm, rapeseed) which is a major drawback for the process as raw materials represent the major part of the final costs of the biofuel. This also brings out concerns about food and feed security. In the last years *Jatropha curca L* oil has obtained attention as an alternative oleochemical feedstock because it is a perennial crop able to produce up to 160 Gallon per hectare-year of a non-edible oil. Even though several authors have reported process conditions and yields on a variety of extraction methods for *Jatropha* oil, few of them present details on the overall process from the industrial standpoint, and also most of them do not assess the impact of the extraction process on the oil quality and its technical requirements for biodiesel production.

In this regard this work describes the development of a *Jatropha* oil production process involving mechanical and solvent extraction. The effect of mechanical extraction parameters such as speed of the screw, diameter and temperature of discharge, each one was analyzed on oil yield and saponification index as a quality parameter. Expeller power consumption and extraction efficiency were studied. In solvent extraction experiments it was evaluated the effect of the nature of solvent by using n-hexano, methanol and ethanol, the seed:solvent mass ratio 1:4, 1:6 and 1:8, and the particle size 1.15, 2.79 and 9.79mm on oil yield and oil properties (saponification index, humidity). The most favorable conditions of each extraction process were established.

An optimization study allowed establishing most favorable conditions for industrial processing involving mechanical extraction of the seeds and solvent extraction of the pressed cake. Oil obtained with this process was characterized and used for biodiesel production.

(800i) CaO-Catalyzed Biodiesel Preparation With a Novel Deep Eutectic Ionic Liquid

Huang, W., Key Laboratory for Green Chemical Technology of Ministry of Education, School of Chemical Engineering & Technology, Tianjin University
Tang, S., Key Laboratory for Green Chemical Technology of Ministry of Education, School of Chemical Engineering & Technology, Tianjin University
Originally presented on: 11/8/2013 14:30:00 - 14:45:00

CaO-Catalyzed Biodiesel Preparation with a Novel Deep Eutectic Ionic Liquid

Wei Huang [a], Shaokun Tang*[a]□Hua Zhao [b]

[a]*Key Laboratory for Green Chemical Technology of Ministry of Education, School of Chemical Engineering & Technology, Tianjin University, Tianjin 300072, China*

[b]*Chemistry Program, Savannah State University, Savannah, GA 31404, USA*

*Corresponding Author. Tel: +86-22-27408578; Fax: +86-22-27408578; E-mail address: shktang@tju.edu.cn

As a promising replacement of fossil diesel, biodiesel has gained proliferating attention in recent years because of its renewable and environmentally benign characters. Biodiesel is commonly produced by homogeneous catalysts such as sodium or potassium hydroxide in industry which is usually accompanied with side reactions and laborious purification steps. Solid base catalysts such as CaO have been used in the transesterification reaction, leading to several advantages such as low byproducts, easy separation, no corrosion and low waste. However, CaO can be easily poisoned by carbon dioxide and water to form a layer of calcium carbonate and calcium hydroxide on the surface of bulk CaO within a few minutes exposure to air. In many cases commercial CaO needs to be calcined at high temperatures to remove the inactive layer before use.

In our work, we identified a straightforward and energy-efficient method to activate commercial CaO for biodiesel production without any pretreatment. A deep eutectic ionic liquid (DEIL) consisting of choline chloride and glycerol (1:2 molar ratio) was prepared and used to activate commercial CaO to be a catalyst during the transesterification of rapeseed oil to produce biodiesel. The yield of fatty acid methyl ester (FAME) with the addition of DES reached 91.90%, while the FAME yield with no DEIL was only 4.03%. The commercial CaO samples before and after reaction were characterized by Powder X-ray diffraction (XRD), Fourier transform infrared spectroscopy (FT-IR) and Scanning Electronic Microscopy (SEM). These characterizations and analyses indicate that the surface of commercial CaO is typically covered by a layer of $CaCO_3$ and $Ca(OH)_2$ which inhibits the activity of CaO, while the addition of DEIL can eliminate the inactive layer to activate CaO bulk due to its strong solvation ability. Moreover, the use of DEIL as co-solvent can also simplify the production separation and purification steps.

Acknowledgment

This project was financially supported by the National Natural Science Foundation of China (21176186).

(800j) Characterisation of Blended Biodiesel Produced By Trans-Esterification of Jatropha Oil

Khan, R. U., Institute of Chemical Engineering and Technology, University of the Punjab, Lahore
Akram, A., Institute of Chemical Engineering & Technology, University of the Punjab
Originally presented on: 11/8/2013 14:45:00 - 15:00:00

Edible and non edible oils can be used for the production of the Biodiesel. Traditionally edible oils such as coconut, soybean, rapeseed, canola, sunflower, corn oil and also palm oil are the most promising sources. This practice is not economical for countries which are not self-sufficient in edible oil. Thus, other alternative, non edible oil, is cheapest and economically more viable for the production of Biodiesel. In the present research non edible vegetable oil, obtained from Jatropha plant, was chosen to produce Biodiesel. Biodiesel produced by this process was mixed with Petro-diesel in different proportions. Results of physicochemical study of blends of Biodiesel and Petro-diesel will be presented.

AUTHOR INDEX

AUTHOR INDEX

AUTHOR INDEX

AUTHOR INDEX

AUTHOR INDEX

AUTHOR INDEX

AUTHOR INDEX

AUTHOR INDEX